From Vines to Wines in Classical Rome

From Vines to Wines in Classical Rome

A Handbook of Viticulture and Oenology in Rome and the Roman West

By

David L. Thurmond

BRILL

LEIDEN | BOSTON

Library of Congress Cataloging-in-Publication Data

Names: Thurmond, David L., author.
Title: From vines to wines in classical Rome : a handbook of viticulture and oenology in Rome and the Roman West / by David L. Thurmond.
Description: Leiden ; Boston : Brill, [2017] | Includes bibliographical references and index.
Identifiers: LCCN 2016041233 (print) | LCCN 2016042781 (ebook) | ISBN 9789004334588 (hardback) : alk. paper) | ISBN 9789004334595 (e-book)
Subjects: LCSH: Wine and wine making--Italy. | Viticulture--Italy.
Classification: LCC TP559.I8 T48 2017 (print) | LCC TP559.I8 (ebook) | DDC 663/.200945--dc23
LC record available at https://lccn.loc.gov/2016041233

Typeface for the Latin, Greek, and Cyrillic scripts: "Brill". See and download: brill.com/brill-typeface.

ISBN 978-90-04-33458-8 (hardback)
ISBN 978-90-04-33459-5 (e-book)

This book is printed on acid-free paper and produced in a sustainable manner.

SODALIBUS ITALICIS

∵

Contents

Acknowledgements

It is indeed a pleasure to acknowledge some of the people most instrumental in seeing this project through to its end. First, I owe a huge debt of gratitude to the people at Brill whose professional acumen is exceeded only by their kindness and *humanitas*. This includes the three readers who offered invaluable suggestions for improving the text, including one who was kind enough to do a close reading. Flawed as it doubtless still is, the text is greatly improved because of their generosity. I owe a special thanks to Ms. Tessel Jonquière and Mr. Pieter van Roon, editors at Brill, for their tireless efforts on behalf of our little brainchild.

I also owe a great debt to Rolando and Filomena Astone and Fabio and Katiuscia Astone, our hosts for several summers of research in the Mezzogiorno. They are no longer merely our friends but our *cara famiglia italiana*. Dott. Fabio Astone has also been an invaluable touchstone for some of the archaeological research. Archaeological work was greatly aided by Aniello Botti, Adriana Zammarrelli, Franco Castelnuovo, Roberto Volpe and Marco Marrone. To my dear friend, Professor Fernando La Greca of the University of Salerno, my Italian twin and partner in archaeological research, I owe deepest gratitude. The work contained herein on the protohistory of the vine in Italy would quite simply have been impossible without his help.

Finally, I offer my deepest thanks to my wife, Sandy Thurmond, my indefatigable photographer and perennial cheerleader.

List of Figures

CHAPTER 1

Prehistory and Protohistory

Biology

Wine in its strictest sense is the product of the alcoholic fermentation of grape sugars to produce, in a reasonably anaerobic environment, a biologically stable liquid that is extremely high in carbohydrates and is thus a rich source of the calories so essential in a pre-industrial diet. Grapes readily ferment because wild yeasts present on the 'bloom', the grayish, powdery film on the waxy surface of the berry, as well as in the air of the vineyard and the winery, are ready and waiting to start this metabolic process as soon as they have access to the sugars. Grapes are one of the few fruits that store carbohydrates primarily as soluble sugars and simultaneously contain significant levels of acids, especially in the form of tartaric acid, which are bacteriostatic and also improve the taste of wine by making it more balanced. It is easy for a modern Westerner to underestimate the value of wine as food in the ancient world, and many of the ancient geoponics are equally myopic. Thus the classical Roman agronomists Cato, Varro, Pliny the Elder, and most notably Columella, speak in some detail about the profitability of wine as a cash crop on the prototype Roman agribusiness, the villa farm, but say essentially nothing about wine as food, perhaps, in their defence, because the fact was so self-evident. They are followed by such modern students of the ancient economy as Duncan-Jones and Carandini who are also silent about this fundamental role of wine. And the modern scholar who has done more, arguably, to advance our knowledge of the prehistory and protohistory of wine, Patrick E. McGovern, can ask, "Why Alcohol and Why Wine?"[1] and then proceed to adduce the role of wine as drug, both therapeutic and psychotropic, as a water purifier, and as a medium for other drugs, but say nothing of wine as a stable source of carbohydrates.

Not to say that these peripheral advantages of wine were and are insignificant. The ancients were well aware of the connection between contaminated water and disease. The fourth-century CE military writer Vegetius, for example, warns soldiers against drinking "bad or marshy water" which he compares to poison, and also cautions that when an army is encamped in one location for too long the water becomes corrupt (*De Re Mil.* 3.2). It has now been

1 Patrick E. McGovern, *Ancient Wine: The Search for the Origins of Viniculture* (Princeton, NJ, 2003): 305-07.

 | DOI 10.1163/9789004334595_002

experimentally demonstrated that wine mixed with contaminated water quickly destroys typhoid, paratyphoid and other typically waterborne pathogens.[2] Water mixed with wine in the proportions typical for Romans (1:1 to 2:1) will have been far safer to drink.

About the medicinal value of wine we are still grossly ignorant, though there is mounting evidence and certainly a great deal of interest. The ancients knew about wine's antibiotic powers as a topical agent and used it habitually for wound-dressing, by itself or as a medium for other bactericides such as enhemes (metal salts), though they were of course ignorant of the operant mechanisms. Otherwise they knew of wine as part of a healthy regimen and as a powerful adjunct in processing other foods.[3] We now know that heavy drinking of wine is implicated in a number of diseases including cardiomyopathy, hypertension, cardiac arrhythmia, cerebrovascular disease, and hemorrhagic stroke, but that light to moderate drinking, especially of red wine, is associated with lower incidence of ischemic stroke, myocardial infarction, coronary heart disease (CHD), and peripheral vascular disease.[4] Wine's role in reduced risk of CHD seems to be linked to the role of flavanoids and other polyphenols as antioxidants in reducing oxidation of low-density lipoproteins (LDLs) to form atherosclerotic lesions. One polyphenol in particular, resveratrol, has generated a great deal of excitement in this regard, and resveratrol is a compound almost unique to the skins of red grapes and the wines they produce.

2 Vernon R. Singleton, "An Enologist's Commentary on Ancient Wines," in Patrick E. McGovern, Stuart Fleming, and Solomon H. Katz, edd., *The Origins and Ancient History of Wine* (Amsterdam, 1996): 75.

3 Luigi Crispino [*Alimentzione mediterranea e stile di vita* (Acciaroli, SA, 2008): 228], for the role of wine in the now-vindicated Mediteranean diet. Martin E. Weisse and Renée S. Moore ["Antimicrobial Effects of Wine," in Merton Sandler and Roger Pinder, edd., *Wine: A Scientific Exploration* (New York, 2003): 299-313] review the scientific evidence and possible operant compounds in treatment of enteric bacteria, especially Salmonella species, of *Staphylococcus aureus* in wound infection, of cholera and *Helocobacter pylori.* Active agents include ethanol, low pH, sulphites, polyphenols and flavanoids. Cf. Luigi Crispino *Virtu terapeutiche del vino* (Acciaroli, 1997); Catherine Cheze, Joseph Vercauteren, and Robert Verpoote, edd., *Polypheols, Wine and Health* (Dordrecht, The Netherlands, 2001 [= *Proceedings of the Phytochemical Society of Europe vol. 48*]; Harry W. Paul, *Bacchic Medicine: Wine and Alcohol Therapies from Napoleon to the French Pardox* (Amsterdam, 2001). The therapeutic use of wine in ancient medicine was pervasive and has generated a vast amount of scholarship. A good introduction is found at Jacques Jouanna and Laurence Villard, edd., *Vin et Santé en Grèce ancienne. Actes du colloque organisé à l'Université de Rouen et à Paris (Université de Paris IV Sorbonne et ENS) 28-30 Septembre 1998* (Paris, 2002).

4 Arthur L. Klatsky, Gary D. Friedman, Mary Anne Armstrong, and Harald Kipp, "Wine, Liquor, Beer, and Mortality," *American Journal of Epidemiology* 158 (6) (2003): 585-95.

Unfortunately, the dramatic *in vitro* effects of the compound have not yet been replicated *in vivo*, apparently because the human gut is so inhospitable to the survival of the compound, and so the jury is still out on this one. But in general the antioxidant role of wine polyphenols, flavanoids, phenolics, flavanols, coumarins, and phytoalexins in reducing the risks of CHD and macular degeneration are now well attested.[5]

All strictly academic, of course, since the ancients scarcely knew of such diseases, much less their etiologies and treatments. And though we are beginning to discover that atherosclerosis may have been more common in antiquity than previously thought,[6] in an age when, we suspect, the average life expectancy was something less than thirty years, it is questionable how serious an issue CHD and macular degeneration will have been in any case for the majority of Romans, since these diseases are so closely associated with aging. But to reiterate: the role of wine as *food* in the preindustrial Mediterranean can hardly be overestimated. The single most critical element of the preindustrial diet is caloric intake, and wine provides calories in spades. But how much per person? The typical Roman subsistence farm, we are told, was two *iugera* (.52 ha = 1.3 acres), and such a farm, even granted the polyculture practiced by Roman farmers, could have easily produced some 2,650 liters (700 gallons) of wine per year. Given the average caloric potential of wine of 875 calories per liter, this amount would provide some 6,350 calories per day, or enough to provide ten people some 635 calories per day. Thus, regardless of how profitable a small vineyard may have been in antiquity, assuming a crop could be made with family labor and perhaps a few slaves, the wine produced would have been a vital subsistence food crop.

In the city of Rome, the food role of wine was, if anything, even more critical. The leading modern authority on the economics of Roman viticulture, André Tchernia, has estimated the average consumption of wine in Rome at some .8 liters (27 oz.) per adult per day, an estimate based on conservative premises about production and importation and perfectly consonant, Tchernia

5 David M. Goldberg, Joseph Yan and George J. Soleas, "Absorption of Three Wine-Related Polyphenols in Threee Different Matrices by Health Subjects," *Clinical Biochemistry* 36(1) (2003): 79-87; J. E. Kinsella, E. Frankel, J. B. German and J. Kanner, "Possible Mechanisms for the Protective Role of Antioxidants in Wine and Plant Foods," *Food Technology* 47 (1993): 85-9; Martin Bobak and Michael Marmot, "Wine and Heart Disease: A Statistical Approach," in Sandler and Pinder (2003): 108-39; Michael Aviram and Bianca Fuhrman, "Wine Flavanoids, LDL Cholesterol, Oxidation and Atherosclerosis," op. cit. (2003): 140-59; Thomas O. Obisesan, "Wine: Protective in Macular Degeneration," op. cit. (2003): 285-98.

6 Adel H. Allam et al., "Atherosclerosis in Ancient and Modern Egyptians," *Global Heart* 9.2 (June, 2012): 197-202.

demonstrates, with comparative evidence from Medieval Europe. Such an amount will have provided some 600 calories per adult per day, roughly one-third the minimum caloric requirement for a typical adult. Whatever else it was in Rome (and it was many other things, as we will see), wine was food.[7]

The success of the European grapevine as a food crop is tied to its physiology and evolution. For one thing, the vine is incredibly productive. Though the European wine grape, *Vitis vinifera vinifera*, takes several years to bear even a passable crop and several more to reach full productivity, thereafter it is remarkably hardy and productive for many years and, ironically, will have been even more so in antiquity, before the ravages of the phylloxera root louse, an accidental exotic from the New World, necessitated the grafting of *Vitis vinifera* scions to American rootstocks which were resistant to these pests. Grafted vines are intrinsically weaker than ungrafted vines. We moderns may suspect an ancient author like Pliny of carelessness or exaggeration when he classifies vines as trees (*NH* 14.9) and illustrates with some examples of prodigious vines. In Sicilian Panormus (Palermo), for example, was an archaic statue of Jupiter made from the wood of a single vine trunk; the Temple of Juno at Metapontum (Metaponto) was supported by columns of vine wood, and the Temple of Diana at Ephesus, we are told, had a staircase made from a single vine trunk. And in Rome itself a single vine shaded one portico of the Porticus of Livia and produced 12 amphorae (314 l. = 83 gal.) of grape must per year. There is no exaggeration here; these were ungrafted vines, obviously of great age. For comparison, Winkler[8] in his seminal work on viticulture has a dramatic picture of the so-called Carpenteria vine, an ungrafted vine planted in 1842 in California, which bore eight tons (7.28 metric tons) of fruit at fifty-one years of age and had at that time a trunk 2.8 m (9′) in circumference. Likewise, an ungrafted vine in Santa Barbara County, California, covered 1115 m² (12,000 sq. ft) and bore a crop of 4.5 metric tons (5 tons) at age 65. The Great Vine at Hampton Court Palace, London, was planted in 1769. The vine is some 1.2 m (4′) in diameter at its base and average yield after more than 240 years is some 270 kg (600 lbs) or grapes per year.[9] Vines on their own rootstocks may live to be 150 years or more and still bear fruit. Furthermore, the vine is hardy in most temperate regions of the northern and southern hemispheres, even in areas where arable

7 André Tchernia, *Le vin de l'Italie romaine. Essai d'histoire économique d'après les amphores* (Rome, 1986): 41-56.

8 A. J. Winkler, James A. Cook, W. M. Kliewer and Lloyd A. Lider, *General Viticulture* (Berkeley, 1974): Fig. 1.

9 Ian Hornsey, *The Chemistry and Biology of Winemaking* (Cambridge, UK, 2007): 73.

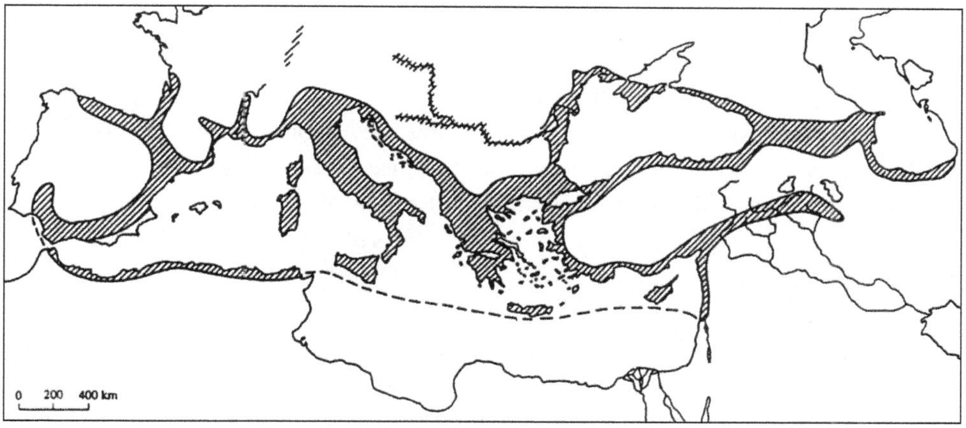

FIGURE 1 *Distribution range of the wild grapevine* Vitis vinifera sylvestris. *From Zohary in McGovern, Fleming and Katz (1996): Map 2.1.*

farming is impractical, and has enormous regenerative powers, even when severely pruned.[10]

The domestication of the vine is in fact yet another example of the co-evolution of early man and the so-called 'founder species'. Grapevines are from the genus *Vitis* in the family *Vitaceae*, whose other members include Boston ivy and Virginia creeper. *Vitaceae* species exhibit a climbing habit, alternate leaf nodes, tendrils which develop from these nodes, and flower clusters opposite the leaves. Flowers may be unisexual or bisexual (hermaphroditic). The ovary has two carpels partly enclosed by a receptacle, the latter of which becomes a two-compartment berry containing up to four seeds. The family is predominantly tropical and subtropical, though the *Vitis* genus of the family is primarily a temperate-zone genus occurring naturally only in the northern hemisphere. The genus is subdivided into *Vitis vinifera,* the progenitor of the European wine grape, and *Vitis muscadinia*, the New World grape. It is now thought that ancestral forms of *Vitis vinifera* were bushy and inhabited sunny locations, but that as humid forests expanded during the Eocene Epoch, the adoption of a climbing habit allowed *Vitis vinifera* to maintain its preference for sunny locations. The genus was well established throughout its range by the end of the Quartenary glacial period (c. 8000 BCE). Glacial periods influenced a southern retreat of the genus as well as a predilection for refuges around the Mediterranean and southern Black Sea and Caspian Sea.

10 Ron S. Jackson. *Wine Science: Principles, Practice, Perception,* 2nd. ed. (San Diego, 2000): 2-3.

Prehistory

The wild progenitor of all European wine grapes of the subspecies *Vitis vinifera vinifera* is *Vitis vinifera sylvestris*, whose preferred habitat was originally the humid forests south of the Black and Caspian Seas and adjacent Transcaucasia, along the fringes of cooler mesic forests of the northern Mediterranean and along the Danube, Rhine and Rhone river valleys of Europe.[11] Thence it spread across temperate Europe and Asia from the Atlantic coast to the Himalayas and southerly to southern Turkey in the east and the African littoral in the west. *Sylvestris* vines are characterized by small, acidic grapes and globular pips; like all other *Vitis* species except North American *Muscadinia* species, it has 2N = 38 chromosomes, and can therefore easily be crossed, these F_1 hybrids being vigorous and fertile. Thus development of new strains via seed propagation is possible, though not necessarily desirable, for reasons to be discussed later[12]

Heroic efforts are currently afoot to ascertain more exactly the time and place of the domestication of *Vitis vinifera vinifera,* i.e., when viticulture, the systematic propagation of the modern wine grape, began. Before we examine these issues, however, it is important to remember that wine can and almost certainly was made from grapes of *Sylvestris* vines and thus vinification, the humanly directed production of wine, indisputably preceded viticulture, perhaps by several millennia. As previously noted, wine in a sense makes itself when grapes are crushed and wild yeasts on the bloom of the berries and in the air are exposed to grape sugars. But in the presence of oxygen, acetic acid bacteria will inevitably proceed to metabolize ethanol into acetic acid and byproducts, and thus wine into a form of vinegar. Thus to speak of vinification in a strict sense we must find evidence not only of the deliberate crushing of grapes, but also of the storage of the resulting fermented must (grape juice) in a reasonably hermetic container.

Did Paleolithic man make wine? This so-called 'Paleolithic Hypothesis' is still precisely that and will likely remain so, but it seems reasonable. According to one scenario, foraging Paleolithic hunter-gatherers happened upon bright, attractive berries of the wild grapevine and, relishing the sweet-tart taste, gathered in a crude, watertight container of some sort, bunches of grapes to eat later. Some of these gathered berries were crushed in the process and the wild yeasts of the bloom established a colony which eventually fermented the

11 Jackson (2000): 13-14.

12 D. Zohary and M. Hopf, *Domestication of Plants in the Old World* (Oxford, 1988): 136-42; Hornsey (2007): 65-113.

remaining grapes to a low-alcohol wine. Intrigued by the taste and the mild psychotropic effect of the ethanol, our Paleolithic wine-bibber might well seek out more of the magical fruit and gradually discover empirically that releasing the juice of the grapes and exercising a bit of patience resulted in even more of the elixir. Alternately, Paleolithic man observed birds and small animals eagerly eating molded grapes; such molds today attack the skin of the grape, simultaneously releasing water from the berry and thus concentrating the sugars therein, as well as permitting access to this rich fermentable for the yeasts of the bloom. Imitating the predilection of his fellow foragers, early man discovered the delights of fermented grapes.

The chances of proving a Paleolithic Hypothesis are very slim, for the simple reason that Paleolithic containers were evidently made from perishable materials such as wood, basketry and leather, and therefore are not available to test for grape residues. Pottery containers date from the very end of the Paleolithic (c. 10,000 BCE) in East Africa and Japan. Stone vessels have been found from the Paleolithic period, though no grape residues have yet been analyzed. Most are open containers not intended to be stoppered. Thus if any did contain wine, it was wine meant to be consumed quickly, before the acetification began.[13]

For Neolithic winemaking, on the other hand, we have more and more evidence, particularly thanks to some of the exciting, emerging analytical branches of archaeology.[14] One such area is paleobotanical archaeology, that is, the examination of plant remains such as seeds, wood and pollen from archaeological contexts. For example, the pips of *Sylvestris* berries differ from those of the cultivated form in having rounder shapes with shorter 'beaks'.[15] Since pips are the primary botanical evidence retrieved in excavation, the relative proportions of length to width of pips was formerly considered reliable evidence of wild vs. cultivated berries. Two problems have emerged, however. First, there is tremendous variation in the proportions of both wild and domesticated seeds and a huge overlap in the middle of the spectrum. Secondly,

---

13 McGovern (2003): 7-11.

14 McGovern [(2003): 16-39] has an excellent, nuanced overview of the main avenus of research along with a balanced assessmnt of the potential promise and limitations of each. My discussion is largely owed to it. J. M. Renfrew ["Archaeology and the Origins of Wine Production," in Sandler and Pinder (2003): 56-69] has a brief but authoritative overview of the protohistory of wine, with emphasis on the gleanings of paleobotany, of which discipline she is one of the founders.

15 Daniel Zohary, "The Domestication of the Grapevine *Vitis Vinifera* L. in the Near East," in McGovern, Fleming and Katz (1996): 26-8 and Fig. 2d.

Smith and Jones[16] have demonstrated experimentally that charring deforms pips to make domesticated pips look more 'wild'. And most Neolithic pips available for analysis are extant precisely because they have been carbonized. Thus, only when a relatively large number of seeds is available for analysis can length-to-width ratios be used to suggest (but not prove) domestication.

For what it is worth, then, numerous finds have been made of grape pips from Neolithic contexts in Europe and the Near East, but almost all from areas that overlap the range of *Sylvestris* grapes and almost all with pips whose morphology conforms fully with wild forms. Thus these finds cannot be regarded as evidence of viticulture and should be regarded as evidence of collection of wild grapes. The first fully reliable paleobotanical evidence of domestication comes from Chalcolithic (ca 3700-3200 BCE) and Early Bronze Age (EBA) (ca. 3200-1900 BCE) sites in the Levant. Specifically, charred pips from Tell esh-Shana in the Jordan Valley (ca. 3200 BCE) as well as grapevine remains from EBA Arad and Lachish in Israel, and Numeria in Jordan, come from contexts which are relatively dry and well outside the range of native *Sylvestris*, thus providing strong circumstantial evidence of domesticated vines by the second half of the fourth millennium BCE in the southern Levant and by the end of the same millennium in the northern Levant. By the late stages of the Bronze Age, archaeobotanical evidence is increasingly common here as are remains of presses, storage vessels, and artistic representations of both viticulture and vinification.

A corollary issue is whether domestication may have occurred simultaneously elsewhere. Archaeobotanical evidence in the form of grape pips of the 'domesticated' type has been found in Neolithic sites in Georgia and Dagestan, but in insufficient numbers and in contexts which cannot prove domestication there. But chemical analysis of a sherd from an early sixth-millennium BCE jar from Shulaveris-Gora (Georgia) found evidence of resinated wine, suggesting to McGovern a more sophisticated viniculture than might be possible by reliance on foraged wild grapes.[17] Nuñez and Walker[18] proposed contemporary

16 H. Smith and G. Jones, "Experiments on the Effects of Charring on Cultivated Grape Seeds," *Jounal of Archaeological Science* 17 (1990): 317-27. An attempt has been made to refine Strummer's original index, but the method is still problematic. Refinement: M. Mangafa and K. Kotsakis, "A New Method for the Identification of Wild and Cultivated Charred Grape Seeds," *J. Arch. Sci.* 23 (1996): 409-18; limitations: Christiane Jacquat and Danièle Martinoli, "Vitis vinifera L.: Wild or Cultivated? Study of the Grape Pips Found at Petra, Jordan; 150 B.C.-A.D. 40," *Veget. Hist. Archaeobot.* 8 (1999): 25-30.

17 McGovern (2003): 37.

18 Diego Rivera Nuñez and Michael J. Walker, "A Review of Paleobotanical Findings of Early *Vitis* in the Mediterranean and of the Origins of Cultivated Grapevines, with Special

domestication of *Vitis vinifera* in the western Mediterranean, specifically in Spain, around 2500 BCE, based on paleobotanical evidence, especially seed morphology. A large pollen rain of *Vitis vinifera* at Laguna de las Madres, Spain, from ca. 3000 BCE was attributed by them to cultivation there; a small group of pips from El Prado de Jumilla had the characteristic 'domesticated' profile, and fragments of vine wood at El Prado and at Cueva del Monte de la Barsella, alongside dried grape skins, are said by them to point to domestication. Other scholars are skeptical. A localized pollen rain in a very small zone proves very little; further, the grape pips adduced as evidence are only a few among many with the 'wild' squatty profile and radiocarbon dating of six of the 'domesticated' pips showed that all were modern. New genetic typing strongly suggests an exclusively Near Eastern origin of the grape, since all vinifera populations in the study were found to be genetically closer to eastern *Sylvestris* than to western *Vinifera* subspecies. The study also supports introgression of imported *Vinifera* with local western *Sylvestris* as the grape was introduced into Western Europe some 2800 years ago.[19]

A second avenue of investigation, the so-called 'Noah Hypothesis', relies on ampelogaphy, the study of grape cultivars, as determined by DNA sampling of paleobotanical remains of both wild and domesticated species, on the analogy of the 'Eve Hypothesis' which purports to trace the human race back to a common mother in East Africa via analysis of mitochondrial DNA. The term 'Noah' is, of course, a clever reference to the Biblical character (Gen. 8:4 and 9:20) who escaped the Deluge and landed on Mt. Ararat, almost certainly modern Büyük Agri Dâgi in the Taurus Mountains of eastern Turkey, and proceeded to husbandry, including viticulture. The basic premise of the hypothesis is that the first center of cultivation of a given plant is the area of greatest genetic diversity.[20] According to this school of thought, the domesticated culture of *Vitis*

Reference to the New Pointers to Prehistoric Exploitation in the Western Mediterranean," *Review of Palaeobotany and Palynology* 61 (1989): 205-37. Cf. Tim Unwin, *Wine and the Vine: An Historical Geography of Viticulture and the Wine Trade* (London, 1991): 59; A. C. Stevenson. "Studies in the Vegetational History of S.W. Spain. Spain II. Palynological Investigations at Laguna de las Madres, S. W. Spain," *Journal of Biogeography* 12 (1985): 293-314.

19 Sean Myles *et al*, "Genetic Structure and the Domestication History of the Grape," *Proceedings of the National Academy of Sciences USA* (2010): 3.

20 A. M. Negrul, "Evolution of Cultivated Forms of Grapes," *Comptes Réndus (Doklady) de l'Academie des Sciences de l'URSS* 18 (1938): 585-88; *idem*, "Proles of Cultivated Grapevines and their Classification," *Ampelography: URSS*, vol. 1 (Moscow, 1946): 133-216; L. Levadoux, "Les populations sauvages et cultivée de *Vitis vinifera* L.," *Annales de l'Amelioration des Plantes* 1 (1956): 31-43 .

vinifera happened ca. 8,000-6,000 BCE in the Transcaucasus and spread, toward the end of the fifth millennium BCE, to the Aegean, Mesopotamia, Syria, and Egypt. But the methodology is fraught with difficulties. For one thing, while claiming DNA evidence as their mantra, Noah proponents tend to rely quite heavily on morphological differences of vines. That is perhaps inevitable since, however promising DNA analysis may ultimately be, it is still in its infancy. Add to that the limited sample of Neolithic DNA and the extraordinary difficulty if not impossibility of extracting usable DNA from ancient remains, plus the fact that 'wild' samples may be escapees, abandoned domesticated vines, or hybrids. Analogy with the Eve Hypothesis is clever but specious, since, as we will see, domesticated vines are propagated almost exclusively by cloning and not by sexual reproduction. The hypothesis gains weight in some circles from a presumed 'real' Great Deluge, namely the sudden infilling of the Black Sea Basin by the Mediterranean after the last ice age, a cataclysm which, it is theorized, forced the migration of the original viticulturalists and the concomitant spread of the technology. Unfortunately, despite the tirelessly self-promoting efforts of one oceanographer, no irrefutable evidence of any such cataclysm has yet been found, not to speak of the mass migration it is said to have precipitated.

Thus a definitive answer to the origins of viticulture is probably still decades away, but a general source region in the area of eastern Turkey (in the environs of Mt. Ararat, be it noted), in Armenia, Georgia, or Azerbaijan south of the Cauasus Mountains in the intermontane valleys—these areas have attracted the most scholarly interest. Noah proponents also point to linguistic research for the homeland of Proto-Indo-European (PIE), the presumed parent language of most of the languages of Europe and India as well as many of those of western Asia. This homeland, once thought to have been the steppes of southern Russia, is now thought to have been this same area of southern Caucasia. In this regard, two Georgian linguists, T. V. Gamkredidze and V. V. Ivanov[21] focused attention on the languages of the Caucasus Mountains, particularly southern Transcaucasia, relying in part on the pioneering work of Russian botanist Nikolai Vavitov who first proposed Transcaucasia as the homeland of *Vitis vinifera vinifera*, and cited a posited PIE word for wine, **woi-no* or **wei-no*, and its presumed connection to a number of IE and non-IE languages such as Hattic *windu*, Hittite **wijana*, Akkadian *înu*, Ugaritic *yu*, Proto-Semitic **wajnu*, Early Hebrew *yayin*, Linear B Greek *wo-no*, and Homeric Greek *woinos*. The Karvelian equivalent is **gwino*, still the spoken form in Georgia today.

21 W. Winter ed., *Indo-European and the Indo-Europeans: A Reconstruction and Historical Analysis of a Proto-Language and a Proto-Culture*, J. N. Nichols, trans, 2 vols (Berlin, 1995).

Gamkredidze and Ivanov proposed that the first speakers of PIE migrated from Transcaucasia and Eastern Turkey about 5,000 BE, heading eastward to Iran, south to Egypt, and west to the Balkans, taking with them their language and the principles of viticulture.[22]

However vexed may be the question of the origins of those principles in time and space, there is no doubt as to the fundamental principles that mark true viticulture. First is selection for hermaphroditic vines. To explain, primitive forms of Vitis from the Tertiary were hermaphroditic, that is, had 'perfect' flowers with both male and female reproductive organs, just as do the modern forms. But for reasons as yet unexplained, the vine became dioecious throughout its range, perhaps during the last Ice Age; flowers became sexually dimorphic, with some vines producing female flowers with developed pistils and ovules and others male flowers with anthers and stamens. Cross-pollination is more difficult in this scheme and thus male vines rarely produce fruit, and female fruit is genetically polymorphic and therefore extremely variable in taste. But in general, *Sylvestris* vines produce small, astringent, many-seeded berries, hardly the raw materials of great wine. Berries are black or dark red, never white. In contrast, the developed forms are large, extremely high in sugar and low in tannins and numbers of seeds, and with variations in form and flavor. Today it is estimated that there are some 10,000 clonal varieties of *Vitis vinifera vinifera*.[23]

Fortunately for early viticulturalists, male vines, planted among female vines for cross-pollination, are sporadically hermaphroditic. Some of their seedlings are hermaphroditic by type, and these are characteristically greater fruit producers with greater numbers and larger sizes of clusters. Thus, simply by selecting for greater productivity, early viticulturalists inadvertently selected for hermaphroditism as well, and exclusively male and female vines were culled.[24]

But if seeds of these vines are used for propagation they exhibit the same sexual dimorphy and lower productivity as the archetypes, and so when seedlings are used the same process of selection must be effected in the next generation. The only way to maintain the desirable traits of the parent vine, in other words, is by clonal propagation, and here again nature may have pointed

22 For those who wish to tread this path, Emilio Sereni ["Per la storia delle più antiche tecniche e della nomenclatura della vite e del vino in Italia," *Atti e Memorie dell'Academia toscana di scienze e letture La Columbaria* 29 (1964): 75-204] has a good review of the linguistic evidence for vitiviniculture in the Mediterranean, though now somewhat dated.

23 McGovern (2003): 11-15.

24 Olmo in McGovern, Fleming and Katz (1996): 43; McGovern (2003): 13-15.

FIGURE 2 *Morphology of the vine: 1) saggita (shoot end), 2) oculus (bud), 3) palma (shoot), 4) uva (cluster), 5) folium (leaf), 6) uva/baca (berry), 7) resex (spur), 8) truncus (trunk), 9) radices (roots).* DRAWING BY THE AUTHOR.

the way. Because it has the habit of vining along the ground if no support is available, vine shoots are not uncommonly covered with dirt by erosion, and these junctures will spontaneously root, a process called layering. Eventually these volunteers are completely independent of the mother vine, yet share all her genetic traits. Obviously, layering was (and is) a perfectly good way to propagate new vines. But eventually some bright spark in the vineyard asked himself the inevitable question, deliberately made cuttings and rooted them and thus discovered that he could mimic nature in a far more efficient way. Once the basic principles of selection for hermaphroditism and clonal propagation were mastered, the slow, empirical process of selection for specific traits such as size of fruit, diminution in numbers of pips, taste, climatic adaptation, and so forth, could begin.

About Neolithic vinification there has been a veritable flood of new evidence in the last thirty years, particularly from the new discipline of molecular archaeology championed by the University of Pennsylvania Museum Applied Science Center for Archaeology (MASCA) and most associated with Patrick McGovern, former Senior Research Scientist in that group. For at least two centuries archaeologists have discovered residues in pottery containers which were thought to be wine residues, and some crude analytical techniques have been used to test some of them. But molecular archaeology took a quantum leap forward in 1988 when residue from jars at the ancient site of Godin Tepe in central western Iran, in the foothills of the Zagros Mountains, were subjected to modern analysis.[25] Godin Tepe was a trade outpost for the lowland Mesopotamian cultures beginning in the fourth millennium BCE. It was the largest site in the area during the period and occupied a strategic position along the Great Khorasan Road, part of the Silk Road. First evidence there of wine and perhaps of winemaking appears in the central citadel of Late Period V (ca. 3,100-2,900 BCE). 'Wine' jars contained a residue of grape juice, as demonstrated by infrared spectrometry of the residue; along with this, the shape and refinements of the jars strongly suggest that the grape juice was in the form of wine. First, a prominent feature of the jars was an appliqué in the

25 Virginia Badler, "The Archaeological Evidence for Winemaking, Distribution and Consumption at Protohistoric Godin Tepe, Iran," in McGovern, Fleming and Katz (1996): 45-56; Patrick E. McGovern, Donald L. Glusker, Lawrence J. Exner, and Mary M. Voigt, "Neolithic Resinated Wine," *Nature* 381.6 (1996): 480-81; Patrick E. McGovern and Rudolph H. Michel, "The Analytical and Archaeological Challenge of Detecting Ancient Wine: Two Case Studies from the Ancient Near East," in McGovern, Fleming and Katz (1996): 57-65; P. E. McGovern, U. Hartung, V. R. Badler, D. L. Glusker and L. J. Exner, "The Beginnings of Winemaking and Viniculture in the Ancient Near East and Egypt," *Expedition* 39 (1) (1997): 3-21; McGovern (2003): 40-63.

shape of a rope across the shoulder and down parallel sides of the jars. Since we know from comparative evidence that such appliqués designated the location of real rope chocks used to stabilize jars in storage, it is obvious that these jars were stored on their sides, just as modern wine bottles, and doubtless for the same reason, to keep the stopper moist and thus prevent cracks in the seal and the consequent invasion of air and acetic bacteria. For the same reason as well, the necks of these jars were quite narrow. Finally, a hole drilled through the side of one vessel after it was fired, about 9.5 cm (4″) above the base, has been interpreted as a means of decanting the wine without broaching the seal or disturbing the lees on the bottom of the vessel once it had been stood upright for several days to allow the lees to settle to the bottom, just as is done with fine aged wines today.

The principle evidence of vinification on site, as opposed to importation of the wine from elsewhere or perhaps decantation here from imported vessels, is a funnel which Badler interprets as a device used along with a textile filter to press juice from grapes. Several jars of the wine type were found in the same room as the funnel, and these are interpreted as empties waiting to be filled. But evidence points to other activities in the room (e.g., large flint cores and a metal chisel), nor have any paleobotanical remains of grapes been found. Thus it is perhaps more logical to see the funnel as precisely that, a funnel used to decant wine from larger bulk vessels such as animal skins into the smaller, more easily shipped or distributed wine jars. But obviously wine was being made somewhere in the vicinity, and at a commercial level difficult if not impossible to sustain using only wild foraged grapes.

Molecular evidence from other sites has emerged. Pottery sherds from the earliest Neolithic levels at Shulaveris-Goa in Caucasia, now housed in the Georgian National Museum at Tblisi, a site which has yielded some of the earliest grape pips of the domesticated profile, were subjected to analysis and were shown to have contained resinated wine. Thus, wine appears in this area from ca. 6,000 BCE. In Anatolia, arachaeobotanical evidence of both wild and domesticated vines is now abundant from sites along the upper Tigris and Euphrates rivers at Korucutepe, Tepecik, Kurban Höyuk, Hacineb, Çayönü, and the Urfa region, ranging from the Aceramic Neolithic (ca. 8,000 BCE) to the Late Chalcolithic (ca. 3,500 BCE). Grape remains gradually increase in frequency and concentration, and by the beginning of the third millennium BCE both viticulture and vinification are well established. Hittite Anatolia yields early wild grape pips from Can Hasan from ca. 7,200-6,500 BCE, and at Cayönü from 9,000 BCE, but McGovern interprets these as a source of fermentative yeasts for fermentation of wheat and barley beers as well as fruits such as hackberries. No other fruits have dependable yeast colonies on their surfaces as do grapes.[26]

To summarize: wine was an enormously popular drink in the Roman world of the classical period for a number of reasons, most notably, I would argue, because it was a relatively cheap, biologically stable and perennially available source of carbohydrates, the high-energy component so critical in the human diet. To wine's food value we should add its value as a medicine, both therapeutic and prophylactic, in both the somatic and psychological spheres. The fact that it can taste delicious and has delightful psychotropic effects when drunk in moderation cannot have hurt.

Wine on a commercial scale implies vine, i.e., vinification ultimately implies viticulture, and the vine as we now have it is the product of a long evolution followed by a short but intensively productive co-evolution with one *Homo sapiens sapiens.* The vine is a member of a large and widely distributed family of bushy plants that developed a vining habit during the Eocene; by the end of the last Ice Age the progenitor of all modern wine grapes, *Vitis vinifera sylvestris,* was scattered across the temperate regions of Europe and western Asia. The vine had become dioecious and had small clusters of small, tangy berries with numerous seeds.

It is entirely likely that 'wine-drinking' per se extends well back into the Paleolithic; some think that Paleolithic man was consuming spontaneously fermented grape juice as far back as 80,000 years ago. It is impossible, given the current state of evidence, to say to what extent Paleolithic man had discovered the rudiments of vinification, but it is not unlikely he had done so. Nor can we say with much precision where and when Neolithic man began to make wine, though a preponderance of evidence points to the Transcaucasus and eastern Turkey at the very beginning of the Neolithic. We can say with assurance that by the Chalcolithic Age (3,500 BCE) wine and viticulture had found its way from this home range to Mesopotamia, Anatolia, the Levant, and perhaps even further afield. A large and increasing body of evidence points to the same area as the homeland of viticulture as well, though there is also tenuous evidence that viticulture may have developed independently in the western Mediterranean, specifically in southwest Spain. The two most momentous steps in the process of domestication of the vine were selection for hemaphrodism and discovery of clonal propagation. Again, by the Chalcolithic there is indisputable evidence of domestication in the eastern range of *Sylvestris*, perhaps a bit later in the west. But at this stage both viniculture and viticulture step onto the stage of history.

26 McGovern (2003): 78-84.

Protohistory

It can be no accident that the regions most likely as the homelands of both vinification and viticulture are located around the periphery of one of the great early centers of civilization, Mesopotamia. Yet another reason for the spectacular success of wine is the fact that it is a valuable commodity which is relatively easy to ship. It is in this role that wine enters the realm of history.[27] Interpretation of residues in vessels at Haji Ferez Tepe, Godin Tepe, and other entrepôts in the foothills of the Zagros Mountains is certainly consonant with the general picture that emerges of trade in this region. During the last quarter of the fourth millennium BCE, the so-called Late Uruk Period in Mesopotamia, the Mesopotamian alluvium underwent dramatic development in urbanization, communication, and material culture. Algaze has demonstrated that this development was largely dependent on the exchange of agricultural products from the lowlands for key commodities required in lowland cities, especially metals and wood, as well as the prestige goods the urbanites demanded, including, we may now say with assurance, wine. The specific vectors for this exchange were a series of entrepôts located along key trade routes around the upland periphery of the alluvium, especially the foothills of the Zagros and Taurus mountains. One such entrepôt was Arslan Tepe in the Malatya region of the southeast Anatolian highlands, where Uruk spouted bottles of a characteristic form have been found. Such vessels were used to ship valuable liquids, especially wine and oil. Analysis of residues from some of these pots has confirmed that wine was indeed one of the commodities, wine which Algaze conjectures may have been repackaged at Arslan Tepe and shipped downstream to the alluvium as well as across the north.

For viticulture and viniculture in the alluvium itself the evidence is not as definitive. Fruit cultivation in the region dates from the early third millennium BCE and includes dates, apples, figs and grapes. Only the date is indigenous, so the others must have ben imported and naturalized in the alluvium. Wine continues to be a rare and expensive commodity here, however, from the third to

27 Guillermo Algaze, "Fourth-Millennium BC Trade in Greater Mesopotamia: Did It Include Wine?," in McGovern, Fleming and Katz (1996): 89-96. Cf. Julian Edgeworth Reade, "The Symposium in Ancient Mesopotamia: Archaeological Evidence," in Oswyn Murray and Manuela Tecuşan, *In Vino Veritas* (London, 1995): 35-56; Marvin A. Powell, "Wine and the Vine in Ancient Mesopotamia: The Cuneiform Evidence," in McGovern, Fleming and Katz (1996): 97-122; Richard L. Zettler and Naomi F. Miller, "Searching for Wine with the Archaeological Record of Ancient Mesopotamia of the Third and Second Millennium BC," in McGovern, Fleming and Katz (1996): 123-31; McGovern, Glusker, Exner and Voigt (1996): 480-1; McGovern (2003): 18-73.

the first millennium. Thus the Sumerian word for 'grapevine' appears in the third millennium, but as late as 2,000 BCE there is no evidence, linguistic or otherwise, of vinification in Babylon itself. Indeed, wine remains an expensive prestige import available only to the rich. Nevertheless, viticulture had a small but significant place in the Babylonian economy. Linguistic evidence makes it clear that grapes were grown as a source for grape syrups and raisins, primarily the latter. Vines were grown in gardens, probably in raised beds to improve drainage and avoid the buildup of mineral salts so prevalent in alluvial soil. There is no evidence that vines were trained to trees, though the practice appears in upper Mesopotamia in the first millennium BCE.

Farther east, the proto-Elamite cities of Shiraz, Tepe Malyan, Susa, and Persepolis are located in the southern Zagros Mountains at elevations amenable to viticulture, and we now know there was a native vine culture there, though evidence of a wine industry remains elusive. Tepe Malyan, ancient Anshan, provides evidence of viticulture in the form of numerous grape pips and actual pieces of grapevine wood dating to the middle of the third millennium BCE. Masses of carbonized and uncarbonized grape pips are best interpreted as remains of grape pomace, the solid remains of pressed grapes, dependable evidence of vinification. This was the period of Elamite hegemony, and we know from cylinder seals and iconographic evidence of the centrality of wine in Elamite culture.[28]

In the north, the Assyrian culture (ca. 2,000-1700 BCE) played a pivotal role in the spread of wine as a commodity, as well as of viticulture itself.[29] Linguistic evidence stretches from ca. 2,200 BCE to the fall of Nineveh in 612 BCE. There is no evidence for either viticulture or viniculture in eastern Assyria in the Kirkuk region, but between the Khabur and Tigris rivers, wine is attested from the eighteenth century, and the name of the region itself may be connected to wine or the wine goddess. It is also here that the appearance of a word for the wineskin occurs, some nine centuries before previously attested. There is also linguistic evidence for a significant wine exportation to Anatolia. From both linguistic and iconographic evidence it is clear that by the ninth century the drinking, possession, and disbursement of wine were powerful symbols of royal authority in both the religious and political realms. Later, the eyewitness account of the Greek historian Xenophon (*Anab.* 1.4.19; 1.5.10) concerning availability of wine in Mesopotamia in 401 BCE agrees remarkably well with

28 McGovern (2003): 165-6.

29 Powell in McGovern, Fleming and Katz (1996): 114-22; David Stronach, "The Imagery of the Wine Bowl: Wine in Assyria in the Early First Millennium BC," in McGovern, Fleming and Katz (1996): 175-95; McGovern (2003): 180-201.

linguistic and archaeological evidence that wine was not only available but relatively plentiful in the area of ancient Assyria but not available further south. In a second account of a march down the Euphrates, wine is plentiful around the Khabur-Euphrates confluence, but is not mentioned thereafter.

The earliest native reference (as opposed to export records from Assyria) to wines of Anatolia date from early second millennium BCE Kültepe in Assyrian Anatolia. Beginning with the Hittite Era (ca. 1600-1200 BCE), written documents from Boğazköy detail the importance of local viticulture. On the other hand, it must be admitted that there is a remarkable lack of archaeological evidence for viticulture and viniculture in the region at this time. While paleobotanical evidence includes both grape pips and peduncles dated to 7200 BCE, the earliest such evidence for the domesticated form derives from the third millennium BCE. Further, not a single grape treading facility or wine press has yet been uncovered. We now know from analysis of residues that at least one of the so-called 'bathtubs' found at Old Assyrian sites functioned in part as a treading vat, and such 'bathtubs' are present in later Hittite contexts as well. Furthermore, almost surely some of the large 'pithoi' found in storage facilities at Boğazköy were used for wine storage, but there are no reports of pips or skins from these jars and no residues have yet been analyzed. Nor have we had much success in locating the source of Anatolian viticulture, though the Biblical tradition, as we have seen, locates it in eastern Anatolia near Mt. Ararat. It is tempting to posit a westward drift in the technologies of viticulture from Transcaucasia to the Assyrian heartland, to Assyrian outliers in eastern Anatolia, and thence to the center and west; ironically, the preponderance of early literary and archaeological evidence derives from the central plateau, where elevation, erratic rainfall, and harsh climatic conditions were (and continue to be) problematic for grape culture. But it appears evident that the Hittites played an important if not key role in the dissemination of the vine and wine culture to other parts of the Near East, especially the Levant.[30]

Elsewhere, the reputation of Cypriote wines was excellent throughout ancient history, and Cyprus seems to have been an early center for both production and distribution of wine. Archaeobotanical remains attest the presence of the vine on the island from the Neolithic, and there is evidence of full-scale production from the Calcolithic (fourth and third millennia BCE). From this era we have an *askos* in the form of a grape cluster from a tomb in

---

30 Ronald L. Gorny, "Viniculture and Ancient Anatolia," in McGovern, Fleming and Katz (1996): 133-74; Sedat Alp, *Song, Music and Dance of the Hittites: Grapes and Wine in Anatolia during the Hittite Period*, trans. Y. Eran. (Ankara, 2000); McGovern (2003): 174-88.

Souskion (Paphos), dating to the third millennium, and most importantly, a ritual vase in ceramic which represents on one side a human figure treading grapes in a rectangular treading vat while bracing himself with both hands by holding supports which are in fact the two *ansae* of the vase. At the front of the raised treading vat is a pipe, below which another figure waits, having placed a bowl beneath the spout. The vase, dating to c. 2000 BCE and discovered in a tomb in the village of Pyrgos (Limassol), is the oldest known representation of the treading process.[31]

Both Mesopotamia and Egypt imported large quantities of wine from the Levant beginning ca. 2500 BCE, so it is evident that both viticulture and viniculture had been established there earlier; the first paleobotanical evidence in the form of carbonized grape pips derives from ca. 3000 BCE. Analysis of several of the 700 or so jars found in the Dynasty 0 (so-called because it preceded the canonical Dynasty 1 rulers) tomb of Scorpion I, dating to 3150 BCE and located in upper Egypt at Abydos, revealed that the jars contained a resinated wine. Further analysis using Instrumental Neutron Activation Analysis (INAA) of the clay fabric of the jars points to the southern Levant as the source of the jars.

By the beginning of the first millennium BCE the Phoenicians had established themselves as a shipping power, and wines from Lebanon were exported extensively. And when Carthage was established by Phoenicians fleeing Tyre at its capture by the Assyrians in 814 BCE, vines were already being grown in this part of North Africa. Viticulture was doubtless transplanted to Spain by the Carthaginian Phoenicians as they colonized from North Africa.[32] Unlike their Phoenician ancestors, the aristocracy of Carthage derived their wealth from agricultural products, notably wine. Carthaginians established trading posts not only in Spain but in eastern Sicily, in Sardinia, and on Motya and other Mediterranean islands.[33]

---

31 Pavlos Floretzos, "La production de vin chypriote à l'époque préhistorique," in Jean-Pierre Brun, Matthieu Poux and André Tchernia, edd., *Le vin. Nectar des dieux, génie des hommes* (Gollion, Switzerland, 2009): 38-9 and Fig. 36.

32 Unwin (1991): 73-7; Carey Ellen Walsh, *Fruit of the Vine: Viticulture in Ancient Israel* [= *Harvard Semitic Monographs* 60] (Cambridge, MA, 2000): 11-41; Albert Leonard, Jr., "'Canaanite Jars' and the Late Bronze Age Aegeo-Levantine Wine Trade," in McGovern, Fleming and Katz (1996): 233-54; Joseph A. Greene, "The Beginnings of Grape Cultivation and Wine Production in Phoenician North Africa, " in McGovern, Fleming and Katz (1996): 311-22; McGovern (2003): 91-106; 201-38. Wine in Ugarit: Michael Heltzer, "Vineyard and Wine in Ugarit (Property and Distribution)" *Ugarit Forschungen* 22 (1990): 119-35.

33 David Abulafia, *The Great Sea: A Human History of the Mediterranean* (Oxford, 2011): 74-82.

As mentioned, Levantine wine was exported to Egypt from early times, but there is incontrovertible evidence of both viticulture and viniculture in Egypt itself from the beginning of the First Dynasty of the Old Kingdom (ca. 3000 BCE).[34] Archaeobotanical evidence includes grape pips from First Dynasty Abydos and Nagada and raisins from Third Dynasty (ca. 2600 BCE) Saqqara. Literary and iconographic evidence includes hieroglyphics and artistic representations of wine presses from Old Kingdom tombs as well as those of viticulture and vinification. And we have wine storage jars by the thousands, complete with vintage labels which include the vintage year, names of winemakers, and descriptions of styles and qualities. Based on this evidence there can be no doubt that both viticulture and viniculture had reached a high level of sophistication already by the early kingdoms. Based on the names of some of the earliest winemakers, it is likely that Old Kingdom monarchs began importing Levantine vintners right along with their wines, in order to establish a local source of the expensive elixir. By the time of the New Kingdoms (ca. 1550 BCE), wine drinking was still not practiced at all levels of society but was far more common than previously.

In the Aegean, there is suggestive evidence of viticulture from Late Neolithic (ca. 4300-2800 BCE) Thessaly and Macedonia, specifically in the form of grape pips of the 'domesticated' form, but the sample is extraordinarily small (32 seeds).[35] The presence of resinated wine is now confirmed analytically for

34 Nicely summarized with extensive bibliography by Mary Anne Murray, Neil Boulton and Carl Heron, "Viticulture and Wine Production," in P. T. Nicholson and I. Shaw, edd., *Ancient Egyptian Materials and Technology* (Cambridge, 2000): 577-608. Cf. Gabriella Scandone Matthiae, "Il Vino nell'Antico Egitto," in Oswyn Murray and Manuela Tecuşan, *In Vino Veritas* (London, 1995): 57-61. Newer analytical techniques in Patrick E. McGovern, Armen Mirzoian and Gretchen R. Hall, "Ancient Greek Herbal Wines," *Proceedings of the National Academy of Science USA* 106 (2009): 7361-66. Cf. H. F. Lutz, *Viticulture and Brewing in the Ancient Orient* (Leipzig, 1922): 46-51; Leonard H. Lesko, *King Tut's Wine Cellar* (Berkeley, 1977): *passim*; Zohary in McGovern, Fleming and Katz (1996): 28-9; T. G. H. James, "The Earliest History of Wine and Its Importance in Ancient Egypt," in McGovern, Fleming and Katz (1996): 197-213; McGovern (2003): 85-106. For Middle and New Kingdoms: Lesko, "Egyptian Wine Production during the New Kingdom," in McGovern, Fleming and Katz (1996): 215-30; McGovern (2003): 105-47.

35 Jane M. Renfrew, "Paleobotanical Finds of *Vitis* from Greece," in McGovern, Fleming and Katz (1996): 255-67; *idem*, *Paleobotany: The Prehistoric Food Plants of the Near East and Europe* (London, 1973); *idem*, "Agriculture," in D. Thocras, ed., *Neolithic Greece* (Athens, 1973): 147-64; Colin Renfrew, *The Emergence of Civilization: The Cyclades and the Aegean in the Third Millennium B.C.* (London, 1972): 278 and Table 5.1; Zohary in McGovern, Fleming and Katz (1996): 28-9; McGovern (2003): 239-78.

pithoi from the Early Bronze Age I (ca. 2200 BCE) palace at Myrtos-Phournon Koryphe, Crete, along with grape stems and pips. The site goes back to the beginning of the third millennium, and wine culture may perhaps be inferred here this far back. In the EBA I context, there is considerable evidence of wine production in the form of *pithoi*, a funnel, and 'bathtub' treading vats. Pressrooms are first attested on Crete at Kato Zakros from Late Minoan I, ca. 1500 BCE, after which they become widespread. By the Late Bronze Age (ca. 1550-1150 BCE) both viticulture and viniculture are well established all over the Aegean, and wine has also become an important article of international trade, witness the 'Canaanite Jars' used for this purpose among others and found all over the Aegean and Eastern Mediterranean littoral.[36]

Already by the Archaic period if not before, Greeks had enthusiastically adopted banquet culture, and it was destined to find its fullest expression in the Golden Age *symposion*.[37] By the Classical period wine was practically ubiquitous in Greece and, as we shall see, prestige wines, not to speak of wine technology and wine culture, were exported extensively to Italy. By the Classical period, the influence of Greek wine culture and technology in all parts of Italy was profound, pervasive, and persistent, and even a thumbnail sketch of that influence here would be totally inadequate. In the event we will have occasion to call attention to that influence as we examine the specifics of wine culture in Italy. Those influences were mediated to the Romans by Magna Graecia in the south of Italy and Etruria in the north, and it is to those areas that we turn in the next chapter.

---

36 Albert Leonard, Jr., "'Canaanite Jars' and the Late Bronze Age Aegeo-Levantine Wine Trade," in McGovern, Fleming and Katz (1996): 233-54. For mixed grogs, Stephanie Pain, "Grog of the Greeks," *New Scientist* 2214 (1999): 54-7 and McGovern, Mirzoian and Hall (2009): 7361-66. For the diffusion of amphorae in ancient Greek trade, Carolyn G. Koehler, "Wine Amphoras in Ancient Greek Trade," in McGovern, Fleming and Katz (1996): 323-37; Virginia R. Grace, *Amphoras and the Ancient Wine Trade* [= *Picture Book 6, Excavations of the Athenian Agora*] (Princeton, 1979).

37 Again, a vast bibliography is available, but two excellent syntheses are François Lissarrague, *The Aesthetics of the Greek Banquet: Images of Wine and Ritual* (Princeton, NJ, 1990) and Marek Węcowski, *The Rise of the Greek Aristocratic Banquet* (Oxford, 2014).

CHAPTER 2

Viniculture in Italy

All of the intriguing possibilities and infuriating ambiguities play out in Italy as well. Thus, Italy is well within the natural range of the wild vine, so there is no reason to assume that viticulture or viniculture were necessarily introduced to Italy at all. The traditional view saw cultural developments in Italy as the direct result of a series of invasions (culture drift), culminating in that of the Italic-speaking 'warrior-hunters' of the early Bronze Age. Our view is now much more nuanced.[1] For one thing, many of the cultural developments in Italy attributed to invaders can more plausibly be explained as organic and internal developments. Secondly, immigration of ideas does not necessarily imply immigration of peoples. Third, differences in settlement patterns and modes of subsistence are largely dictated by environmental factors such as the necessity for transhumance for hunting and stock-rearing in Mediterranean Italy, rather than by ethnic divisions. Indeed, the same ethnic groups may have had two or even three separate seasonal settlements in a given year. Finally, there is mounting if not definitive evidence that the indigenous wild grape was discovered in the Mesolithic in Italy and was maintained, at least at a casual level, until the late Bronze Age saw developed viticulture there. The presence of grape pips in human contexts is attested sporadically from this Mesolithic,[2] and there is an ever-increasing sample from the Neolithic.[3] All are morphologically firmly in the 'wild' category. But doubtless the most exciting evidence comes from the Neolithic village of 'La Marmotta', now submerged at the bottom of Lake Bracciano about thirty kilometers north of Rome. Not only does evidence suggest that this was a colony established from the Neolithic Near Eastern or Aegean world, but some of the evidence suggests to the longtime

---

1 Graeme Barker [*Landscape and Society: Prehistoric Central Italy* (London, 1981)] has the best discussion.+

2 L. Costantini, "Semi e carboni del mesolitico e neolitico della Grotta dell'Uzzo, Trapani," *Quartenaria* 23 (1981): 233-47. Gaetano Di Pasqulae and Elda Russo Ermolì ["L'ambiente della vite nell'area tirrenica e le evidenze della sua coltivazione in epoca etrusco," in Giovanni Di Pasquale, ed., *Vinum Nostrum. Arte, scienza e miti del vino nelle civiltà del Mediterraneo antico* (Florence, 2010): 62-5] have a good review of the archaeobotanical evidence.

3 Filippo Delpino, "Viticoltura, produzione e consumo del vino nell'Etruria protostorica," in Andrea Ciacci, Paola Rendini and Andrea Zifferero, edd., *Archeologia della vite e del vino in Etruria* (Siena, 2007): 134.

 | DOI 10.1163/9789004334595_003

director of this excavation, Maria Antonietta Fugazzola Delpino[4] that wine was being made at the site from local wild grapes. Since the village dates from c. 5700 BCE, this evidence of Neolithic winemaking may actually predate Hajji Firez Tepe. Elsewhere in Tuscany, at Gran Carro near Lago di Bolsena and at Cures, carbonized seeds have been interpreted as proof of cultivation of the fruit, and thus developed viticulture, here from the Neolithic.[5]

There is another form of archaeological evidence which is, in my opinion, highly relevant here. All over the Mediterranean, beginning in the Levant in the second millennium but spreading quickly westward, are rock-cut installations for the processing of grapes. They occur in all parts of Italy, including the large islands, but on the mainland they tend to cluster along the Apennine ranges in the natural range of the grape, from Tuscany to Calabria. In the north they are today most commonly called *pestarole*, in the south *palmenti*.

A bit of clarification is necessary. In the first place the more common term, *palmento*, is used in various senses; in various locales and times the term *palmento* may refer to a vat or grouping of vats either cut into native stone or constructed of masonry, or for the pressroom where such vats are located. In Sicily, for example, the term is used generically for any rustic winery. But in the strict sense used here it refers to vats excavated in native rock, what we might designate as *palmenti rupestri*. The typical form has a larger, usually higher vat, used for treading the grapes, which communicates by means of a channel cut into the party wall with a smaller, usually lower vat. The term itself has been variously explained, e.g., from L. *palmit-*, 'vine shoot', but clearly it derives immediately from Latin *pavimentum*, 'pavement' and ultimately from L. *pavire*, 'to pound'.[6] Functionally *palmenti* have been described by various authors as vats for dying or cleaning wool, for the tanning of hides, for the extraction of dyes from plants, for the retting of flax or hemp, even for the fermentation of green manures. There is no reason such vats cannot have been used in these ways, at least incidentally or subsequently, but it is quite clear that their original purpose was always for the processing of grapes. In fact, in some of the

4 Cf, e.g. Maria Antonietta Fugazzola Delpino, Andrea Pessina and Vincenzo Tiné, edd., *Civiltà dell'argilla. Le prime communità del Neolitico* (Dragoni, CE, 2004).

5 L. Costantini and L. Coastantini Biasini, "Cures Sabini-paletnobotanica. Nota preliminare" *Archeologia Laziale* 7.1 (1985): 86-90; F. Delpino, "L'ellenizzazione dell'Etruria villanoviana: sui rapporti tra Grecia e Etruria fra il IX e VIII secolo a.C.," in *Atti del 2° Convegno Internazionale Etrusco* (Rome, 1985): 105-16.

6 Cf. CIL 6.219: *paimentum*.

FIGURE 3 Palmento *at Novi Velia.* PHOTO COURTESY OF PROF. FERNANDO LA GRECA.

FIGURE 4 *Upper and lower vats of the* palmento. PHOTO COURTESY OF PROF. FERNANDO LA GRECA.

remoter parts of Italy there has been a continuous or nearly continuous history of their use for this purpose right down until the 1960s.[7]

Thus the function can be recovered from actual practice: the upper vat, usually square, rectangular or roughly circular, is filled with water which is allowed to remain there for several hours or overnight in order to make the vat more impermeable. Then the water is drained and the harvested grapes are dumped into the vat and trodden with naked feet to extract the must. This must is allowed to remain on the grape solids, the pomace, for several hours or overnight to extract color, aroma and flavor compounds and to clarify as gross lees settle to the bottom of the liquid. The interconnecting channel, several centimeters above the floor of the upper vat, has been plugged with clay and this is now removed to allow the clarified must, now probably rapidly fermenting, to drain into the lower vat where it is allowed to further clarify before being decanted into fermentation vessels. The pomace in the upper vat, meanwhile, may be pressed to extract residual must; many of the attested *palmenti* have cuttings in the surrounding rock which can be interpreted as those for the framework of a press, apparently either a simple lever press or a lever and screw press or a lever and screw with counterweight (cf Ch. 4, 'Presses'). But many do not, suggesting either a simpler kind of press such as a bag press or no pressing at all. It is important to note that some 80% of must can be extracted from grapes by thorough treading and that pomace that is not totally exhausted is excellent fodder for farm animals. In Calabria[8] two other refinements appear. First, smaller perimeter holes suggest the fittings for some sort of simple roof, and second, there are sometimes vertical channels cut into the lateral

---

7 The bibliography is meager but growing. In general, see Aniello Botti, David L. Thurmond and Fernando La Greca, "Un Palmento ben conservato a Novi Velia ed altri palmenti nel territorio del Cilento. Osservazioni ed ipotesi," *Annali Storici di Principato Citrà* 9.2 (2012): 5-52; and A. Masi, "Un esempio di archeologia dell'agricoltura: i palmenti," in Andrea Ciacci and Andrea Zifferero edd., *Vinum: un progetto per il riconoscimento della vite silvestre nel paesaggio archeologico della Toscana e de Lazio settentrionale* (Siena, 2005): 83-95. Best documentation for those in other parts of the Mediterranean will be found in L. Quilici, "Opifici rupestri dell'Italia centrale in età antica e medioevale," *Atti del XXI Convegno di Studi Maceratesi (Matelica, 1985)* (Macerata, 1988): 41-66 and *idem*, "Segni del paesaggio agrario nell'Etruria rupestre. Impianti per la viticoltura," in Marina Martelli, ed., *Tyrrhenoi Philotechnoi* (Rome, 1994): 183-93. For ancient Etruria and northern Latium, Quilici but also A. Fatucchi, "Palmenti romani da uva dell'Etruria nord-orientale," *Rivista di Storia dell'Agricoltura*27.2 (1987): 9-29. For Campania, Botti, Thurmond and La Greca. For Calabria, O. Sculli, *I palmenti di Ferruzzano.Archeologia del vino e testimonianze di cultura materiale in un territorio della Calabria meridionale* (Florence, 2002). For Sicily, S. Puglisi, *La valle dei palmenti. Archeologia vitivinicola e rupestre in Sicilia* (Messina, 1990).

8 Sculli (2002): 57-64.

walls of the upper vat, usually about halfway along its length. Into these channels is inserted a board (Sculli uses the dialect term *consù*) whose width is equal to the depth of the vat and in which numerous holes have been cut. In this case, after the grapes have been trodden, the *consù* is inserted and pomace placed in the part of the vat farther from the channel, a horizontal board is placed on top of the pomace enclosed in the box created by the *consù* and above this 'follower' spacers allow the press beam to gain purchase. The residual must is pressed from the pomace and flows through the holes in the *consù* into the empty part of the upper vat where it clarifies and is decanted. Another refinement shows up in some Tuscan *palmenti*, namely a circular hole in the floor of the upper chamber, a 'sump'', as it were, for the collection of pomace.[9]

The problem of the 'mysterious' *palmenti*, of course, is not their function but their chronology. How does the archaeologist date a cutting in a native rock? Especially when its chisel marks have weathered away with age and when we have ample evidence, not only of the continuous use, but of the manufacture of these facilities for more than three millennia? Obviously, short of some new molecular archaeological technique, archaeological context is everything. For example, two *palmenti* at Pian Conserva, Tolfa (RM) may be connected to two nearby vineyards which can be dated by context to the second century BCE, though the excavators thought it more likely they were medieval.[10] But this area is rife with wild vines and so pre-Roman use is by no means excluded. Elsewhere in Etruria, a *palmento* at San Giovenale, Regello (FL) is interpreted as deriving from the sixth century BCE, but the archaeological context here goes back to the late Bronze Age.[11] In Sardinia many *palmenti* doubtless derive from a Phoenico-Punic stratum, but again this island is a natural home to wild vines and there is ample evidence of human habitation well back into the Mesolithic.[12] In Calabria many of the hundreds of *palmenti* have carved into them Latin, Byzantine and Armenian crosses and Sculli thinks the majority date from the Byzantine era, but much of this iconography was manifestly carved into preexisting structures.[13] In the wider Mediterranean context, treading facilities that are practically identical have been found in many parts

9 Fatucchi (1987): 14-20.

10 M. Giannace, A. Masi and F. Vallelonga, "Il 'Progetto Vinum': vite silvestri e siti archeologici," in Ciacci-Zifferero (2005): 39-45.

11 Quilici (1994): 41-65; Y. Backe-Fosberg, *Crossing the Bridge: An Interpretation of the Archaeological Remains in the Etruscan Bridge Complex at San Giovenale, Etruria* (Uppsala, 2005).

12 P. van Dommelen, C. Gomez Bellard and G. Pérez Jordà, "Produzione agraria nella Sardegna punica fra cereali e vino," *L'Africa Romana* 18 (2010): Vol 2, 1187-1202.

13 Sculli (2002): 55-6.

of the Levant, and some are clearly from a Bronze Age context,[14] as are they in Minoan Crete.[15]

Finally, there is the whole issue of the sophistication of Italic cultures, especially the so-called 'Apenninic' cultures. As the greatest scholar of the last eighty years in this area has aptly said, "The old image of a primitive Italy populated and civilized by incursions from the north has been turned upside down by discoveries showing the existence of developed, sophisticated and enduring Neolithic cultures in the Italian Mezzogiorno and the islands, of Bronze Age civilizations of the type dubbed 'Apenninic' spread across the entire peninsula, ... of marked and repeated influences from the eastern Mediterranean Orient, which culminated in the presence of Mycenaean traders, who may even have been in some sense colonists, in Apulia and the Sicilian region as well as along the Tyrrhenian coast and in Sardinia."[16] Early settlements in the Mezzogiorno were stable and prosperous and were evolving toward a proto-urban form as, for example, at Coppa Nevigata in Apulia or Castello di Lipari in the Aeolian Islands. Here there is indisputable evidence of a developed economy of agricultural and artistic production, a well-defined social structure and political organization, and deep and enduring participation in Mycenaean culture.

Some of the *palmenti* in the Cilento region of Campania may be taken as test cases here. This is an area where Paleolithic, Neolithic and Bronze Age settlement is not uncommon, and where the forms of several of the *palmenti* show obvious affinities with Minoan and Mycenaean forms. Several are located on the flanks of Monte della Stella at altitudes above 1100 meters (3600′), an altitude at which the vine has not been cultivated in living memory. But if viticulture here is inexplicable, given the current climate, there was a period between 1000 and 1350 CE when the climate was significantly warmer, the 'Medieval Warm Period' and, predictably, one of these *palmenti* is within the ambit of the convent of Sant'Arcangelo di Perdifumo. Case closed? By no means. There is another climatic possibility, namely the end of the 'Holocene Climatic Optimum' when the climate was gradually cooling toward something similar to that of today, namely, at the end of the second millennium and the

14 G. W. Ahlström, "Wine Presses and Cup-Marks of the Jenin-Megiddo Survey" *BASOR* 231 (Oct., 1978): 45-6.

15 Leftéris Platon and Katérina Kopaka, "Ληνοί Μινωικοί. Installations minoennes de traitement des produits liquides," *Bulletin de correspondence hellénique* 117.1 (1993): 35-101.

16 Cf. Massimo Pallottino, *A History of Earliest Italy* (Ann Arbor, 1991): 25-55. Elisabetta Setari ["La cultura del vino nel mondo italico. Forme, immagini, rituali," in Di Pasquale (2010): 110-19] for the Mycenaean currents in the Mezzogiorno. Salvatore Bianco ["Vino e simposio nelle communità dell'Enotria," in Di Pasquale (2010): 120-31] has a good review of the ceramic evidence.

beginning of the first BCE. At this time, wild grapes will have grown at this level, and systematic viticulture is entirely possible as well.

Surely then, we must assume with Puglisi[17] that some of our *palmenti* are older than the Greco-Roman epoch and that a few at least date to the beginning of the first millennium if not before. If so, we may assume that systematic viniculture as well, at least in southern Italy and the islands, is much older than previously thought. Viniculture on this scale would almost surely imply some form of viticulture as well, probably a type of paradomestication of sylvatic grapes by transhumant pastoralists.[18]

Thus it is likely that invaders did not so much introduce as mediate the use of the grape. Forni, for example[19] thinks that Mycenaean traders did not introduce viticulture to Italy but may well have stimulated its development, particularly by way of the introduction of better cultivars. And this may have been purely unintentional. Consider that raisins and brine-cured olives will have been among the most common foods of Mycenaean/Minoan sailors trading in the western Mediterranean, and that the pips of these two fruits will have been left among the detritus of their forays onto land for the purpose of trade. Over a period of time such trading posts develop middens and such middens, fertilized as they so often are with the organic remains of human occupation, most prominently human excrement, could well create so-called 'spontaneous gardens' from which local people could have propagated superior cultivars. Alternately, grapevine cuttings are among those organic materials used as dunnage in lading ancient ships, and in the cool, damp hold of a ship such cuttings might well have survived, perhaps to be discarded as

17 Puglisi (2009).

18 La Greca in Botti, Thurmond, La Greca (2012): 42-6.

19 Forni in Ciacci, Rendini and Zifferero (2007): 73-6. Cf. Gras (1985): 256-60. Research into the protohistory of Roman wine is being radically advanced at the time of writing by Progetto Vinum, a consortium of academic and vinicultural interests whose purpose is to use emerging technologies to advance the study of the prehistory and protohistory of Italian wine. Cf. Attilio Scienza, "Interesse e attualità dello studio della vite selvatica nella Toscana e nel Lazio settentrionale: una comune area di ricerca tra archeologia e biologia," in Ciacci and Zifferero (2005): 11-14; Andrea Ciacci and Andrea Zifferero, "Introduzione al 'Progetto Vinum': un progetto per il riconoscimenta della vite silvestre nel paesaggio archeologico della Toscana e del Lazio settentrionale," op. cit., 15-40. In the same volume, DNA research: Gaetano Forni, "Tappe e tracce della domesticazione della vite," 41-6; Attilio Scienza, Osvaldo Failla and Roberto Anzani, "Consistenza e valutazione del patrimonio viticolo spontaneo," 47-55; Serena Imazio and Massimo Labro, "I marcatori molecolari e il loro impiego nello studio della domesticazione della vite, attraverso il confronto con la vite selvatica," 57-66. Archaeobotany: Miria Mori Secci, "Archeobotanica della vite," 67-74.

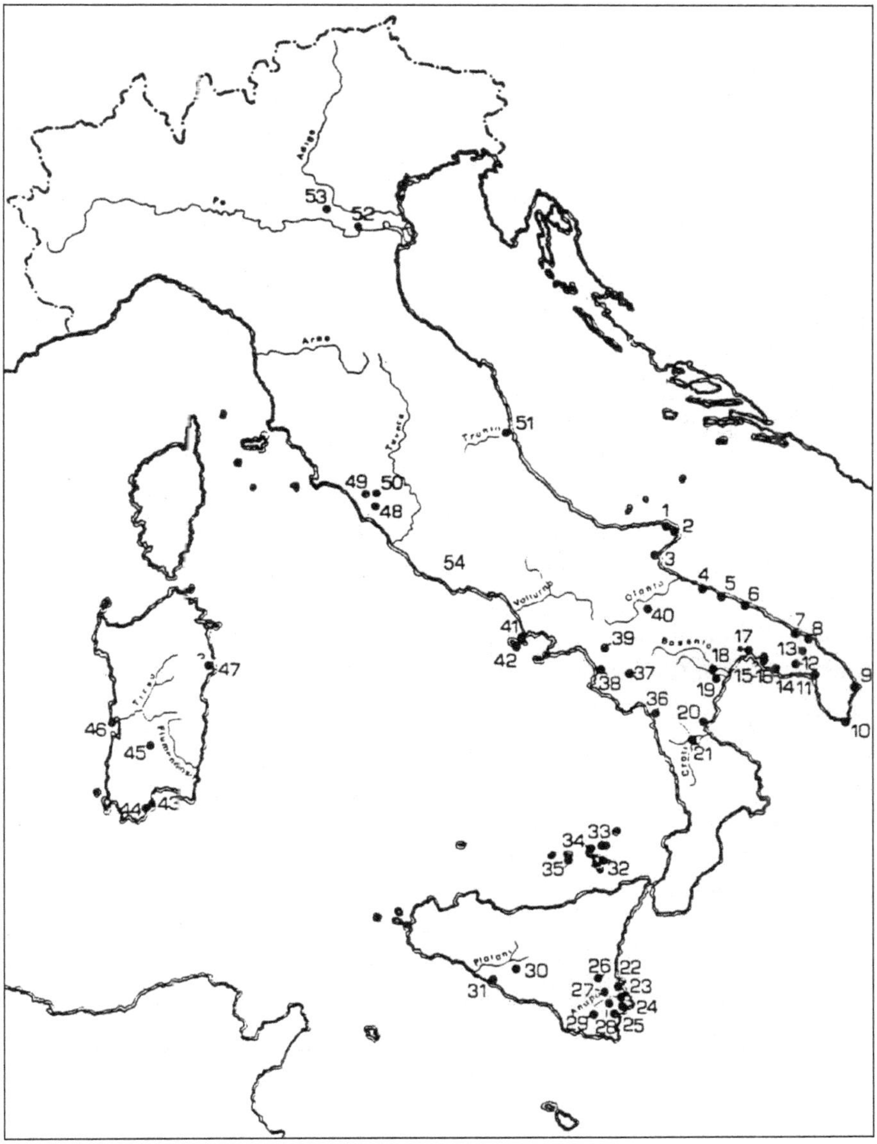

FIGURE 5 *Distribution of Mycenaean ceramics in Italy. From Gilda Bartoloni,* La cultura villanoviana. All'inizio della storia etrusca *(Rome, 1989): Fig. 3.11.* COURTESY OF THE AUTHOR.

ships were unburdened of their cargo at trading outposts. There, given the right circumstances, they could put down roots and flourish.

One such Mycenaean outpost occurs at Vivara, a small island in the Bay of Naples between Ischia and Procida. The site of Punta Mezzogiorno on the island attests frequent and continuous contacts with the Mycenaean/Minoan world as early as the seventeenth century BCE, and at another island site, Punta d'Alaca, comparable evidence derives from the sixteenth century. Here a Vaphio cup (late Helladic IIA, i.e., 16th-15th c.) shows evidence of an herb-flavored, unresinated wine.[20] Such Mycenaean ceramic is attested all along the Italian littoral and up many of the rivers.[21] And the flow of trade and culture—and perhaps colonization—was in both directions, we now know.

With the collapse of Aegean cultures in the late Bronze Age our evidence becomes exiguous, in Italy as elsewhere. But as one scholar so aptly states, "Europe at the height of Mycenaean power in the fourteenth and thirteenth centuries BC presented a foretaste of what was to come ... It is almost as if the Mycenaean system was a trial run for the fully fledged Greco-Roman world."[22] Beginning with the tenth century, the Early Iron Age, new cultures begin to emerge in Italy, mediated by contacts with the great trading cultures of the archaic Aegean, notably Phoenicia and Greece. Obviously, Italy is ideally located geographically for maritime trade with both eastern and western Mediterranean. The fact that Mediterranean sailors eschewed the open waters, preferring to island hop or skip along coastlines (*cabotage*) made the Italian Aegean, Ionian and Tyrrhenian seas a focus for such trade.

A case can be made that traders from the Phoenician *entrepôt* of Carthage were at least as important in the spread of advanced viticulture in the west as the western Greeks. Carthage, traditionally founded in 814 BCE, a date in accord with archaeological evidence, was, like other Phoenician outposts, made wealthy by trade with the western Mediterranean, but the real key to Carthaginian success lay in her agricultural hinterland where wheat, olives and wine were produced for local consumption and export. Carthaginian outposts, in turn, were established along the Mediterranean and Atlantic coasts of Spain, in Western Sicily at Motya, and in Sardinia. In this regard it is interesting that

---

20 M. Marazzi, C. Giardino and C. Pepe, "Vivara, Settlement of Punta d'Alaca, Naples, Italy Late Helladic I-II, 16th-15th Centuries B.C.," in Holley Martlew, ed., *Archaeology Meets Science: Biomolecular Investigations in Bronze Age Greece; The Primary Scientific Evidence, 1997-2003* (Oxford, 2008): 273-80.

21 Cf. Gilda Bartoloni, *La cultura villanoviana: all'inizio della storia etrusca* (Rome, 1989): p. 75, Fig. 3.11; K. A. Wardle, "The Palace Civilizations of Minoan Crete and Mycenaean Greece," in Barry Cunliffe, ed., *The Illustrated History of Prehistoric Europe* (Oxford, 1994): 237-43.

22 Barry Cunliffe, "Iron Age Societies in Western Europe and Beyond, 800-140 BC," in Cunliffe (1994): 336.

there was clearly a Phoenician contingent at Pithekoussai itself, the earliest western Greek outpost.[23] McGovern conjectures that it was Phoenician and Carthaginian traders who mediated true wine culture in Protovillanovan Etruria beginning around the ninth century. A Greek-style 'mixed grog' can be inferred from evidence at Murlo dated before 575 BCE, at Casale Marritimo and at Verucchio from the eighth and seventh centuries. McGovern believes that Etruscan, like many other European cultures, had a tradition of mixed fermented beverages in which grapes were used primarily as a source of yeasts, but that this 'grog' culture was supplanted by a pure wine culture under the influence of arriving Phoenicians and Greeks.[24] Be that as it may, Etruria itself, driven by her agricultural fertility and fabulous wealth in metal ores, quickly became a major exporter and was already expanding her territory and influence by the ninth century. In the eighth, proto-urban centers evolved from villages and orientalizing influences became profound, attesting extensive contact with eastern Mediterranean cultures.[25] That there were Etruscan traders/pirates in the eastern Mediterranean is amply attested by archaeological evidence from Lemnos, though Gras' insistence that the famous Homeric Hymn in which the new god Dionysus' kidnap by Etruscan pirates is really a tale of how they tried to dominate the Mediterranean wine trade[26] is rightly characterized as "excessively ingenious."[27]

This same scenario—a casual form of viticulture made more systematic under the influence of eastern contacts—may also be seen in Sardinia. Earliest evidence of wine culture in Sardinia is a cache of wine pips from the nuraghe Genna Maria at Villanovaforra dating from the ninth or early eighth century, perhaps not coincidentally the traditional date of the first arrival of Phoenicians at Nora. At this period archaic Greek amphorae are almost nonexistent. Yet we know that there was contact with some wine culture in Sardinia at least as early as the Mycenaean period and it seems clear that both Nuragic and Phoenician cultures practiced viticulture on the island.[28]

Meanwhile the Mezzogiorno saw the relatively rapid development of archaic Greek trading outposts followed by systematic and extensive coloniza-

23 Abulafia (2011): 74-9.

24 Patrick E. McGovern, *Uncorking the Past: The Quest for Wine, Beer and Other Alcoholic Beverages* (Berkeley, 2009): 189-97.

25 Pallotino (1991): 52-3.

26 Michel Gras, *Trafics tyrrhéniens archaïques* (Rome, 1985): 648-9.

27 Abulafia (2011): 105-7.

28 Gras (1985): 217-27.

tion.[29] In a word, the Mezzogiorno became Magna Graecia. The first attested Greek trading outpost in the west was established by Euboaean Greeks at Pithekussai on the island of Ischia, not far from modern Vivara, dating from c. 770-760 BCE.[30] The outpost was a mixed ethnic trading post established for the transference of raw iron ore from Etruria to the eastern Mediterranean or, as now seems more likely, for trade in iron artifacts, manufactured on the island from Etruscan ore, to the Campanian mainland. Thus in a sense the 'commodity' traded was a new technology—iron—for which there was high demand, made even more profitable here by the high cost of transport and the proximity of Ischia to that demand. If so, the products of developed Greek viticulture may have been another such technology which found a natural *point d'apuis* on the island. The outpost had been founded by Euboean Eretrians and Chalcidians, according to Strabo (5.4.9), the same groups who had also founded an *entrepôt* at Al Mina on the Levantine coast. Strabo says that their prosperity at Pithekoussai was assured by "the fertility of the soil and by the activities of their goldsmiths." Ridgway surmises that Strabo's *eukarpia* is fertility in the restricted sense of viticulture, given the island's ideal hilly terrain, porous and mineral-laden soil, and mild climate. Since there is no gold on the island but archaeological remains of foundries have been found, Strabo's reference to goldsmiths may refer to working of precious metals from the Etrsucan ambit, but the word may also be emended to *chatreia*, 'clay beds,' or 'potteries,' and thus a source of pottery vessels including pithoi for shipment of wine. Such a surmise would also perhaps resonate wth the name of the island, otherwise inexplicable as 'Monkey Island'. Could this trade have been a factor in the establishment of a true colony with an agricultural *chora* at Cumae?[31] In any case there followed a huge influx of Greek colonists all along the Ionian and southern Tyrrhenian littoral as well as on Sicily. Their influence on Italic cultures was profound. But, to reiterate, these Greeks almost surely found a well established wine culture.

For example, tradition says (e.g., Herodotus 1.167; Strabo 6.1.1) that the Greeks who entered the area that would become Calabria and the Cilento, the home of so many of our *palmenti*, encountered Oenotria, 'The Land of the

---

29 The best general discussion is Christian Vandermersch, *Vins et amphores de grande Grèce et de Siciles IVe-IIIe s. avant J.-C.* (Naples, 1994).

30 A. G. McKay, *Ancient Campania,vol* 2 (Hamilton, Ont., 1972): 75-88; David Ridgway, *The First Western Greeks* (Cambridge, 1992): 31-7; Ridgway, "Nestor's Cup and the Etruscans," *Oxford Journal of Archaeology* 16.3 (1997): 325-44; R. Ross Holloway, *The Archaeology of Ancient Sicily* (London, 2002): 136-54; Abulafia (2011): 89-94.

31 Holloway follows S. C. Bakhuizen, *Chalcis-in-Euboea: Iron and Chalcidians Abroad* (Leiden, 1976).

Vine'. Their culture is well enough defined that we can identify Oenotrian artifacts as far away as Etruria; for example, an Oenotrian cup at Capena, a fibula at Veii, and ornamental objects from Bisenzo, dating from the eighth century.[32] An alternate tradition says that the country took its name from Oenotrus, mythical Arcadian colonist of the generation before the Trojan War. But one is always skeptical of such euhemerizing etiologies. One tribe from the region, the Serdaioi, are clearly historical; they are in fact the tribe which established treaties with Sybaris guaranteeing the security of the Sybaritic subcolony at Poseidonia (Paestum). The Serdaioi were the first Italic people to mint, beginning at the end of the sixth century, and their coins, beautifully executed, have the head of Dionysus on the obverse, a cluster of grapes on the reverse.

The first coins minted on Sicily derive from Naxos, c. 530 BCE, and are to be associated with the port's commercial activities. The oldest series bears the head of Dionysus on the obverse, a cluster of grapes on the reverse, an apparent reference to the importance of viticulture here in an area ill-suited to growing cereals.[33]

Clearly, then, the two centers for systematic vitiviniculture in Italy, starting from the early Bronze Age, were the future Etruria in the north and the future Magna Graecia in the south. And clearly in both cases the development of vitiviniculture resulted from the interplay of immigrant and native populations. Attempting to disentangle the roles of specific cultures in the development of Italian viniculture only does violence to the evidence;[34] Italian culture in the early Iron Age is very much a 'vigneto miscello', a mixed vineyard with a variety of ethnic cultivars, and the cultural ferment in that vineyard will bear fruit spectacularly some two hundred years later in the form of three brilliant classical cultures. One can hardly imagine a less apt description than 'Dark Age'.

---

32 Bartoloni (1989): 186.

33 Luca Cerchiai, Lorena Jannelli and Fausto Longo, *The Greek Cities of Magna Graecia and Sicily* (Los Angeles, 2004): 158.

34 For example, the trellised vine is supposed to be typical of viticulture in Magna Graecia and the *arbustum* that of Etruria. But the arbustum was known in Magna Graecia from early on. The famous cultivar Byblinos, one of the first acclimated in Italy, traditionally at Rhegium, was a married vine (Gr. *anandendrás*), and perhaps also at Metapontum. The cultivar Aminea had the reputation of being especially adapted to the arbustum (Pliny *NH* 14.22) and this is the most famous of the *cépages* of Magna Graecia. Cf. Vandermersch (1994): 54-5.

Etruria

Evidence of the grape in Bronze Age Italy is attested at Gran Carro, Le Pozze and Blera, in Lazio, Cures Sabini, Radda, and Siena in Tuscany; Rieti in Umbria; Santa Margherita, Ancora; Campobasso, Molise; Pontecagnano and Roccagloriosa, Campania; and San Giovannili Ruoti, Basilicata.[35] More recently, Marta Mariotti Lippi and colleagues have documented remains of wild grapes at Pontedra, Pisa; Livorno Stagno; San Lorenzo a Greve and Pian d'Alma in Tuscany. The domesticated grape is now attested at Scanzo, at Cala Colombo and Madonna delle Grazie (BA) Puglia. In fact, at almost every site examined by Meriotti Lippi's team, both *Sylvestris* and *Vinifera* were found.

The first evidence of grapes in Etruria in clearly human contexts occurs initially along coastal regions and thereafter moves inland and further afield. For example, *Sylvestris* appears from the Neolithic in a human settlement at Pienza (SI).[36] From the late Neolithic comes a grape pip at Podere Casanuova near Pontedera (PI) dated to the end of the fourth millennium, again, consistent with the wild vine. Analysis of pollen from a site near Lago di Massaciuccoli (LU) dating from the early Bronze Age (c. 2000 BCE) suggests that the vine there was the object of protection but not cultivation and was thus at the level of paradomestication. From the Middle Bronze Age (c. 1500 BCE) evidence of consumption of the fruit in some form multiplies, as it does in Magna Graecia, perhaps as a result of contact with the Aegean world. But only toward the end of the second millennium is there evidence of systematic culture of the vine, and this is also the period when pips of an unmistakably domesticated type are consistently found.[37]

First evidence of widescale wine production in Etruria comes from the famous shipwreck off Languedoc, France, where a large proportion of the lost cargo was wine of Etruscan provenance. In fact, wherever Mediterranean people settled or traded, they brought, not just wine, but a 'civilization of wine'. For example, the Etruscan bucchero vases found in Provence and Languedoc are 95-100% drinking cups.[38] Amphora studies in southern France, particularly

35 Forni in Ciacci, Rendini and Zifferero (2007): Table 1.

36 L. Castelletti, "Rapporto preliminaire sui resti vegetali della serie Neolitico-Bronzo di Pienza (Siena)" *Rivista Archeologica dell'Antica Provincia e Diocesi di Como* 156-7 (1974-5): 243-51.

37 di Biancamaria Aranguren et al., "L'avvio della coltura della vite in Toscana: l'esempio di San Lorenzo a Greve (Firenze)," in Ciacci, Rendini and Zifferero (2007): 88-92; Delpino in Ciacci, Rendini and Zifferero (2007): 134-5.

38 Jean-Paul Morel, "Greek Colonization in Italy and in the West," in T. Hackens, Nancy D. Holloway and R. Ross Holloway, edd., *Crossroads of the Mediterranean* (Providence, RI, 1981): 130.

along the southern reaches of the Rhone, indicated the terminus for much of the trade as well as the use of wine in banqueting rituals. Later field surveys in the Albegna valley in the 1990s began to reveal the source of this wine. An excavation at sixth/fifith century Podere Tartuchino in this area is our first evidence of a winery discovered anywhere in mainland Italy. Here the center of a room interpreted as a pressroom was dominated by a *dolium defossum,* a large terracotta fermentation jug buried up to the shoulders to provide thermal stability. Analysis of its interior revealed that it had been smeared with pitch to make it waterproof. Around this jar were the footings of three treading platforms which fed must into the dolium.[39]

Excavation on the island of Giglio, as well as the port cities of Orbetello, Lagune and up the valley of the Albegna reveal the complexity of viniculture as early as the Middle Bronze Age. Hoard after hoard of wine vessels reveals the prevalence of Etruscan wares, especially of the type classified as Py 1.2 to Py 3, but also the presence of Punic and both Eastern and Western Greek wares. And already by the Late Bronze Age there was a well attested trade route for Etruscan wine with termini at the Golfo Campese, Giglio and at Tramonte on the Lagoon of Orbetello, as well as a north-south route along the coast and an easterly route along the river and perhaps even a connection via the Tiber valley with central Etruria.[40]

There can be little doubt that it was the Etruscans who introduced viniculture on a systematic level into the Po Valley at the beginning of the Iron Age, by way of the Padane Gauls or the Raetian Gauls or the ethnic Veneti, if not all three. Perhaps as early as the tenth century BCE, pruning hooks and other viticultural tools in bronze begin showing up in the archaeological record in the environs of Bologna.[41] At Verucchio, Romagna, there is evidence of the

39 J. P. Joncheray, "L'Epare grecque, ou etrusque, de Bon Porte," *Cahiers d'Archéologie Subaquatique* 5 (1976): 5-53; F. Py and M. Py, "Les amphores étrusques du Vannage et de Villevielle, Gard" *MEFRA* 86 (1974): 141-254; Philip Perkins, *Etruscan Settlement, Society and Material Culture in Central Coastal Etruria* [= *BAR Int Ser.* 788] (Oxford, 1999); I. Attolini and P. Perkins, "The Excavation of an Etruscan Farm at Podere Tartuchino" *PBSR* 60 (1992): 1-76; Philip Perkins, "Production and Distribution of Wine in the Etruscan Albegna Valley," in Ciacci, Rendini and Zifferero (2007): 185-90; Andrea Zifferero, "La produzione e il commercio del vino in Etruria," in Ciacci and Zifferero (2005): 100-04.

40 G. Ciampoltrini and P. Rendini, "Vie e porti del vino nella valle dell'Albegna in età Etrusca (VI-V secolo a.c.)," in Ciacci, Rendini and Zifferero (2007): 176-84. Cf. C. Ampolo, "Per un studio dell'alimentazione dell'Etruria e di Roma arcaica," in *L'Alimentazione nel Mondo Antico* (Rome, 1987): 9-13; M. Cristofani, "Duo sunt liquores" op. cit. (1987): 37-40.

41 Larissa Bonfante, Giovannangelo Camporeale, Alexandra Carpino and John R. Clarke, *New Perspectives on Etruria and Early Rome* (Madison, WI, 2009): 23-4.

domesticated vine from the seventh century, at Veronese Castelrotto a century later, and a bit later still at Trentino and in Friuli. Viticulture spread thence, first to the Insubrian Gauls, the original inhabitants of Lombardia, then to the Senones, Boi and Cenomani beginning from the fourth century. We have archaeobotanical evidence from the end of the seventh century at Castelletto Ticino, where training of vines was by so-called 'long pruning', i.e., training to trees, the classical *arbustum Gallicum*,[42] in contrast with the traditional *arbustum Italicum* associated with the Etruscans (Pliny *NH* 17-201-11; 14.12; Col. *DRR* 5.7).[43]

By the middle of the second century BCE there was a virtual explosion of wine exports from Etruria, largely centered on the Cosan plain and the Albegna valley. But the trade ends with startling abruptness in the last two decades of the first century. It now seems likely that this sudden demise is in part real, in part illusory. Real because the wine trade, developed almost exclusively for export to Gaul, found other markets or was abandoned for more profitable pursuits such as the rearing of livestock. Apparent, because this is precisely the period when we begin to see, in the shipwrecks which provide so much of our information, evidence of shipment *en vrac,* that is, use of bulk dolia in the main cargo holds and of amphorae only in the smaller fore and aft holds. Add to this possible shipment in wooden *foudres* and *tonnes*, as well as the development of local Gallic viticulture. Transpadane Etruria seems to have opened up an alternate trade route through the emporium at Spina, at the head of the Adriatic, and thence into eastern Gaul and the Danube regions.[44]

Magna Graecia

Sicily is typical of the difficulty of unraveling Italian wine history in the Mezzogiorno. First, the wild grape is native even here as on the mainland, so it is entirely possible that Paleolithic and Ceramic Neolithic peoples used the wild grape as a source of fermentable, perhaps in a mixed grog, perhaps in true wine. Secondly, there is ample evidence of Mycenaean contact with the island, including trade in ceramic pieces associated with wine culture and perhaps in

---

42 R. Chavallier, *La romanisation de la Celtique du Pô* (Rome, 1983): 238; E. Sereni, *Terra nuova e buoi rossi* (Torino, 1981): 169; Myriam Giannace, "Archeologia della vitivinicoltura: metodi di coltivazione e lavorazione dell'uva nell'anitichità preromana," in Ciacci and Zifferero (2005): 75-81.

43 Cf. Forni in Ciacci, Rendini and Zifferero (2007): 76-9.

44 Jean-Paul Morel, "Les trafics maritmes de la Grande Grèce," in Francesco Prontera, ed., *La Magna Grecia e il mare: studi di storia marittima* (Taranto, 1996): 143-72.

the form of our *palmenti*. Third, it is now evident that acculturation during the age of Greek colonization (735-650 BCE) of the eastern and southern parts of the island was far from unilinear and that, furthermore, there is as much continuity in native, especially Sikel, culture as there is Hellenizing. Finally, Punic outposts on the northern and western coasts will have provided another element to the cultural mix. That dynamic can be attested by wine wares. Not only are there both native and Greek-style wares attested in Both Sikel and Greek settlements, but Greek-style colonial wares are also common and it is impossible to determine the ethnicity of the manufacturers of the latter.[45]

We have similar problems of interpretation on the mainland of Magna Graecia. Excavations at the site of Pizzica Pantanello near the Greek colony of Metapontum (BA) confirms the literary notices that viticulture was practiced here from the ninth century, well before Greek colonization. Archaeobotanical remains, specifically in the form of carbonized, mineralized and mummified seeds, indicate the relative prevalence of grapes in the diet, the presence at this early date of the domesticated grape vine, and the possibility of at least three different cultivars.[46] At nearby Incoronata, both Greek and indigenous levels yielded plant remains including grapes from contexts dated securely to the late eighth-early seventh century.[47]

And yet, clearly the period of Greek colonization can be associated with a development in the material culture in such coastal colonies as Metapontum, Siris-Policora, Sybaris, Croton, and Locris, not to mention of Elia/Velia and Poseidonia/Paestum on the Tyrrhenian littoral. And Greek influence can clearly be tied to the systematization of agriculture, including viticulture, in these same areas.[48] The fourth century was a chaotic time politically in Magna Graecia as Italic peoples of Samnitic origins such as the Lucanians, Pelasgians, Iapygians and others conquered one after another of the Italiote Greek cities. However, far from being a disaster for Italian viticulture, again and again we see an intensification of agricultural settlement and production. For example,

45 Tamar Hodos, "Wine Wares in Protohistoric Eastern Sicily," in Christopher Smith and John Serrati, edd., *Sicily from Aeneas to Augustus: New Approaches in Archaeology and History* (Edinburgh, 2000): 43-54.

46 Lorenzo Costantini, "Archeologia della vitivinicoltura in Basilicata. Un bilancio delle ricerche archeobotaniche a Pizzica Pantanello, Metaponto," in Ciacci, Rendini and Zifferero (2007): 98-108.

47 Lortenzo Costantini, "The Origin of the Mediterranean Diet in Italy," *Rivista di Antropologia* 76 Suppl. (1988): 7-15; J. C. Carter, *Discovering the Greek Countryside at Metaponto* (Ann Arbor, 2006): 78-9.

48 E. Sereni, "Per la storia delle più antiche tecniche e della nomenclatura della vite e del vino in Italia," 29 (1964): 128-31.

at Lucanian Roccagloriosa, remains of grapes during this period represent 31% of the floral remains so far analyzed and 85.7% of the fruit remains. Recent analysis of archaeobotanical remains from the central plateau of the city show a boom in arboriculture, particularly of the vine, between the fifth and third centuries, concomitant with "an unprecedented expansion of rural settlements" here as at Metapontum and Poseidonia.[49] At Roccagloriosa local 'pseudo-Chian' amphorae may be evidence of the export of wine from this area (Bussento/Buxentum) destined to produce a *grand cru* in the classical period.[50]

But the period of greatest expansion of viticulture, in Italy as a whole as in the south, corresponds with the end of the Second Punic War. Factors include the stability of the political environment, especially in Latium, Campania and Lucania; the importation of grain from Sicily and North Africa, freeing up large tracts of land for viticulture; the rise in wealth of Italian peoples as a result of Eastern conquests, wealth manifested especially in the form of slave labor; the institution of the highly efficient villa farm (agribusiness); the exploitation of commercial markets in the western provinces; and the transfer of viticultural expertise from Magna Graecia, Sicily and Carthage— for example, the translation into Latin of the Carthaginian Mago's agricultural treatise. The heyday of Italian viticulture was the second half of the second century BCE through the second century CE.[51]

Archaic Rome

And what of Rome herself?[52] The Romans themselves thought that Campania was the source of their viticultural practices (Cato in Pliny, *NH* 14.46-7), but, again, it is viticulture on a commercial scale to which they refer. The hills of Rome were inhabited, we now know, long before the traditional date of the

49 S. Bökönyi, L. Costantini and J. Fitt, "The Farming Economy," in M. Gualtieri.ed. *Fourth Century B.C. Magna Graecia: A Case Study* (Partille, Sweden, 1993): 281-307. Cf. E. Greco, "Poseidonia entre le VIe et la IVe siècle avant J.C.: quelques problèmes de topographie historique," *Rev Arch* (1979.2): 219-34.

50 Gualtieri (1993): 340-1.

51 Flaminia Arcuri, "Il vino in età romana," in Luigi Rossi, ed., *Il vino nel Cilento* (Acciaroli, Sa, 1994): 38-9.

52 Riccardo Santangeli Valenzani and Rita Volpe, "La restituzione del paesaggio agrario della vitvinicoltura a Rome e nel suburbio," in Ciacci, Rendini and Zifferero (2007): 48-54; C. Fayer, *Aspetti di vita quotidiana nella Roma arcaica: dalle origini all'età monarchica* (Rome, 1982) 151; Bartoloni (1989): 50-1; 188-9.

city's founding, and grape pips of the *Sylvestris* type have been found in human contexts on the Palatine from the Neolithic and Early Bronze Age. Already by the eighth century, probably under the influence of Etruscan neighbors north of the Tiber, the early Romans had adopted the banquet culture and the accoutrements of aristocratic burial with the full panoply of imported banqueting vessels. The tradition which has King Numa interdicting wine in certain religious rites has been used to suggest that the earliest Romans did not know the vine, but this flies in the face of archaeology and the tradition itself; elsewhere we hear that when Aeneas fought the Etruscan king Mezentius he vowed the whole Latian vintage to the gods in exchange for victory. What the tradition does indicate is that at this early stage wine was strictly an expensive prestige product; King Numa's law in fact is the first in a long series of Roman laws attempting to curb conspicuous consumption. There are trenches for vineyards in the Roman *suburbium*, dug into the tufa plateau, as early as the fifth to fourth centuries BCE. This was definitely a time of agricultural intensification here; remains of villas have been found only 700 meters from each other.[53]

Thus it is not so much a question of when and how viticulture came to Rome as it is when and how Rome subsumed the viticultural regions of central Tyrrhenian Italy and thus when Italian viticulture became 'Roman' perforce. The heartland of Italian viticulture during the classical period was not Latium but Campania. Republican wine amphorae, especially Greco-Italic and Dressel 1 styles, point to Campania as the source of an organized commerce in wine well before the Hannibalic War. During the Samnitic Wars (343-290 BCE) agriculture, viticulture and sheep farming were the principal economic activities in the region, but by the end of the Republic, viticulture on a commercial scale was widespread and highly profitable, as evidenced by the intermarriage of noble Roman *gentes* with local aristocrats so engaged, such as the Lasii and the Holconii, the Eumachii, Numistri and perhaps the Allei of Pompeii. By the Augustan Age these local *gentes*, along with the Gelii, also wine producers, were the most prominent families in Pompeii.[54] Wine was, in fact, the chief crop in the fertile Campania. Johnson's[55] characterization of Pompeii as the

53 Annalisa Marzano, *Roman Villas in Central Italy: A Social and Economic History* (Leiden, 2007): 103-04.

54 Paavo Castrén, *Ordo Populusque Pompeianus* (Rome, 1975): 38-95; J. Day, "Agriculture in the Life of Pompeii," *Yale Classical Studies* 3 (1932): 166-208; Wilhelmina F. Jashemski, *The Gardens of Pompeii, Herculaneum and the Villas Destroyed by Vesuvius* (New Rochelle, NY, 1979): 218.

55 Hugh Johnson, *The Story of Wine* (London, 2004): 36.

'Bordeaux of Roman Italy' may be a bit enthusiastic; after all, Pompeii is the only commercial city in the region which has been extensively excavated. But it is certainly true that the overwhelming majority of the villas so far excavated in the region have wineries.[56]

Purcell[57] has exploded the myth of a first-century CE collapse in Italian viticulture. It is in fact a time of enormous expansion in demand, coupled with a huge expansion of viticulture in the Roman suburbium and further afield in the Tiber Valley.

Wine in the Western Provinces

We now turn to those areas where Rome provided the impetus for the introduction of vitiviniculture. But it should be remembered that the Eastern world, most especially Greece and Egypt, continued to supply large quantities of wine to Rome. Strictly speaking, however, that is not our purview. About the influence of these parts of the Roman world on the Roman perception of wine as a status symbol we shall have much more to say in a later chapter.

Gaul

The same complexities which surround the transmission and spread of viniculture in Italy recur in the provinces as well, unsurprisingly. Roman wine culture spread first to Gaul, initially to the Transpadane Gauls of northern Italy, as we have seen, and then to the Gauls of southern France. Here it is the role of the Romans themselves which is hard to ascertain, at least in particulars.[58] The wild vine is indigenous here too, as all over the western Mediterranean climate zone, and grape remains have been found all along the southern littoral dating from the Paleolithic. Such remains become common in the Mesolithic and Neolithic. Exclusively wild grape pips are found up until the Bronze Age, but firm evidence for regular cultivation dates from the middle Iron Age, more precisely from the fifth century. Analysis of pip morphology suggests an intermediate stage of domestication and is taken as amelioration of local indigenous

56 N. Purcell, "Wine and Wealth in Ancient Italy" *JRS* 75 (1985): 6-9.

57 Purcell (1985): 10-19.

58 Roger Dion, *L'histoire de la vigne et du vin en France* (Paris, 1959); Tchernia (1986); Anthony King, *Roman Gaul and Germany* (Berkeley, 1990): 102; Françoise Formenti and J. M. Duthel, "The Analysis of Wine and Other Organics Inside Amphoras of the Roman Period," in McGovern, Fleming and Katz (1996): 79-86; Jean-Pierre Brun and Fanette Laubenheimer, edd., "La viticulture en Gaule," *Gallia* 58 (2001).

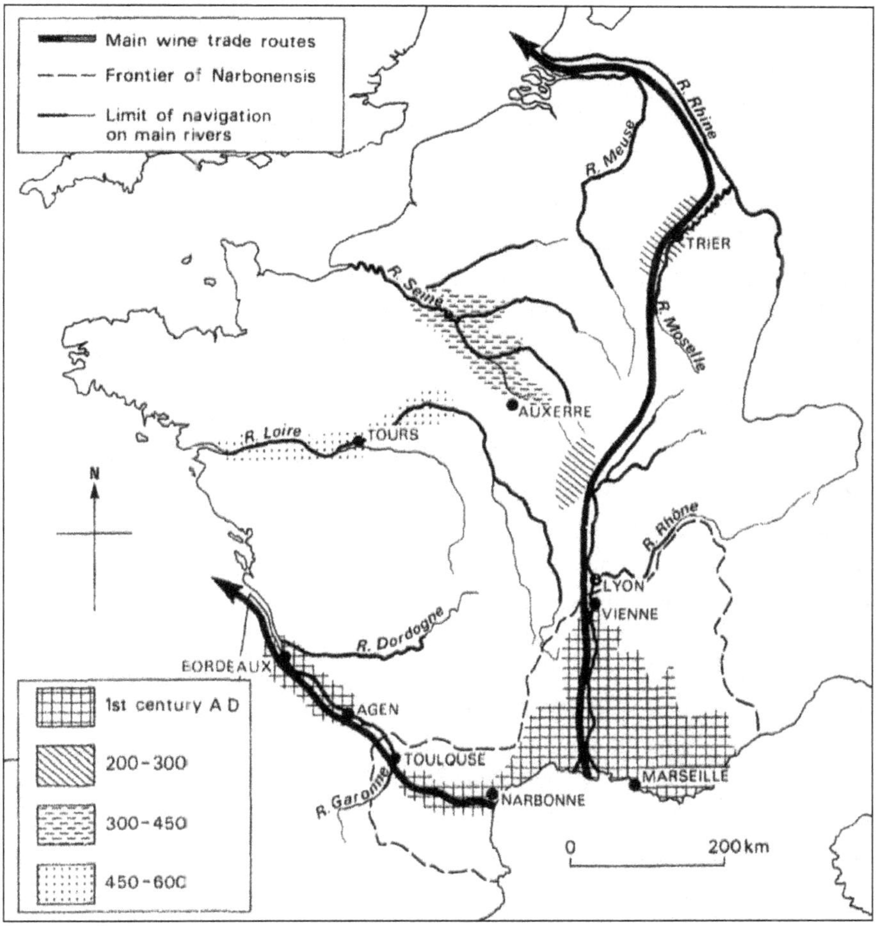

FIGURE 6 *Spread of vitiviniculture in Roman Gaul. From Tim Unwin,* Wine and the Vine *(London, 1991): Fig. 20.* COURTESY OF TAYLOR AND FRANCIS.

cultivars. As we will see, DNA evidence suggests that some of our most important modern cultivars are products of this incrossing.

In pre-Roman mainland Gaul, viticulture was concentrated around Marseilles, ancient Massilia, a Phocaean Greek emporium founded in 600. Here it is only after the capture of Phocaea itself by the Lydians and after the Battle of Alalia (between 540 and 535) that the influx of large groups of colonists created a *chora* sufficient to create viticulture capable of supplying not only the local demand for wine but excess for exportation. The expansion of this vignoble continued right up until the middle of the first century CE.[59] Thus, wine production at a commercial level was derived from Greek practice and preceeded Roman influence.

Still, the influence of mainland Italy in the transmission of viniculture in France is undeniable. That influence is most accessible to us by way of the systematic study of the amphorae in which wine was shipped. The pioneer in this methodology is André Tchernia,[60] but the bibliography is now vast.[61] More recently, molecular studies of wine residues and the clay fabrics of amphorae are beginning to radically refine these same studies.[62] Greek, Etruscan and Punic ceramic wine containers begin to show up in southern France as early as the seventh century, especially in Languedoc. Thereafter, increases of local pottery production around Marseilles suggest the beginnings of viticulture there as well. During the fifth and fourth centuries, a virtual monopoly of the Gallic wine trade by Marseilles is evident. Italic wine trade with Gaul is attested from the second century. Greco-Italic amphorae are found in shipwrecks and on the mainland as far north as Toulouse and Limoges. Dressel 1a-c amphorae, an Italic type especially associated with wine from Campania and Etruria, begin showing up in the first century. It has been estimated that an average of 60,00 hectoliters (160,000 gal.) of wine per year were exported to Gaul from about 130 to 30 BCE. The vast majority probably originated in southern Italy, especially in Campania, and in Etruria. There were diffused in Gaul between c. 200 and 30 BCE several millions of amphorae of the style Campanian A, a typical style of Magna Graecia.[63] At the end of the first century other Italic types, Dressel 2-4, displace Dressel 1; because of a more compact shape they can carry significantly more wine per unit of container weight. But overall numbers of such vessels are smaller than for Dressel 1, perhaps, to reiterate, due to ship-

59 F. Py and R. Buxó i Capdevila, "La viticulture en Gaule à l'Âge du Fer," *Gallia* 58.1 (2001): 29-43; Sereni (1964): 124-6. For Liguria: Emilio Sereni, *History of the Italian Agricultural Landscape* (Princeton, NJ, 1997): 131-6. For a concise and authoritative summary of historical developments in Gaul from perhaps the greatest scholars in the field, see Jean-Pierre Brun and Fanette Laubenheimer, "Le vignoble gaulois," in Brun, Poux, Tchernia (2009): 217-29.

60 Tchernia (1986).

61 S. Martin-Kilche, ["Amphorae in the Roman West: Discussion and Research since 1989" in J Eiring and J. Lund, edd., *Transport Amphorae and Trade in the Eastern Mediterranean* (Athens, 2004): 263-72] has an extensive bibliography to that date, but see also J. Theodore Peña, *Roman Pottery in the Archaeological Record* (Cambridge, 2007).

62 For example, Formenti and Duthel in McGovern, Fleming and Katz (1996): 81-6.

63 Morel in Francesco Prontera (1996): 143-72.

ment *en vrac*. Marie-Brigette Carre[64] points out that one limitation of the technique was the depth of holds of ancient ships. For ships of some 20 meters and 50 tons capacity, the technique was efficient, but for smaller ones impossible because of the height of the dolia and for those larger inefficient because their deeper holds allowed for three tiers of interlocking amphorae. In any case this proved to be a failed technology since shifts in only a couple of these large vessels could easily cause the whole cargo to shift and make the ship turn turtle. But clearly by this time Roman trade had completely eclipsed Massiliote. Also at this time a local type with the same 'squatty' profile, the so-called 'Gauloise' type, appears. Simultaneously, chemical analysis of clay bodies of Dressel 2-4 amphorae indicates local manufacture of this type at Lugdunum (Lyon), indicating a return to or extension of local wine production in southern France.[65]

Romanization of France in the first centuries of the common era is associated with a virtual explosion in viticulture. It is in the first century, for example, that the Roman province of Narbo (Narbonne) develops commercial viniculture capable of exportation, even back to Italy herself. During the reign of Nero and Vespasian, archaeology supports a multiplication of large vinicultural installations here along with massive development in local installations for production of amphorae. Under Vespasian Gallic amphorae represent 40% of such containers found at Ostia. This growth in exports is clearly to be associated with a veritable profusion of Roman-style villas found on the Mediterranean littoral as well as in the valleys of the Hérault and Rhone. During the second half of the first century many such installations show evidence of enlargement, and viniculture begins to move into more marginal areas. In the second century some of these marginal installations and more modest ones in highly productive areas are abandoned, but larger installations in productive areas continue. Is this crisis or mere consolidation? The expansion of the villa system continues elsewhere in the first half of the third century, but the situation changes radically after 260 CE when a large number of villas are abandoned, especially east of the Rhone. In part this seems a response to changes in commercial and climatic conditions, in part a change to more profitable stock breeding.[66]

It is also during these first three centuries that viniculture begins to move northward from the Gallic littoral. A famous case involving Fonteius, governor of Narbonese Gaul, charged with extortion by the locals when he imposed

64 Marie-Brigette Carre, "Transport en vrac'," in Brun, Poux and Tchernia (2009): 290-1.

65 Fromenti and Duthel in McGovern, Fleming and Katz (1996): 79-86.

66 Jean-Pierre Brun, *Archeologie de l'huile e du vin en Gaule romaine* (*Paris*, 2005): 68-76.

import fees on Roman wine making its way to Tolosa (Toulouse) and other parts of the interior, suggests considerable competition between local wines and imports and also suggests that wine made its way via Tolouse down the Gironde River to the emporium at Bituriga (Bordeaux) and thence to the British Isles.[67] Strabo (4.1.11) famously regarded the *massif central* as a sort of northern limit for viticulture and oleoculture, except along the central Rhone, but this same writer attests that viticulture had surely reached the area of Roman Vienna (Vienne), tribal center of the Allobroges, by Augustan times, and by the time of Martial it had become "Vienne the vine-clad" (Ep. 13.107). Thus by the end of the first century the wines of the area were famous.[68]

Lyon (Lugdunum) was an important êntrepot for the dispersal of wine to and from Gaul from at least the second century BCE. There *negotiatores vinarii* were among the most prominent citizens and wines made their way through Lyon from Italy, the Iberian peninsula, southern Gaul, the eastern Mediterranean (Greece, Asia Minor, the Levant and Egypt) and from North Africa.[69] During the Antonine period there is still a rich corporation of *negotiatores vinarii* at Lyon, and a famous address to Emperor Constantine of 312 CE from the city of Augustodunum (Autun) signals thast the Côtes de Nuits and Beaune, the region known as the *pagus Arebrignus*, had been a notable vignoble already for many years. Autun had replaced Bibracte as capitol of the Aeduans and Langres became that of the Lingones. But how early was this vignoble established? Certainly the complaints that the use of the propagation system known as *provignage* had made the soil unworkable suggest that vines had been established there for many decades. Dion[70] compares this to a similar situation in 1875 and surmises we are talking about less than a century for entanglement of roots to occur. But certainly before 250 the Côte d'Or was noted for its vineyard.

In Aquitania there is ample literary evidence for the development of viticulture from the first century forward, evidence recently augmented by archaeobotanical and other archaeological evidence (dolia, amphorae, viticultural utensils, villa farms), which clearly marks the development of viniculture here from the second quarter of the century. The cessation of manufacture of

67 Dion (1959): 95-116 discusses the importance of the case.

68 Cf. Matthieu Poux, "Le 'vin du triumvir', premier cru lyonnais," in Brun, Poux, Tchernia (2009): 232-3; Hugues Savay-Gueraz, "*Allobrogica*, le vin du Allobroges," op. cit. (2009): 234-5.

69 Armand Desbat, "Le commerce des vins à Lugdunum," in Brun, Poux and Tchernia (2009): 264-72.

70 Dion (1959): 148-66.

amphorae here in the second century is clearly a result of the transition, typical by then, to wooden *foudre* and barrels.[71]

Three centuries after Strabo had seen the Cézennes as a natural barrier to viticulture, it was clearly established in Bordeaux, near Paris,[72] and in the Moselle valley. Obviously the greatest difficulty overcome in these northerly regions was the intolerance of the vine for winter cold, far more than the lack of summer sun. The key was the development of new cultivars via incrossing, such as the famous Allobrigica of Dauphiné. This area of the central Rhone was a perfect 'laboratory' for the development of new varieties, since in the space of less than 100 kilometers (62 miles), Mediterranean flora give way to central and northern species. Perhaps the Allobrogica achieved its famous cold-hardiness on the lower slopes of the Alps and then was brought north with the tribes of the Allobroges, eventually becoming the Pinot grape, and thus explaining the grape's nickname *'Montagnard'*.[73] Elsewhere, in the Loire basin sites are first mentioned near Orléans in the last decade of the fourth century (Sidonius Apollinarius *Ep.* 4.21.5).

In the northeast, the central Rhone became a natural conduit for the transmission of viticulture as it had been for centuries for wine. From the nexus at Lyon, wine made its way up the Saône to its headwaters and thence to the Moselle, then down this river to the Rhine. Viticulture traveled in its wake. Viniculture was well established in the microclimate of the Moselle valley before the third century.[74] By the time of Ausonius in the mid-fourth century (*Mosella* 161-7; 335-48) we hear of the riverside "planted from ridge top to river's edge, while bargemen shout crude jokes at vinedressers." And even here in this Roman outpost we find Roman-style villas comparable to those at Cumae and Baiae.

Spain and Portugal

As we have seen, the wild grape is indigenous on the Iberian Peninsula and though it now seems unlikely that the vine saw early domestication here as well as in the East, there is at least a remote possibility that viniculture can be

71 Brun (2005): 105-21.

72 Cf. Christian Vernou, "Le vin des *Parisii* et des cités septentrionales," in Brun, Poux and Tchernia (2009): 241.

73 Dion (1959): 117-33.

74 M. Lachiver, *Histoire des Vignobles Français* (Paris, 1988); Dion (1959): 147-50; Rheinisches Landesmuseum Trier, *2000 Jahre Weinkultur am Mosel-Saar-Ruwer* (Trier, 1987); Edith Mary Wightman, *Roman Trier and the Treveri* (New York, 1971): 189-92; S. Loeschke, *Denkmäler vom Weinbau aus der Zeit der Römer-Herrschaft am Mosel, Saar und Ruwer* (Trier, 1933).

posited here from the Neolithic. In the pre-Roman period, a locally produced amphora patterned after Punic prototypes is used widely in the Iberian peninsula for transport and storage of liquids, including wine. But in any case once again it is the coming of Roman influence which sees viniculture established on a commercial level. In Spain that history begins in Baetica, roughly modern Andalusia, granted provincial status in 197 BCE and the object of strong Italian immigration. Thus we may point to viticulture here from this time if not before.[75] Numerous indications from coins and literary notices show viticulture concentrated along both sides of the Baetis (Guadalquivir) River. Thence it was extended during the first two centuries before the Augustan period, by which time it was extensive in this province. Strabo (3.2.6) says of Turdetania (Baetica) that it had gained a reputation for both the quantity as well as the quality of its wine exports in his day. Varro has only general references to Spanish viniculture but Columella has numerous specific references, as might be expected from this native of Gades (Cadiz), and speaks of his uncle as the most expert vigneron of the province. He laments that in his own day Roman men stock their wine cellars, not with the products of their own labors, but with wines from the Cyclades, Gaul and Baetica (*I. Praef.* 20).

This growing exportation of Baetic wines is clearly traceable in amphora studies as well. Several distinct styles of Baetic amphorae are identifiable. One, the so-called Haltern 70, is firmly tied to Baetica by inscriptions on the necks of several from the Port-Vendres II shipwreck. From Belo (Bolonia, province of Cadiz) come Dressel 1C amphorae dating from the second half of the first century BCE, though these were also used for the famous Spanish garum. Some Dressel 2-4 amphorae probably also derive from Baetica, though in this case the evidence is thin. Haltern 70 amphorae, almost surely Baetic in origin, have been found in several places in Gaul as well as at Ostia. Finally, a few Dressel 28 amphorae derive from Baetica.[76]

The other large province of Roman Spain was Hispania Tarraconensis, Taragon and northern Spain. Literary sources are difficult to use here because they are notoriously imprecise in their geographical designations. 'Hispania' per se seems to designate Baetica. But Pliny (*NH* 14.30), in speaking of the grape cultivar *cocolobis*, "which they call *balisca* in Hispania," seems to mean Hispania Tarraconensis. Elsewhere (*NH* 14.71) he speaks of the excellent quality of Tarragonese wines, a sentiment shared by Silius Italicus (*Punica* 15.178, 369).

75 The definitive work on Iberian viniculture is Robert Étienne and Françoise Mayet, *Le vin hispanique* (Paris, 2000). Cf. Fabienne Olmer, "Le vin hispanique," in Brun, Poux and Tchernia (2009): 276-7.

76 Étienne and Mayet (2000): 73-97.

Martial (*Ep* 4.14.2-4), a native of Bilbilis (Cerro de Bambola, near Calatayud), compares them with the wines of Etruria. In 102 CE Florus (*Vergilius Orator an poeta* 2) praises the climate and fertility of Tarragona and compares the wines favorably with those of Italy. Martial (*Ep*. 1.49.21-22) and Pliny (*NH* 14.71) both speak of the wines of Laietania (Barcelona) but both praise her vines for yield rather than quality. Pliny the Younger (*Ep*. 2.13) mentions Sagonte where Voconius Romanus has a vineyard and also praises the wines of Lauro, as had his uncle. Though the location of this town is disputed, it is most likely in northeastern Spain near modern Llerona. We also have several amphora stamps of a *vinum lauronense*. In general, it would seem that Tarraconensis was noted more for the fertility of the vines along the coast and for the quality of those in the hills behind, just as we should expect.[77]

In this province as well, amphora studies have been helpful. We have imitations of Dressel 1b amphorae from local sources from Augustan times. Variations are numerous here, especially of Dressel 2-4 types. Many are stamped, or rather imprinted, so their provenience can be definitely established. Several vineyards and ateliers in Tarraconensis belong to well known Italian families, including C. Massidius Nepos, father of L. Massidius, proconsul of Sicily under Augustus, and perhaps M. Porcius Cato, named suffect consul in 36 BCE. Find spots for Tarragonese vessels include Gaul, the Gaulish coast (shipwrecks) and Ostia. Amphorae with flat bases, that is, Galloise 4 and Oberaden 74, are common from the Augustan age, and those with pitched interiors may be assumed to be wine jars. Centers of production for these jars are now well established, lying all along the eastern littoral from Ampurias (La Escala) to Valencia.[78]

Diffusion of Spanish wines was extensive. Baetic wines were shipped all over western Europe's littoral and along the major rivers. Find spots in Spain occur along the Ebro River, the eastern littoral, Majorca, both sides of the Straits of Gibraltar, the western littoral, and up the Tagus (Tejo) and Duoro rivers. In France they occur all along the Mediterranean littoral as well as along the Rhone and Moselle. In Germany they occur along the Lower Mosel and Rhine. Several have been found in southeast England. In Italy they occur in Sardinia, Elba, along the Tiber, especially in Rome, and around the Bay of Naples. Tarragonese wine amphorae occur occasionally in Spain and North Africa but are concentrated in Gaul along the Garonne and Gironde, along the Mediterranean littoral, along the Loire, Rhone and Sâone. In Germany they are found along the Rhine. They are widely dispersed in southern England. In Italy

77 Éteinne and Mayet (2000): 101-210.

78 Étienne and Mayet (2000): 109-210.

they occur in Liguria, Ostia, Rome and Sicily. Those stamped from M. Porcius alone are widely diffused in Gaul. In general Baetic wines are widely diffused but never frequent, whereas, beginning in the Augustan period, the frequency of Tarragonese wines in Gaul and along the Rhine is quite stunning.[79]

Finally, although the Balearics administratively were attached to Tarraconensis, they are habitually spoken of separately in the literary sources (Balearides or Gymnesies). Colonized beginning in 122 BCE, the Balearics soon saw the cultivation of the vine and the olive. Cicero (*de Rep* 3.19) and Pliny (*NH* 14.71) speaks glowingly of the quality of the wines.[80]

Wine was also grown on a commercial scale in Portugal, Roman Lusitania. Athenaeus (8.330e) cites one of the lost books of Polybius' history that wine was a native product in Lusitania in his own day (c. 200-118 BCE) and was in fact quite cheap. Since Polybius was an eyewitness, Portuguese viniculture from at least as far back as the second century BCE is securely established, and may have been introduced with the first colony in 197. We can tell quite a lot about Lusitanian viniculture from tomb monuments constructed in stone in the shape of wine barrels. The fact that these also date from the second century suggests that viniculture on a commercial scale (transport in barrels) was early as well. Viniculture apparently followed Roman colonization along the course of the Tagus River. The majority of these funerary barrels have been found in the vicinity of Beja, that is, along the central Guadiana River which formed the frontier between eastern Lusitania and southwestern Baetica.[81]

Further Afield

As we have seen, one of the most important treatments of viniculture in Rome was written by the Carthaginian Mago (date uncertain), whose advice on agriculture was considered so important that his treatise was translated into Latin at public expense. Ironically, there is little archaeological evidence of wine production in *Africa vetus* (i.e., the area of Carthage), but considerable in *Africa nova,* that is, Roman Africa, especially in the valley of the Bagradas River (today the Medjerda in northwest Tunisia) and especially under the auspices of the imperial *conductores* after the imperial fisc arrogated huge plots of North African land.[82]

---

79 Éteinne and Mayet (2000): 213-45.

80 Étienne and Mayet (2000): 243-45.

81 Étienne and Mayet (2000): 15-58.

82 Arnaldo Marcone, *Storia dell'agricoltura romana: dal mondo arcaico all'età imperiale* (Rome, 1997): 185. Cf. Michel Bonifay, "Le vin d'Afrique," in Brun, Poux and Tchernia (2009): 288-9.

In the east, Probus' famous edict to the Gauls 'to plant vines and grow rich" (Eutropius *Brev.* 9.17; Flavius Vopiscus, *Probus* 18; Aurelius Victor, *Lib. de Caes.* 37.3) has been taken, not as an extension of the rights of non-Roman citizens to grow vines, but as a geographical extension of viticulture, not only in Gaul but in the Danubian provinces of Illyrium and Panonia. Recent research which establishes the Heunisch cultivar as one parent of a number of western European cultivars has given rise to the theory that Probus instructed the legions to disseminate this hardy 'Hunnish" native of his own Panonia as they made their way into western France and Germany.[83] If true, it was hardly the revolutionary development the authors of the book in which the theory appears would have us believe, but it still illustrates well how viticulture has been extended, namely by the intercrossing of a cold-hardy variety with one or more of the classed Roman ones.

Western, so-called 'Rough' Cilicia probably developed viticulture under Roman influence, though of course viticulture had been known in Anatolia for millennia. Rough Cilicia was noted by such writers as Pliny (*NH* 14.81-2) as a source of *passum*, raisin wine, and this wine was still highly regarded in the mid-tenth century (*Expositio Totius Mundi* 39). There was even a kosher form exported to Roman Palestine for Passover. And we now have considerable archaeological evidence of viticulture here.[84] Finally, wine culture if not viniculture was exported by Rome even further afield, to Arabia Felix and India.[85]

---

83 Attilio Scienze, "La storia della civiltà europea racconta attraverso l'origine dei suoi vitigni," in Giovanni Negri and Elisibetta Petrini, *Roma, Caput Vini* (Milan, 2011): 181-202.

84 Nicholas K. Rauh et al., "Viticulture, Oleoculture and Economic Development in Roman Rough Cilicia" *Münstersche Beiträge zur antiken Handelgeschichte* 25.1 (2006): 49-98.

85 Caroline Singer, "The Incense Kingdoms of Yemen," in Peacock and Williams (2006): 10-25; André Tchernia, "The Dromedary of the Peticii and Trade with the East," in F. DeRomanis and A. Tchernia, edd., *Crossings: Early Mediterranean Contacts with India* (New Delhi, 1997): 238-49; Grant Parker, *The Making of Roman India* (Cambridge, 2008): 147-202; Elizabeth Lyding Will, "The Mediterranean Shipping Amphoras from Arikamedu," in Vimala Begley and Richard Daniel De Puma, edd., *Rome and India: The Ancient Sea Trade* (Madison, WI, 1991): 151-6; Kathleen Warner Slane, "Observations on Mediterranean Amphoras and Tablewares Found in India," op. cit. (1991): 204-15.

CHAPTER 3

Viticulture

A bad wine can be, and sometimes is, made from good grapes, but an excellent wine can never be made from bad grapes. A wine can be no better than the grapes from which it is made. Winemakers, even today, freely acknowledge that great wines are made largely in the vineyard, not the winery. Thus viticulture is a critical element of our story.

Fortunately for us, we are in a particularly good position to know a great deal about Roman viticulture. The Romans were heirs to several millennia of viticultural experiment, much of which had been codified and transmitted in written form by the first century of the common era. And most of that lore was codified and transmitted by the Romans themselves, both within Italy and abroad in the western provinces where she had taken the vine. One impetus more than any other drives that dynamic, namely, the establishment of commercial villa farms in the Italian heartland, beginning in the second century BCE, an inchoate agribusiness, so to speak.

We should be clear that good wine has always been made, even up to this day, with the simplest of operations. All that is essential to produce good wine is a small plot of fertile, well situated land, good vines, favorable climatic conditions at harvest, a small tub to tread the grapes in, and a reasonably anaerobic container in which they can ferment, and hermetically sealed containers in which the wine so produced can age and develop its wonderful nuances. These and obsessive care.[1]

But to produce wine at a commercial level requires considerably more land and equipment, and the Roman *villa rustica* answered that need. Romans had discovered that viniculture could be enormously profitable. Combined with the fact that it was considered *déclassé* for members of the senatorial class to be engaged in commerce, with the singular exception of farming, and there was a strong incentive for capital to flow into this outlet. Thus the famous Mediterranean Triad of wheat, olives and grapes, grown for sale and export.

We are triply blessed to have three agricultural manuals specifically focused on villa farming for gentlemen farmers, the earlier *De Agricultura* of Marcus Porcius Cato from c. 180 BCE, the *De Re Rustica* of the Augustan scholar Varro,

1 For camparative practices in Egypt: Clotilde Ricci, *La cultra della vite e la fabricazione del vino nell'Egitto Greco-Romano* (Milan, 1924): 31-46; in the modern Vesuviana, Silvestro Sannino, *Civiltà agricola vesuviana* (Salerno, 2009): 181-205.

 | DOI 10.1163/9789004334595_004

and the *Res Rusticae* of the mid-first century agronomist Lucius Junius Moderatus Columella. Columella was an immigrant from Spain where his family owned estates and he therefore gives us considerable insight into the phenomenon of villa farming in the provincial hinterlands as well. The encyclopaedist Pliny the Elder, whose monumental natural science touches on almost every branch of human knowledge of the time, has yet another vinicultural *vademecum* embedded in his manual on plants. Add to these the hundreds of incidental references to grapes and wine. We are especially fortunate that these literary efforts span the better part of three centuries so that we can, to an extent, witness the birth, childhood and early youth of commercial viniculture in the Roman world.

Establishing the Vineyard

The vine is a remarkably adaptable species, as we have seen, but a healthy environment in which to grow can go far towards ensuring the success of commercial viniculture. In establishing a vineyard the ancient vigneron as his modern counterpart had three options: he could assume the care of an existing vineyard, he could establish a new vineyard on farmland previously used for other crops, or he could establish a new vineyard on virgin land. In all these cases he was required to rigorously evaluate the elements which collectively we moderns refer to as *terroir:* exposure or aspect, slope, soil quality, climate, and varietal choice.[2] Of the three options, Columella (*RR* 3.11.1-6) advises that the best site is one previously untilled rather than one with grain fields or a plantation of trees or vines. The worst possible site is an old, neglected vineyard; not only is the soil exhausted but it is a tangle of old roots (vine roots can grow to depths of 9 m / 30′) and still has the 'poison' of old age in it. Elsewhere (*De Arb.* 3.5) he recommends that an old vineyard not be replanted to vines for at least ten years. Wild land is best, easily grubbed out since bushes and wild trees have only surface roots. Next best land is plowed land free of trees, then a plantation of vines on widely spaced trees, the *arbustum* training system, or an olive grove, but preferably old olive trees that have not been wedded to vines (the *arbustum* system again). But if the choice of an old vineyard is necessary, all old roots must be laboriously grubbed out and burned, the ground generously manured and new topsoil brought in from bramble thickets (i.e., soil amenable to vining plants).

2 A good introduction to these elements, accessible to the layman, is James E. Wilson, *Terroir* (Berkeley, 1998): 8-49.

The ancients were perfectly aware of the elements of good terroir, if not the science behind them. Columella recommends a site with moderately loose (good drainage), moderately rich soil, on a gentle slope, in a moderately moist climate, a plot with good drainage and water retention but not too many springs, with water that is neither bitter nor brackish, located in a moderate climate. Too much rain is far worse than too much sun. All perfectly true. Thus the famous taxonomy of Falernian wines where the highest quality of these prestigious wines was grown on the upper slopes, so steep, in fact, that the land had to be terraced. "Traces of terracing can still be seen along the slopes of Monte Massico, and although these have not been dated it seems probable that many go back to the Roman period. Certainly the slopes must have been terraced by the Romans for vine cultivation to have been possible here. These terraces have now been abandoned and vine growing is practiced only on a small scale down on the plain."[3] This brings up an interesting question, namely, which came first, the terraces or the *premier crú*? Did the reputation and consequent higher price of Falernian justify the expense of terracing the higher slopes, or had the ancients stumbled on one of the secrets of growing great wine, namely that wine grows well on slopes, a commonplace, but that grapes grown on upland slopes and, by necessity, on terraced land, produced wine of superior balance and complexity? Something like that was also happening in Magna Graecia from the fourth century BCE, as campestrine areas along the Ionian coast were progressively abandoned to cultivation of grain, and viticulture moved inland into the hills and foothills of mountains.[4]

As to the size of the farm, again the ancients knew well that grapes could be grown for crop size or for quality, and that the higher price commanded by outstanding grapes could offset limitations on crop size. Thus a small plot with excellent terroir was ideal for growing high-quality grapes to produce prestige wines, whereas larger plots in lesser terroirs could still be profitable if they produced in abundance. Columella (*RR* 4.3.6) cites an example from his contemporary, Julius Graecinus, who wrote a manual of viticulture which, unfortunately, has not survived. Graecinus tells of his neighbor, Paridius Veterensis, who had two daughters and gave one-third of his vineyard to the elder as dowry but made the same revenue from the remainder by tending it more carefully, then repeated this success when he gave a second third to his younger daughter and still suffered no loss of revenue.

Poor terroir has both poor soil quality (what modern agronomists call poor pedology) and/or poor climate (Columella *RR* 4.22-89). Poor soil is infertile,

3 M. Aylwin Cotton, *The San Rocco Villa at Francolise* (London, 1985): xx.

4 Vandermersch (1994): 44.

salty, acidic, and/or poorly drained. Columella includes elements that we would call aspect, namely, soil on a steep or rugged slope, or in a valley too dark from lack of sun exposure. Specifically, poor soil has too much sand, tufa, lean gravel or hard gravel with no organic matter. If, on the other hand, an old vineyard is vigorous but unproductive because of bad stock (*mali generis*) or shoddy culture, it may be profitably rejuvenated by ingrafting better vines on old rootstocks.

The ancients strove mightily to develop hard-and-fast rules on aspect or exposure, all of which Pliny ultimately rejects, correctly, as it happens, since good aspect depends on prevailing climate and winds as well as the nature of the soil, the vine variety, and the poximity of the vineyard to bodies of water. (Pliny *NH* 17.19-24; Cf. Columella *RR* 3.12.5). Columella recalls the great disagreement as to aspect among authorities such as Saserna, Tremelius Scrofa, Vergil, Democritus, and Mago. The first two are Romans who had written agricultural treatises, and we know that Scrofa had, along with his wife, estates in Campania and Sabine territory. Vergil, of course, is from Cisalpine Gaul; Democritus is Greek and Mago is Carthaginian. Little wonder that they can offer no definitive advice on aspect. But in general, says Columella, in colder northern climates one looks for southern aspects, in moderate climates eastern ones, unless, that is, prevailing winds from the south or southwest preclude this as they do in Baetica, but in hot climates such as in North Africa (Mago) one looks strictly for northern aspects.[5] Pliny (*NH* 18.334) says that vines here should be planted facing Boreas, the northeast wind.

Recommendation of gentle slope is correct since it improves drainage and prevents so-called 'cold-damming', the tendency of cold air to pool in low-lying pockets. As it happens, the recommendation can now be demonstrated archaeologically: the large vineyard in Pompeii in the so-called 'Forum Boarium' of the northeast sector which was famously re-excavated by Wilhemina Jashemski[6] had a definite slope away from the *triclinium* and pressroom in the northwest corner, in places as much as 5%. The loose, volcanic soil will have been able to absorb or drain most rainfall, but a depression down the center of the north-south footpath suggests that it was also used for

5 Cf. Joseph A. Greene, "The Beginnings of Grape Cultivation and Wine Production in Phoenician/Punic North Africa," in McGovern, Fleming and Katz (1966): 314-15. The areas suitable for viticulture in North Africa are strictly the northern coastal zones where the mountain ranges are oriented southwest to northeast and north-facing slopes catch almost all the available moisture coming from the Mediterranean. Ergo, both Mago and Columella are perfectly correct. Not only are southern slopes in the 'rain shadow' but their southerly aspect allows too much radiant heat and drying.

6 Jashemski (1979): 214-15.

drainage of heavy rainfall. The small portion of the east-west footpath excavated also showed evidence of drainage toward the north-south footpath. Contour of the soil surface in undamaged areas showed that the vineyard was hand cultivated, not plowed. For reasons to be discussed below, this method was more effective in creating good tilth; loose soil tilth conserves water and there is barely enough rainfall in this section of Italy today to raise grapes without irrigation. On the other hand, too much moisture is actually a worse problem since the vine cannot stand to have its 'feet' in the water for very long, and thus Columella gives specific directions for ditching land cleared for cultivation which is too moist and/or has poor drainage (*RR* 2.2.9-11).

Another truism is that vines, especially in cooler climates, love to look at bodies of water, a dictum that the ancients obviously knew well. Reflected sunlight from bodies of water can raise both temperatures and sun exposures significantly. Pliny (*NH* 17.29) mentions an instance in Thessalian Larissa where the draining of a lake resulted in such a dramatic lowering of ambient temperatures on the surrounding slopes that olives could no longer grow there and vines were often frostbitten (*aduri*), a problem that had never happened before. Likewise Aenos (in modern Bulgaria) experienced a rise in temperature when the Maritza (the modern Cerna) River was diverted near to it. We are reminded of those areas along the Moselle and Rhine at the northernmost limits of grape culture where successful viticulture is possible only on steep, south-facing slopes which overlook expanses of the river, especially in the bends.

Grapes are remarkably adaptable when it come to soils, growing happily in every type from sand to almost pure gravel. The key is to tailor cultural practices to the specific type of soil. The Romans were aware of this but also knew that the expense of some cultural practices made viticulture on some soils unprofitable. Thus they made concerted efforts to classify soils on a scientific basis. An effort doomed to failure, of course; as White adduces in his masterful disquisition on the subject,[7] without the rudiments of chemistry any such effort must rely on purely empirical criteria and these sometimes lead to gross blunders and even superstitions. Still, it is quite impressive how well our agronomists did, given the limitations of their theoretical frame of reference.

Four basic elements define soil quality: prevailing texture, i.e., proportions of sand and clay with other constituents; soil structure, i.e., the size of various particles in the soil; moisture status, the ability of the soil to retain and drain

7 K. D. White, *Roman Farming* (Ithaca, NY, 1970): 86-109. The testimonia are found in the appendices.

water, not to be confused with moisture content at any given time; and soil temperature.

Soil structure and texture combine to create so-called tilth and it is obvious that the Roman classification system begins to 'leak' when they make distinctions beyond this level. Varro (I.9) has the most complete taxonomy. He describes twelve types of particles present in soils: rock, marble, rubble, sand, loam, clay, terra rossa, volcanic dust, chalk, ash, gravel and *carbunculus*. He further characterizes soil texture as slightly or moderately or heavily rocky, etc. The elusive *carbunculus* illustrates the limitations of the empirical system. The term, literally 'glowing ember', suggests red soils, and red soils prevalent in central Italy are typically either red ocher, clays tinted with iron oxide to give them their red color, or terra rossa, decayed limestone also tinted by iron oxides. It is clear from the testimonia that our authors either use the term interchangeably for both soils, or use it for one or the other but in the opposite sense from other writers. Thus the confusion as to the meaning of the term. But it is equally clear that, in the matters of most import such as how the soil must be tilled they were perfectly capable of astute distinctions despite the apparent confusion. Hence Pliny (*NH* 17.33) advises that some red soils are difficult to work and have poor drainage. Precisely. Red ocher is so permeated with clay that it absorbs and retains moisture and thus stubbornly clings to agricultural implements. Other red soils (terra rossa), Pliny adds, are difficult to break up but once reduced to the proper tilth provide excellent drainage and produce good crops. We are reminded that terra rossa soils of the Coonawarra region produce some of Australia's finest wines.[8]

Soil structure is very loosely classified by the agronomists as dense (*densus*, *spissus*) or loose (*rarus*, *solutus*). The ancients tested for density and friability by kneading a small amount of soil in the hand with the addition of a small amount of water to see how readily it clumped and then broke apart when thrown to the ground (though note that Vergil confuses structure with 'fatness' or fertility: *Geo* 2.248-50), or by digging a trench and refilling it to see how much superfluous soil remained afterward. Such tests are crude, but reasonably effective.

8 On red clay soils, White [(1970): 56, n. 23] says of Varro's *carbunculus* (1.9.2) where he defines it as a soil condition rather than a type of soil that "the name may have been attached to this soil by reason of the curious 'carbuncular' effect produced on the surface when exposed to hot sun, causing protuberances to rise." White cites Professor E. R. Orchard of the Department of Soil Science of the University of Natal, who reported the condition in parts of the former Orange Free State.

Again, the ancient sources are obviously incapable of making distinctions between moisture status and moisture content but knew perfectly well that dense soils with poor drainage were unsuitable for vines. The aspect of temperature they treated only cursorily and only in the baldest terms of cold (*frigidus*) and warm (*calidus*). Otherwise they evaluated soils, in addition to the simple tests mentioned, by color (red, black); by smell when newly plowed; by taste after passing water through soil in a wine strainer and then sampling the liquid for acidity and salinity; and most important by far, by observation of the type of ambient vegetation it supported and the types of healthy crops it supported elsewhere in the same area.

For the vineyard, Columella (*RR* 3.12.1-5)[9] recommends specifically black, moderately loose and easily worked soil, so-called *pulla*. This is specifically the black volcanic soil of Campania. Even such soil mixed with tufa stones (*tofum*) can be used, since the stones when brought to the surface will be weathered by storms into gravel. Gravel and pebbles in soil are good if they occur in 'fat' (rich) soil. Flinty soil (*silex*) is also friendly to vines. We are reminded that some of the finest wines in the world come from soils that are dominated by gravels; perhaps the most famous vignoble in Bordeaux, for example, Graves, is named for its gravel soils. Clayey soil is serviceable unless it is predominantly clay (*argilla*), in which case it is decidedly hostile to the vine, as are pure gravel and excessively wet, salty, 'sour' (acidic) or dry soil. Black or reddish sand is suitable when mixed with loam, but soil with a preponderance of 'carbuncle' (terra rossa in this case) must be heavily manured. Red ocher (*rubrica*) is heavy and offers little purchase for vine roots, but once a root system is established is nourishing. On the other hand it is difficult to work when wet because it is sticky, as well as when it is completely dry because it is too hard. Soil constitution and texture can be modified, says Columella, though this is obviously labor-intensive and potentially expensive and thus to be avoided if possible. Columella mentions his uncle, "a very learned and careful farmer," who modified the tilth of his soil by mixing clay in gravelly soil and gravel in stiff, clayey soil (*RR* 2.15.4). About such modifications we will have more to say under the heading of fertilizers.

Finally, some general considerations. The best site for a vineyard should have access to osier willows and cane thickets as well as to mixed forests or at least chestnut groves (*RR* 4.30-33; Pliny, *NH* 141-53). These provide the ties, props, stakes and trellises essential for training vines, as well as for any number of other farm items and procedures. Elsewhere Columella gives specific directions for the establishment and/or cultivation of each. Very arid environments

9 Cf. Francesca Piccoli, *Il vino del nord Italia in epoca Romana* (Verona, 2004): 43-6.

which do not permit cane and willow thickets can be planted with broom (*genista*) which can be bundled to make vine trellises and can be used individually as ties. Otherwise bramble (*rubus*) may be used in the same way. Considerations also include the relative safety of the neighborhood and the nature of local transport (Varro *RR* 1.16). The relative cost of transport in antiquity was high and that of land transport was exponentially dearer than riverine and especially sea transport. Thus availability of affordable transport could easily determine the profitability of a vineyard. And archaeology now richly documents the propensity for villa farms to be located along navigable bodies of water and along primary and secondary roads.[10]

Finally there was the question of which vines to plant in a new vineyard, which scions to graft in an old one. The ancients at least as early as the Greeks were aware that specific microclimates and terroirs were suited to specific varietals (Theophrastus 2.3.2; Cato *Agr* 24 & 51; Columella *RR* 3.2; 3.21.8; 12.44; Palladius 3.18.1-2). As early as Cato's treatise, for example, best soil and aspect demanded the small Amminean (*amminium minusculum*: *Agr*. 6.4), the twin Eugenum (*geminum eugeneum*) and the small Helvolan (*helvolum minusculum*). Heavier soil or terroir subject to fogs called for the large Amminian (*Amminium maius*) or Morgantina (*Murgentinum*), the Apician or the Lucanum. Other varieties, especially hybrids (*miscellae*) are adaptable to almost any site. By the time of Columella some 260 years later, it seems the Amminian had fallen out of fashion and the agronomist was at pains to defend the noble vine (*RR* 3.6-7). Palladius, writing in the fifth century CE but carefully respecting the elder statesmen of viticulture, especially Columella, asserts that the Amminian alone returns a noble wine wherever it is planted.

By the first century CE there were well over two hundred varieties named, though many of the names are doubtless synonyms. Varieties are rated for their productivity, flavor, aging potential, ability to reliably set fruit, 'dregginess', affinity for specific training regimens, tolerance of adverse weather conditions such as winds and storms, time of budbreak and flowering and therefore length of maturation, affinity for soil fertility, tilth and moisture, and healthfulness of the wine produced (Columella *RR* 3.2.14-29). Columella has a detailed description of a number of varieties (*RR* 3.2.3-13) and asserts the general rule that if the site will produce a noble wine it should be planted to noble grapes, even if only moderate producers, since the higher prices of the wine so produced will compensate, but if the site will not support a noble varietal it should be planted to the most prolific vine it can sustain. In general relatively

10 Annalisa Marzano, *Roman Villas in Central Italy: A Social and Economic History* (Leiden, 2007):156-65.

flat campestrine terroirs yield greater crops of lower quality whereas sloping hilly land produces lower yields of higher quality. The dictum is as true today as it was two millennia ago, incidentally. In a temperate climate like coastal and piedmontese Italy, hills with a northern aspect are more productive, while those with a southern aspect yield less but of higher quality. In general, we are told, the most adaptable vine of all, except in cold climates, is the Amminian, the Chardonnay of its age.

Ergo, for level terrain the ancient vigneron is urged to choose a cultivar tolerant of mists and fogs, for a slope one tolerant of drought and wind (*RR* 3.1.3-7; cf. Palladius 3.18.1-2). Fat, fertile soils require a 'thin' vine which is not overly productive, lean soils a more productive vine. Heavy soils (poor drainage) demand a vigorous cultivar which produces much wood and foliage (greater transpiration) while loose, rich soils demand the opposite. Moist soil demands a variety with small, thick-skinned berries, dry soil the opposite. Cold or foggy microclimates demand an early ripening variety or one with firm, hard-skinned berries, as do windy and unsettled climates. Warmer, gentler climates demand vines with more delicate berries. Dry climates will sustain cultivars which are habitually rotted by too much rain or heavy dews, dewy climates can be planted to varieties that tolerate dry conditions. A climate subject to hailstorms demands cultivars with tough, broad leaves capable of protecting the fruit.

Ultimately (*RR* 3.2.31) Columella approves the general dictum laid down by Cornelius Celsus and Cato before him that no variety should be planted unless approved by local consensus and none retained unless proved in the laboratory of practical experience. And in any case it makes good sense to grow several cultivars on the same farm since adverse weather affects them differently and they mature at different rates. But, be it noted, Columella (*RR* 3.21.10) would have us grow them in different plots, not in the mixed plots so popular in many parts of Italy, so that they can be vinified and stored separately. Then at the time of bottling they may be sold separately to maintain their varietal character or blended to achieve a more balance product, just as is done today.

As previously noted, by the first century of the Common Era a bewildering number of cultivars are named by our sources.[11] Heroic efforts have been made to identify some of these varieties with modern counterparts. Older scholars

11 Those interested should consult Raymond Billiard, *La Vigne dans l'antiquité* (Lyon, 1913): 304-17; Jacques André, "Contribution au vocabulaire de la viticulture: les noms de cépages" *Revue des Études latines* 30 (1953): 126-56; Dion (1959): 121-6; Tchernia (1979): 91; Piccoli (2004): 24-30; Mario Fregoni, Le varietà di vite a Pompei all'epoca di Plinio il Vecchio," in Di Pasquale (2010): 192-3; Scienza (2011).

relied heavily on linguistic evidence. Thus Dion[12] followed André in deriving the French cultivar Ugne from Greek *eugeneia* and conjectures it may have been introduced during the seventh century BCE, during the Greek colonization of Provence. If true, the unassuming little vine has done well for itself; the Ugne itself is so pedestrian a grape that several attempts have been made to ban its use in French viniculture, but recent DNA studies have confirmed[13] that Ugne is parent, along with the Pinot, of the world's most versatile white cultivar, Chardonnay. Elsewhere, the first mention of the region of Bordeaux in Strabo (4.2.1) says nothing of viticulture here but mentions the tribe who occupied it, the Bituriges. After the establishment of the Pax Romana they had successfully initiated viticulture by establishing a cultivar, the *biturica*, capable of sustaining this northern climate. This vine, according to Columella (*RR* 3.2.19) was imported from Spain where it was called *cocolobis*. Pliny (*NH* 14.27) remarks its ability to withstand cold without shatter (Fr. *couler*) and its tolerance for cold, wet, stormy terroir, as does Columella. Dion associates this vine with the famous Cabernet, one of whose dialect names is *vidure*, thought to be a deformation of *biturica*. The impetus for the adoption of the vine will have been the huge profits to be made in the wine trade with the British Isles, especially Ireland (Tacitus, *Agricola* 24.1)[14] The same DNA studies previously mentioned have now established that the parents of the world's most famous red grape, Cabernet Sauvignon, are the Cabernet Franc, perhaps the original *biturica*, and the local *Sylvestris* grape. Elsewhere Dion again follows André[15] in seeing the *duracina* grape of Cato and others as the source of the Narbonnaise *duraze* (Pamiers), *duras* (Gaillac), *durazaine* (Saint-Peray) and *duret* (Annonay). Elsewhere Piccoli[16] finds the source of the noble Piedmontese grape, Nebbiolo, which is "nourished by the moisture of the clouds," in the famous Gallo-Roman *spionia/spinea*, a grape originally peculiar to Ravenna. Is this the same as the *uva Gallica* cited by Pliny (*NH* 14.34) as characteristic of Cisalpine Gaul? If so, *spionia* will be a Gallo-Latin term to designate a vine of northern Italian origin, a product of the intercrossing of local wild vines and cultivated varieties, the

12 Dion (1959): 78; André (1953): 53. Vandermersch [(1994): 44-52] tries his hand as well.

13 Carole P. Meredith,"Science as a Window into Wine History" *Bulletin of the American Academy of Arts and Sciences* 56.2 (Winter, 2003): 54-70. Attilio Scienza ["L'origine dei vitigni coltivati. Una storia interdisciplinare della cultura europea," in Di Pasquale (2010): 25-31] has a good discussion of the theoretical mechanics of varietal diffusion and hybridization.

14 Dion (1959): 121-6, 134-66; Etienne and Mayet (2000): 7; Frédéric Berthault, "*Biturica*, le vin des Bituriges," in Brun, Poux and Tchernia (2009): 236-8.

15 Dion (1959): 109-17; André (1953): 365.

16 Piccoli (2004): 21-4.

name deriving from *spinus*, 'wild plum', due to the shape of the berries. We may compare Novarese *spanna* / *sponia* / *spolnia*, synonyms of Nebbiolo. Connection with the plum is evidenced by the Ossolan name of the *spanna*, namely *pruneto*.

So goes this line of inquiry. We are actually on the cusp of being able to make some definitive determinations about some of these issues, thanks to DNA mapping of microsatellites. For one thing, we now know the mechanism by which red-berried varieties have become white (all wild grapes are more or less red),[17] a mutation which is thought to be responsible for many other elements of varietal character. Secondly, the grape genome was early on completed and DNA sequencing gives us a powerful new tool, especially useful since vines are propagated clonally.[18] But, of course, such analysis depends upon the availability of 'ancestral' DNA as well as that of 'offspring' (strictly speaking not a proper term for clones), and that is a luxury we rarely have. Otherwise we are thrown back on etymology for the 'ancestral' element, and that is hardly satisfactory because of the dizzying amount of homonymy (the same name given to different cultivars) and synonymy (different names given to the same cultivar) that we find in varietal nomenclature.

DNA analysis relies on analysis of so-called 'microsatellite markers', genes which carry discriminating traits. For example, it has been thought that the Pinot grape has been in Burgundy since before the Roman conquest, and it is now known that this cultivar is the progenitor, via a simple genetic mutation of the sort previously described, of Pinot Noir, Pinot Gris and Pinot Blanc as well as Meunier. And as we have seen, through Pinot blanc and the ubiquitous Heunnisch, which appears in Burgundy as Ugni Blanc or Gouais Blanc, of Chardonnay as well. But for the attribution of Heunnisch as the vine which the emperor Probus brought to France as a gift, we have only conjecture if not outright speculation.[19] Another example: Pliny and others identify Raetica as an outstanding white grape of Raetia, the Alpine province. Etymologically and geographically, the white Rèze of Valais, Switzerland, should be its modern form. Authors of an important study[20] have analyzed a number of cultivars using genotype databases and genetic markers and have found that twelve

17 Shozo Kobayashi, Nami Goto-Yamamoto and Hirohiko Hirochika, "Retrotransposon-Induced Mutations in Grape Skin Color," *Science* 304 (14 May 2004): 982.

18 Pinder and Meredith (2003): 260-73.

19 Pinder and Meredith (2003): 260-73; Meredith (2003): 61-4.

20 José Vouillamoz, Anna Schneider and M. Stella Grando, "Microsatellite Analysis of Alpine Grape Cultivars (Vitis vinifera L.): Alleged Descendants of Pliny the Elder's *Raetica* Are Genetically Related," *Genetic Resources and Crop Evolution* 54 (2007): 1095-1104. Cf. Marguerite Gagneux-Granade, "Retica, le vin des Rhétes," in Brun, Poux and Tchernia (2009): 239-40.

cultivars show first-degree (parent-offspring or full-sibling) relationship or second-degree (grandparent-grandoffspring, uncle-nephew or half-sibling) relationships with Rèze. Four had parent-offspring relationship, namely Cascarolo Bianco (Piemonte, Italy), Arvine Grande (Valais), Gropello di Revò (Trentino, Italy) and Nosiolo (Trentino). Unfortunately, analysis of ancient DNA from grape pips derived from Roman archaeological sites was unsuccessful, so there is as yet no genetic evidence that any of these is related to the famous Raetica.

Particularly promising on this front, however, is the 'Progetto Vinum' cosponsored by the University of Sienna and the Associazione Nazionale Città del Vino.[21] Though strictly preliminary, the results do not look good for some of the linguists: For example, it was Della Porta as early as 1592 who first made the association Aglianica-hellenica-helvola. Thus the southern Italian cultivar Aglianico was presumed to be the ancient Helvola. But earlier documents had associated the name with Aragonese Spanish llano, 'plain', and thus 'the grape of the plain (Campania)'. Recent DNA researches, as reported by Attilio Scienza,[22] confirm a great genetic distance between Aglianico and any Greek species.

The Farmstead

Finally, if he is establishing a new vineyard, our vigneron must build a facility where his vinicultural activities will transpire. The size and style of the farmstead will be proportional to the size of the farm, of course, but here our literary sources focus exclusively on the commercial vineyard and thus on the famous *villa rustica*. We should be reminded, however, that farmsteads, like all other vinicultural elements, doubtless ran the gamut from simple to elaborate. Viticulture was organized in one of four main ways in Rome. There were the simple peasant smallholders who practiced viticulture as part of mixed

21 Cf. Andrea Masi, "'Progetto Vinum': Archeologia del siti campionati e primi risultati," in Ciacci, Rendini and Zifferero (2007): 201-16; Serena Imuzio et al., "'Progetto Vinum': Metodi analisi del genoma e primi risultati' (op. cit.): 238-48, an excellent and concise synthesis of the botany of the vine from earliest beginnings to modern molecular archaeology; Andrea Ciacci and Andrea Zifferero, "Il 'Progetto Vinum': Prime considerazioni conclusive," (op. cit.): 249-72.

22 Attilio Scienza, "Per una storia della vitcoltura Campana," in A. Scienza and M. Boselli, edd., *Vini e vitigni della Campania. Tremila anni di storia* (Naples, 2003): 56. Cf. Giuseppe Guadagno, *Campania antica: storia, archeologia, vita quotidiana* (Benevento, 2009): 15 and n. 25.

farming in order to eke out a subsistence for their *familiae* and perhaps a bit more; there were the peasants who no longer owned the land but continued on the ancestral plot as tenants to wealthy landlords (*coloni*) or as sharecroppers (*partiarii*), and there were the freeholders who practiced commercial farming, from a modest level, where the landlord was resident in the farmstead and supervised and perhaps worked along with the slave laborers, to the owners of huge plantations (*latifundia*) only periodically in residence if at all, such plantations being worked by large slave gangs overseen by *vilici*.[23]

As usual in such matters, we are grossly ignorant of the farmsteads of smallholders.[24] Archaeological evidence is exiguous, usually no more than post holes and trenches, and thus our best conception of the *casa* or *tugurium* (Pliny *NH* 16.14; Vergil *Ecl.* 1.69; Col. *RR* 12.15; Festus, s.v. *tugurium*) can be gained from the little archaic 'hut urns' used to house the ashes of the cremated from the Late Bronze Age and Early Iron Age. This is a round or ovoid or roughly rectangular hut with a hip or gabled roof, wattle-and-daub walls, clear evidence of roof trusses, a thatched roof, and a perimeter trench at the fall line of the eaves to channel away rain water, whose infiltration is deadly to clay walls. The pediments at front and back are left partially open as smoke holes for a circular hearth, typically a low platform of hardened clay found in the middle of the floor. The most carefully surveyed part of Italy outside the environs of Pompeii is doubtless southern Etruria, and here evidence for smallholdings with *tuguria* has been found in contexts which suggest polyculture that includes vineyards. For example, from Monte Forco in the Ager Capenus comes a *tugurium* 10.95 × 5.15 m in size with an enclosed farmyard outside and a projected size of the estate of ten *iugera* (2.5 ha/6.3 acres). Likewise at Crocicchie on the Via Clodia comes a small *tugurium* associated with another building interpreted as a barn, constructed in early Imperial times. Hence, the same basic form after a millennium. It may be significant that the *tugurium* at Crocicchie later received a bath complex, suggesting that the proprietor was

23 White (1970): 384-414.

24 Cf. Carla Fayer, *Aspeiti di vita quotidiana nella Roman arcaica* (Rome, 1982): 221-40; Timothy W. Potter, *Roman* Italy (Berkeley, 1987): 113-16; Gilda Bartoloni, *La cultura villanoviana. All'inizio della storia etrusca* (Rome, 1989): 110-18; Alfonsina Russo Tagliente, *Edilizia domestica in Apulia e Lucania: ellenizzazione e società nella tipologia abitativa indigena tra VIII e il III secolo a.C.* (Congedo, 1992): 25-34; J. Rasmus Brandt and Lars Karlsson, edd., *From Huts to Houses : Transformations of Ancient Societies ; Proceedings of an International Seminar Organized by the Norwegian and Swedish Institutes in Rome, 21-24 September 1997* (Stockholm, 2001); cf. Vitruvius, *de Arch.* 3.1.2-3.

FIGURE 7 Tugurium. *From Alfonsina Russo Tagliente,* Edilizia domestica in Apulia e Lucania *(Galatina, LE, 1992): Fig. 2a.*

thriving.[25] Nor does the archaeology of this region suggest that such smallholdings were being gobbled up by plantations; in fact, they are common here well into imperial times, despite the fact that a great deal of land was indeed held by a wealthy elite, perhaps weekend 'hobby farmers' from Rome. One such simple structure contained a number of luxury goods, suggesting that the inhabitants were hardly at the subsistence level.[26]

25 G. D. B. Jones, "Capena and the Ager Capenas *PBSR* 30 (1962): 116-208; and 31 (1963): 147-58; T. W. Potter and K. M. Dunbabin, "A Roman Villa at Crocicchie, Via Clodia," *PBSR* 47 (1979): 19-26.

26 L. Camin and W. McCall, "Settlement Patterns and Rural Habitation in the Middle Cecina Valley between the Hellenistic to Roman Age," *Etruscan Studies* 9 (2002): 19-27.

It may also be significant that an Etruscan longhouse from Luni sul Mignone has a rectangular stone socle on which the gabled, thatched roof was placed directly to create in effect an A-frame house on a stone foundation. Again, the purpose seems clear, namely, to prevent the infiltration of water into the walls of the house by simply eliminating the walls. Perhaps this is a transitional phase to true masonry construction. The first attested masonry 'house' in archaic Rome is the Iron Age Regia[27] and it incorporates a courtyard. We may perhaps posit an evolution from wattle-and-daub *tuguria* facing a simple courtyard formed from dry rubble or hedge walls, followed by small stone cottages with thatch or terracotta tile roofs and a courtyard.

The Latin *villa* is nothing more or less than a 'farmhouse', but by the time of our focus the term was no longer strictly adequate, as the redundancy of the term *villa rustica* will attest.[28] Modern confusion over the exact significance of the term has a fine ancient pedigree, witness the discussion among Varro's dialogists over the term in the prologue of Book III. It is one of the etymological jokes of which Varro is so fond that the dialogists, who never reach anything approaching consensus, are seated at the time at the Villa Publica of the Campus Martius, used to house the offices of the censors, to receive foreign ambassadors, as a base for army levies, and as the site where victorious generals waited to hear if they had been granted a triumph. In other words, for all sorts of things which have nothing to do with either *fructus* or *luxus*.[29]

Rostovzev,[30] based on extensive experience with Campanian villas, especially in the Vesuviana, had early on created a typology of villas in which three basic types emerge: the true *villa rustica,* usually a fairly modest facility in

27 Fayer (1982): Fig. 47.

28 Best synthesis in Nicola Terrenato, "The Auditorium Site in Rome and the Origins of the Villa" *JRS* 14.1 (2001): 5-32. In general see John H. D'Arms, *Romans on the Bay of Naples* (Cambridge, MA, 1970): 9-17; Andrea Carandini et al, *Schiavi e padroni nell'Etruria romana: la Villa de Settefinestre dallo scavo alla mostra* (Bari, 1979); Stefano De Caro, *La villa rustica in località villa Regina a Boscoreale* (Rome, 1994); J. Edwin Frazer, ed., *The Roman Villa. Villa Urbana* (Philadelphia, 1998); A. G. McKay, *Houses, Villas and Palaces in the Roman-World* (London, 1975); H. Mielsch, *Die römische Villa: Architektur und Lebensform* (Munich, 1987); *idem, La villa romana* (Florence, 1990); Piccoli (2004); F. Reutti, ed., *Die römische Villa* (Darmstadt, 1990); J. T. Smith, *Roman Villas: A Study in Social Structure* (London, 1997).

29 On the the literary and political agendas of the sources and the danger of using them uncritically, Carin M. C. Green, "The Shepherd of the People: Varro on Herding for the Villa Publica in *De re rustica* 2," in Becker and Terrenato (2012): 32-44, and Brandon Reay, "Cato's *De agri cultura* and the Spectacle of Expertise (op. cit.): 61-8.

30 Michael I. Rostovtzeff, *The Social and Economic History of the Roman Empire,* 2nd. ed. (Oxford, 1957): pp. 551-67.

which the proprietor, his family and slaves lived and worked; a second type which might or might not include productive facilities (a *pars rustica*) and which the proprietor is presumed to have visited periodically, turning over the daily management of the facility to an overseer (*vilicus*); and a third facility which was strictly productive, an industrial farm, as it were. A specific subset of the second category was the *villa maritima,* a seaside villa to which the wealthy, especially those of the city of Rome itself, retreated to enjoy the pleasant weather and spectacular sea views. Although it has been generally assumed that early *villae maritimae* on the Bay of Naples were productive as well, as were clearly their inland counterparts, D'Arms makes a convincing case that, as early as the second century BCE they were almost exclusively for leisure and escape. In D'Arms' own words, "It is *delectatio* and *amoenitas*, rather than *fructus*, which concerned wealthy owners of *villae maritimae*." But by the time of the late Republic we have Cicero's explicit evidence (*De lege agraria* 2.78) that wealthy Romans were eager to buy farms in the Campanian hinterland so that the profits derived from them could be used to finance their lavish coastal estates at Cumae and Puteoli.[31]

Be that as it may, it is with Types II and III that we have most to do, since we have every reason to believe, from both literary and archaeological evidence, that it was such facilities that produced the majority of wine consumed in the Roman world from the second century BCE. Cato's exquisitely detailed description of the facilities and equipment of a modest, so-called 'Catonian' villa leads us to believe that the form as the center of commercial agriculture was a development of his own period. Unfortunately, we are not as well provided with examples from the field as we could wish.[32] The classic form is the '*cortile*' villa,

31 But on a *pars rustica* as part of the *villa maritima*, cf. Marzano (2007): 44-6.

32 The following in no way purports to be complete but does list some of the more accessible and representative samples. For the older excavations, reported (poorly) in *Notizie degli Scavi,* see White (1970): 416-45 and Appendix. Late Republican Campania: P. von Blanckenhagen, M. A. Cotton and J. B. Ward-Perkins, "Two Roman Villas at Francolise, Prov. Caserta," PBSR 33 (1965): 55-69; M. A. Cotton, *The Late Republican Villa at Posto Francolise* (London, 1979): 7-18; M. A. Cotton and G. Métreaux, *The San Rocco Villa at Francolise* (London, 1985): 11-34. Etruria and northern Latium: M. Moretti and A. Syubini Moretti, *La Villa Volusii a Lucus Feroniae* (Rome, 1977); Gianfranco Gazzetti, *La villa dei Volusii a Fiano Romano* (Rome, 1994); Potter and Dunbabin (1979) 19-26; I. Attolini and P. Perkins, "The Excavation of an Etruscan farm at Podere Tartuchino" PBSR 60 (1992): 1-76; T. W. Potter, *Excavations at the Mola di Monte Gelato: A Roman and Medieval Settlement in South Etruria* (London, 1997); C. E. Östenberg, "Villa Sambuco," in Axel Boëthius, *Etruscan Culture. Land and People* (Mälmo, 1962): 313-20; A. Small, ed., *The Excavations at San Giovanni di Ruoti* (Toronto, 1994). For the provinces: J. Percival, *The Roman Villa* (London, 1976).

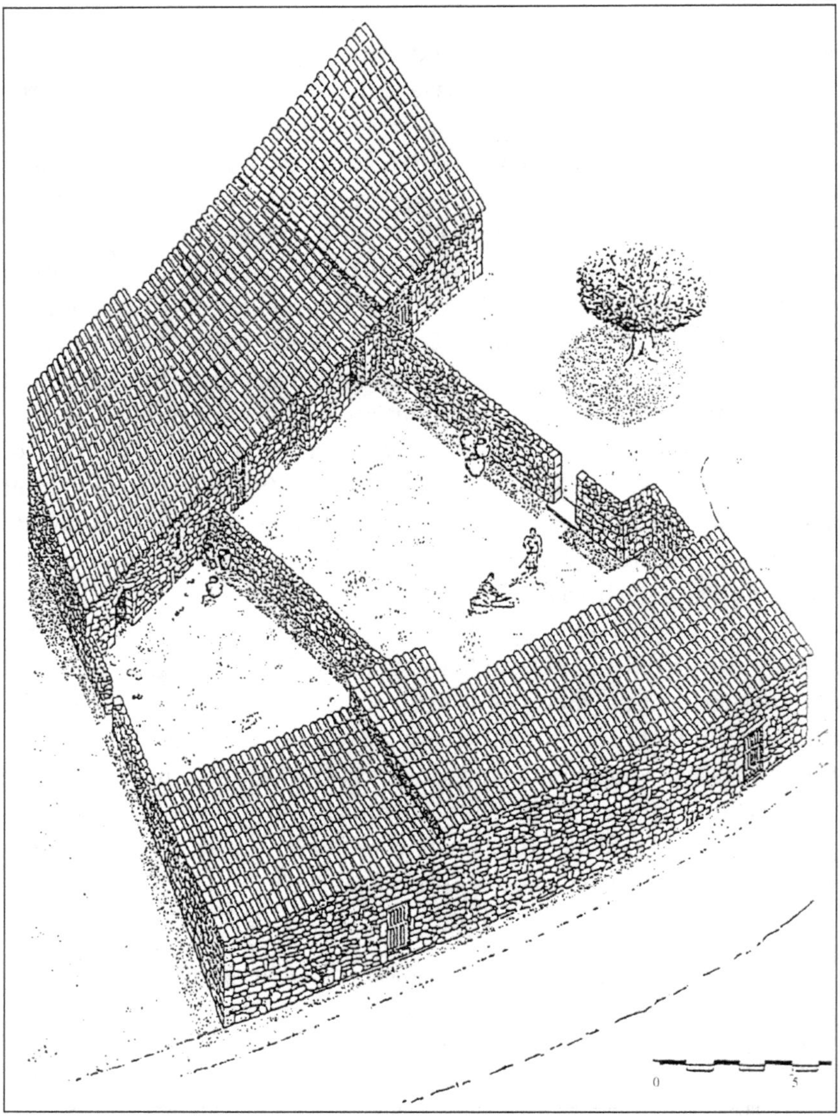

FIGURE 8 *Prototype* cortile *villa. From Russo Tagliente (1992): Fig. 106.*

that is, one which encloses a farmyard around whose perimeter cluster residential rooms, stables, and rooms for various productive processes. Already in 1902 Patroni[33] had described the classic design, and though we needn't credit

33 G. Patroni, "L'origine della *domus*" *Rendiconti Reale Accademia dei Lincei, Classe di Scienza* 11 (1902): 467-507.

his claim that the atrium of the *domus* was an evolved form of the *cortile* (Latin *chors* or *cohors*, as in a fragment of Varro preserved by Nonius (p. 83L): "cohortes sunt villarum intra maceriam spatia."), the archaeology clearly confirms the basic premise.[34] The form has obvious Greek and Punic analogues,[35] for example, a farmhouse at Metaponto and another near Olynthos on the Greek mainland. In Messapia the *cortile* villa first appears in an Italic context from the sixth century BCE at Cavallino (Lecce), at first a polygonal enclosure before a range of three rooms, then with the courtyard subdivided into two plus the addition of more rooms at the south.[36] Examples are also found in the Tolve.[37] Further north and west in Lucania, a pattern of small farms with no farmsteads has been interpreted as evidence that farmers lived in nucleated settlements and walked to their fields. A settlement requires a certain amount of infrastructure such as a source of water, transportation routes, defensive structures, access to equipment and storage, etc. Here there is a trend toward smaller settlements further afield and small farms (less than 2000 m2). Such settlements can only work when there are defensible settlements within reach to which smallholders can retreat when invaded.[38] The next step is a 'settlement' complex consisting of one farm, and courtyard villas with regular masonry walls and tiled roofs begin to appear in Lucania already in the fourth century BCE. At Lavello (Potenza), a house from the second half of the fourth century includes a large courtyard, an area interpreted as a residential complex, and a large room at the south of the courtyard intepreted as the base of a tower, though in this case one made of adobe and wood, given the fact that we don't see the typical thickening of the foundation walls. By the end of the fourth century

34 On the presumed evolution: Nicola Terrenato, "The Enigma of 'Catonian' Villas: The *De agri cultura* in the Context of Second-Century BC Italian Architecture," in Jeffrey Becker and Nicola Terrenato, edd., *Roman Republican Villas: Architecture, Context and Ideology* (Ann Arbor, MI, 2012): 69-93.

35 Potter (1987):111-12; H. Lauter, "Hellenistische Vorläufer der Römischen Villa," in Frazer (1998): 211-7; Mario Torelli, "The Early Villa: Roman Contributions to the Development of a Greek Prototype," in Becker and Terrenato (2012): 18-31. Villa culture in the Middle Republican Roman Campagna: Rita Volpe, "Republican Villas in the *Suburbium* of Rome," in Becker and Terrenato (2012): 94-110.

36 Alfonsina Russo Tagliente, *Edilizia domestica in Apulia e Lucania: ellenizzazione e società nella tipologia indigena tra VIII e III secolo A.C.* (Lecce, 1992): 56-62.

37 E. Greco, "Poseidonia entre le VIe e la IVe siècle avant J.C.: quelques problèmes de topographie historique" *Revue Archéologique* (1979:2): 261, who takes it as the model of the Catonian villa: Russo Tagliente (1992): 173-80.

38 Alistair M. Small, "Changes in the Pattern of Land Settlement and Land Use Around Gravina and Monte Irsi," in Elio Lo Cascio and Alfredina Storchi Marino, edd., *Modalità insediative e strutture agrarie nell'Italia meridionale in età romana* (Bari, 2001): 35-53.

such *cortile* villas become increasingly common in Lucania, for example at Muro Tenente (Brindisi), Banzi (Potenza), Serra di Voglio (Potenza), and Montegiordano (Cosenza).[39] At Montegiordano, the villa complex dated to the second half of the fourth century was at least two stories tall and around the *cortile* were service rooms with a press and pithoi. This villa also had a tower and stuccoed walls.[40]

In the Ager Faliscus, the area studied by the South Etruria project, some 35% of scatters were interpreted as belonging to huts of the *tugurium* type. Small farms with masonry villas of well formed tufa blocks represented 43% of scatter, and large villas 28%.[41] Here a number of cases show that villa owners actually built private roads to connect to Roman roads such as the Via Clodia and Via Amerina. An example is the villa at San Giovanni à Pollo where such a *diverticulum* connected with the town of Sutri, over two kilometers away on the Via Cassia.

But all 'evolutionary' schemes must now take account of a dramatic discovery in the immediate environs of the city of Rome itself, a fully developed masonry villa some 1.5 km outside the Aurelian walls along the Via Flaminia in the Tiber floodplain, discovered in 1996 during construction of a new concert hall and thus called the Auditorium Villa.[42] We are blessed that the villa underwent successive reconstruction, apparently because of periodic floods here, and the same siltation which prompted rebuilding has beautifully preserved the older foundations and some of the appurtenances. But the most remarkable fact about the site is that the earliest phase can be securely dated from both ceramic studies and building techniques to roughly 550 BCE, centuries before the Catonian form 'emerged'. And yet, here we clearly see a large, rectangular *cortile* surrounded by smaller rectangular rooms, one of which preserves part of an oven or furnace. In the southwest corner of the *cortile* is a large, trapezoidal room which suggests a storage area. In the second phase, about a half century later, the villa has more than doubled in size by the addition of much larger courtyard to the north and the addition of several more rooms. Now the sequence of rooms on the north side has become the residential quarter, there may be the postholes for a wooden portico along the east side which contains a beautifully preserved pressbed for an olive press, one of

---

39 Russo Tagliente (1992): 100-02.

40 Helle W. Horsnaes, *The Cultural Development in North Western Lucania c. 600-273 BC* (Rome, 2002): 48-52.

41 Timothy W. Potter, *The Changing Landscape of South Etruria* (London, 1979): 108-22.

42 Terranato (2001): 5-32 and A. Carandini, ed., *La fattoria e villa dell'auditorium nel quartiere Flaminio di Roma* (Rome, 2007).

the earliest ever found, and the former *cortile* has now become a *cella*, to judge by a large number of dolium fragments found there.

Two early sites in Magna Graecia near Francolise are instructive. That at Posto is valuable as an example of a modest 'Catonian' villa which was later expanded and refined with the addition of a bath suite. The villa is built on an artificial terrace. Along one side of the *cortile* is a portico, while a residential section and bath suite are on the opposite side flanking the entryway. Exterior access to the *cortile* was provided by a large postern gate at the northeast corner of the platform. Though the farmhouse, especially in the later early imperial phase, was used mostly for oil production, the type is universal enough and would have been perfectly serviceable for a vineyard or mixed farm as well. A multistory tower is conjectured for the southeast corner, where the walls of the foundations are significantly thicker than elsewhere, as they are at Sambuco and as we see in Pompeian wall paintings.[43] The second villa, at nearby San Rocco, is a more developed type with a distinct *pars urbana* and *pars rustica,* in this case on two adjoining terraces but at different levels. This site dates to the early part of the first century BCE. Here as well there was a radical enlargement, complete with a bath suite and a discrete oil production facility (*oletum*) dating to the early empire. Either the owners were prospering wonderfully or, as seems more likely, the property ahd been bought by someone of substantial means.[44]

At the other end of the peninsula we have a late Republican villa, the so-called 'Villa Sambuco,' near San Giovenale.[45] The villa is firmly dated to the late first century BCE. Along the north side of the *cortile* are rooms for storage and processing of agricultural products, including a large open area at the northeast which could well be a *cella vinaria.* Along the south are a stable, the main entrance, living quarters, the basis of a tower, and a tool shed.

From the Augustan period on we have a profusion of villa sites, partly due to the fortuitous preservation in the wake of the eruption of Vesuvius but also no doubt a true reflection of the intensification of commercial viticulture at this time. Sadly, there have been some 147 villas excavated in the vicinity of Vesuvius alone which have some vinicultural facilities attested, but of these only two have been excavated and documented to modern archaeological standards. Sadder still, well over half were destroyed after they were excavated (most were found in the process of modern building schemes) and the overwhelming majority received only cursory notices in the literature. Most of those which

43 Cotton (1979).

44 Cotton and Métraux (1985).

45 Östenberg (1962).

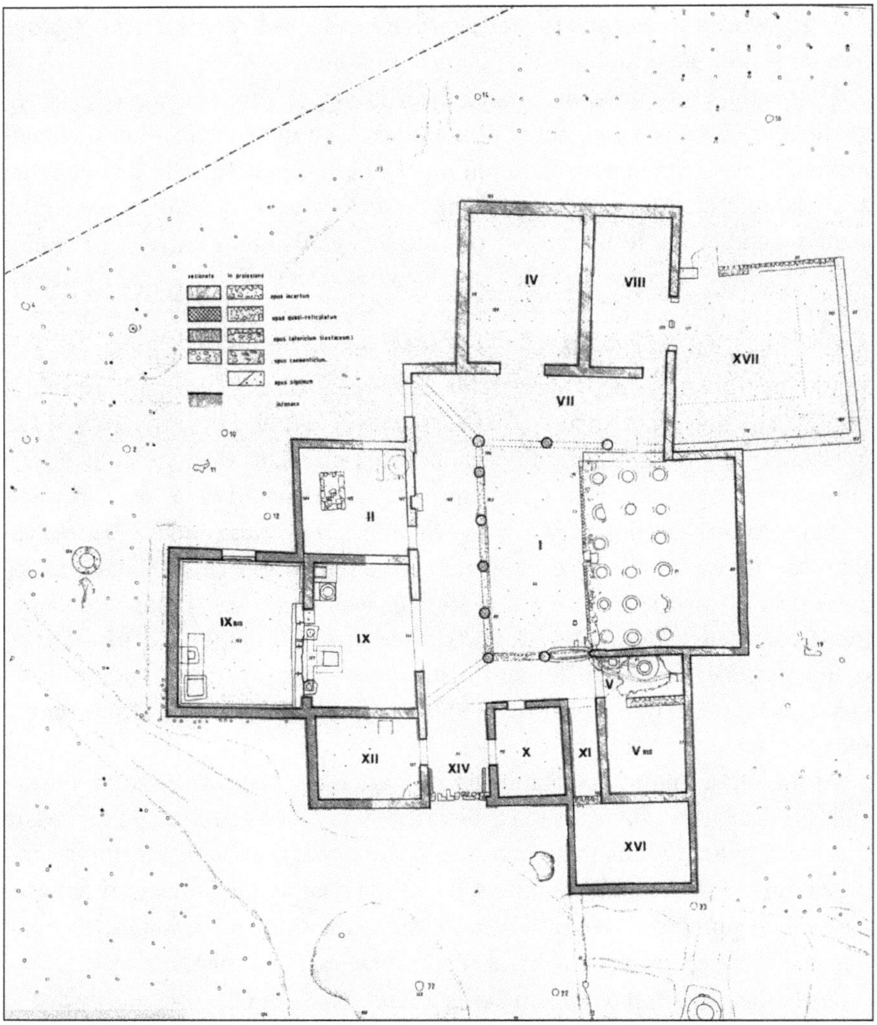

FIGURE 9 *Plan of the Villa Regina. From Stefano De Caro,* La villa rustica in località Villa Regina a Boscoreale *(Rome, 1994).*

were not destroyed were reinterred, so that today only eight are still visible. The good news is that the eight are now documented.[46] And one, the so-called Villa Regina at Boscoreale, has been scientifically excavated and extensively documented, in this case by the masterful work of Stefano de Caro.[47] Again we

46 Greta Stefani, "Le ville rustiche del territorio Vesuviano," in *Ministerio per i Beni e le Attività Culturali Sorprintendenza Archeologica di Pompei* (Naples, 2000): 13-56.

47 De Caro (1994).

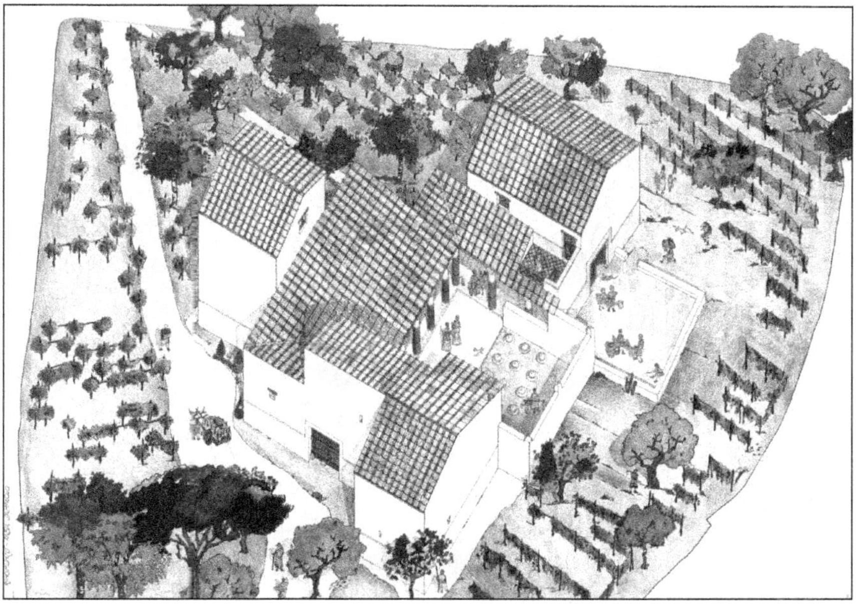

FIGURE 10 *Reconstruction of the Villa Regina. From De Caro (1994).*

have the classic form, with access to the *cortile* by way of a large passageway from the exterior on the south, through which farm wagons could be driven. The portal was closed by a heavy gate facing the public road and capable of being barred. There appears to be a second postern gate at the northwestern corner giving access to vineyards, orchards and other farm fields. On the west side we find the productive facilities, specifically a *torcularium* or pressroom, actually two rooms, toward the west, a large treading vat (*calcatorium*) which also contained the press bed of a large lever press.[48] The eastern half of the *torcularium* is actually an extension which creates an ell, and on the northern wall of this extension is a large window with shutters through which harvested grapes could be delivered directly into the *calcatorium*, as is true at the Villa of the Mysteries as well. The *torcularium* is dominated by a large lever press, operated from the east side in the staging room. Between the truncated walls which separate the two parts of the *torcularium* is a low socle on which the press beam (*prelum*) could rest when not in use. In front of this socle is a sump, 65 × 60 cm, surrounded by a frame for a trapdoor, some 2.5 m deep and accessed by three steps carved into the stone. The function of the sump is easily interpreted as a receptacle for the must of pressing. Along the same side of the courtyard

48 De Caro (1994): 27-46.

FIGURE 11 *Villa Regina. The* torcularium *is at right. Note the window through which harvested grapes could be passed directly to the treading vat/press.* PHOTO BY SANDY THURMOND.

are a kitchen and a storeroom as well as the postern previously mentioned. To the north of the *cortile* is a two-story work area which may also have contained quarters for slaves. The upper story, a later addition, was a granary and hay loft. To the east of this complex, abutting the northern end of the *cella vinaria,* is an open-air threshing floor (*area*). The south, west, and north perimeter of the cortile are an impluviate portico, and the eastern side is dominated by the open-air wine cellar (*cella vinaria*) dominated in turn by 18 *dolia defossa,* i.e., dolia buried up to their shoulders in sand to act as a natural refrigerant. In the southeast sector of the villa is a two-story complex of rooms that constituted the living quarters of the family.

Such an arrangement is repeated, with variations, again and again in the archaeological record of the period.[49]

49 e.g., Villa del fondo Prisco: Stefani (2000): 45-7; Villa del fondo Imperiale in località Civita di Nitto (Stefani (2000): 49-53; the Villa del fondo Agricoltura in località Pisanella : Stefani (2000): 59-64 (not to be confused with the more famous Villa Pisanella); the Villa

Likewise, a villa rustica at Pisanella excavated in 1906-08 by Giovanni Di Palma, some 90 meters from the Villa of Numerius Popidius Florus, had two entrances, one toward the road and the other toward the farmfields, both of which led to a *cortile*. The villa had a *torcularium* with a large window through which grapes could be transferred at harvest, a large *calcatorium* in cocciopesto with holes for insertion of *arbores* and a dolium for reception of pressed and trodden must. On the northern wall of the *torcularium* on a white stucco ground was a fresco of Bacchus. To the north of the production area was the residential wing with an *excedra*, *cubicula*, and a small bath suite. At the southern end of a portico to the west of the *torcularium* was an *apotheca*, a storage magazine. The residential quarters were served by a combination *cucina/ praefurnium*.[50]

Indeed, the only particular in which the Villa Regina is somewhat atypical is in its *al fresco* bathing facilities; a number of other *villae rusticae* of the period have small bath complexes, typically next to a furnace room (*praefurniam*) which also heats water for an adjacent kitchen (*culina*). For purposes of comparison, the wine produced from the dolia in the *cella vinaria* of the Villa Regina, some 10,000 liters, will have brought in from 5,000 to 7,500 sesterces and may have produced a profit of 500-750. Of course the farm produced a number of other crops as well, so perhaps we are not too far wrong in suggesting an annual income of roughly 1,000 HS. Subsistence income is reckoned at 500 HS, the annual income of a Roman legionary of the period being 900 HS. Our farmhouse will have supported a family comfortably but by no means luxuriously. And the *familia* will have had the benefit of a healthful, varied diet, a delightful climate, and some spectacular surroundings. All in all, a very

---

Pisanella : Angiolo Pasqui, "La villa pompeiana della Pisanella presso Boscoreale," *Monumenti antichi* 7 (1897): 397-554; Stefani (2000): 65-8; the Villa di Ioannis near Aquileia: Piccoli (2004): 90; D. Scagliarini Cortaita, "Le 'villae' romane nell'Italia settentrionale," in E. Roffia, ed., *Ville romane sul Lago di Garda* (Brescia, 1997): 53-81; the Villa Russi near Ravenna: Piccoli (2004): 89; Scagliarini (1997): 53-81. It is easy to forget that villas developed over time not only in the appurtenances of *luxus* and *amoenitas* such as mosaic floors and bath suites, but in their productive facilities as well, witness a settling vat for olive oil, inscribed with the consular date of its 'inauguration', so proud was the owner of his new technological improvement and so eager was he, apparently, to make it a public icon of his success. Cf. Marzano (2007): 92.

50 M. Della Corte, "Altra villa rustica, esplorata da Signore Giovanni di Palma, nel fondo 'Agricoltura' di sua proprietà in contrada Pisanella, commune di Boscoreale, l'anno 1906 (giorni 15) e l'anno 1908 (mesi 6)" *Notizie delli Scavi* (1921): 46-7.

pleasant place to live.[51] This in poignant counterpoint, of course, to its sudden, violent end.

A short distance away in Boscoreale lay the so-called 'Villa Pisanella', now, sadly, reinterred.[52] Here a greater effort has been made to separate the *pars urbana* from the working parts of the farm, with bedrooms, dining facilities and a complete bath suite clustered at the west end of the courtyard. Yet we see the same standard features: press rooms, *cellae*, an open-air threshing facility, slave quarters, a porticoed *cortile*, all on a more grandiose scale. Calculations indicate that dolia here could hold some 93,800 liters (24,800 gal) of wine, which would indicate an estate of some 100 iugera (25 ha/ 65 acres). Obviously this was the estate of a wealthy landowner, as is indicated as well by the lavish decoration, a hoard of silver tableware, and more than a thousand gold coins.

The classic example of the industrial farm is in the Ager Cosanus on the northern coast of Etruria, the famous 'Villa Settefenestre'.[53] Here the *pars urbana* is composed of a large *domus* with a peristyle court, a loggia the length of one whole side, even a communal *latrina* that accommodated twenty. The *pars rustica* had facilities for grain and olives but the emphasis was clearly on wine production. We can make some estimate of production based on a hypothetical estate of some 500 iugera (125 ha/310 acres). According to the figures suggested by Roman agronomists this would suggest a production in excess of 100,000 liters (26,400 gal.) per year. There is evidence that the estate was owned by the senatorial family Sestii, and amphorae from their Tuscan estate or estates were shipped up the northwest littoral of Italy to southern France and northeastern Spain as well as far up the Rhone valley. We know that the Sestii also owned ships, and it has been estimated that the profits from the export of wine alone will have amounted to some 60,000 HS per year. The *domus* was extensively refurbished in the late first century CE, but a dramatic decline in wine production occurred soon after, perhaps as competition from overseas forced a move toward cereals and livestock (the estate has a huge piggery). Whatever the new emphasis, it was apparently not economically viable, since

---

51 Mary Beard, *The Fires of Vesuvius: Pompeii Lost and Found* (Cambridge, MA, 2010): 157-60 for the calculations.

52 Pasqui (1897): 397-554; White (1970): 422; Arnold De Vos, *Pompei, Ercolano, Stabia* (Rome, 1982): 242; Potter (1987): 96-7; A. Carandini, "La villa romana e la piantagione schiavstica," in Giulio Einaidi, ed., *Storia di Roma,* vol. 4 (Torino, 1988): 127:166-8.

53 A. Carandini and A. Ricci, edd, *Settefinestre: Una villa schiavistica nell'Etruria romana* (Modena, 1985). Careful reassessment of the 'Villa Schiavistica' model popularized by Carandini, including the famous courtyard surrounded by 'slave cells' at Settefinestre: Marzano (2007): 129-53.

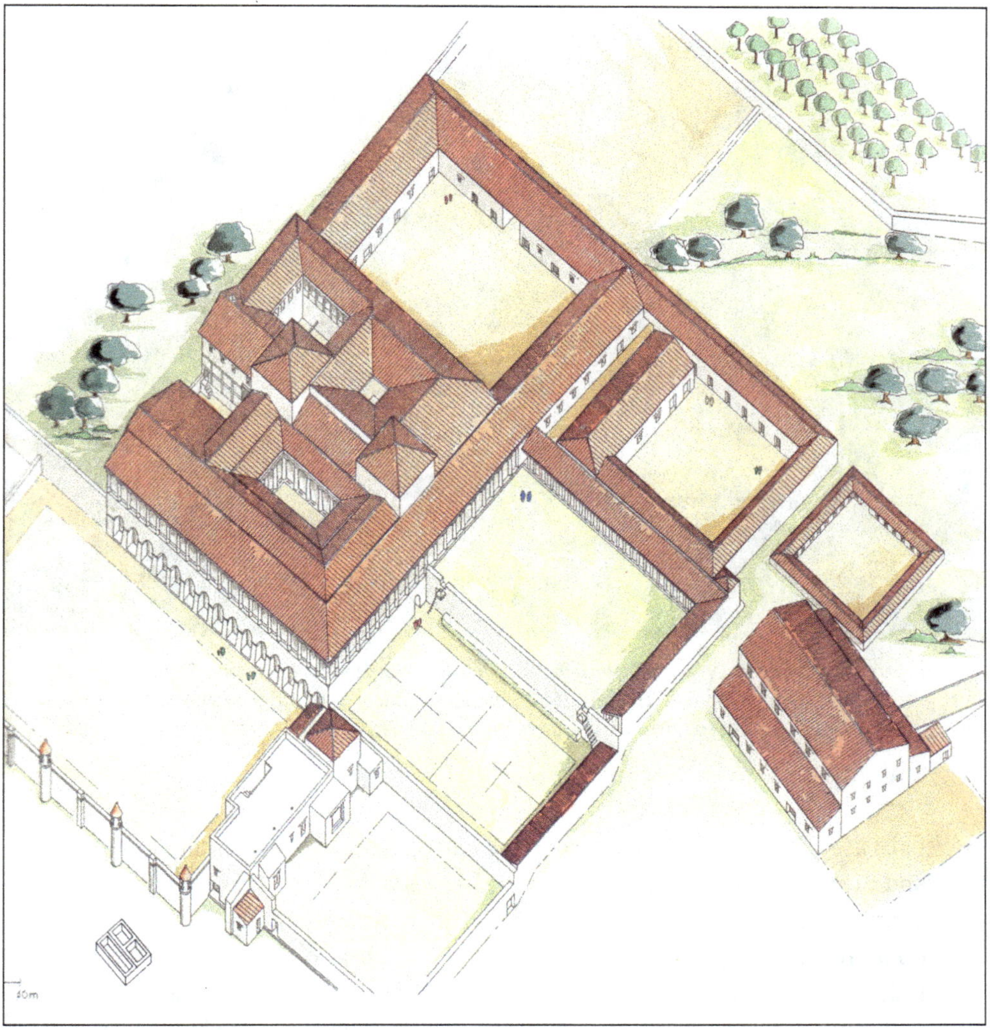

FIGURE 12 *Reconstruction of the Villa Settefinestre. From Andrea Carandini,* Settefinistre. Una villa schiavistica nell'Etruria romana, vol. 1 *(Modena, 1985): Fig. 151.*

the villa was abandoned within the next fifty years and was derelict by the second century.

In the northern provinces, the form of the villa was modified somewhat. For example, the classic 'Gallic' villa is best represented at the site of Molard, 15 km northwest of Augusta Tricastinorum (Donzère), where the living quarters are in a separate, square complex and the *pars rustica* is a long, rectangular building (15 × 67 m) with a clerestory level at the back, truncated with two rectangular

FIGURE 13 *Winged corridor villa on a mosaic from Tabarka, Tunisia. From John Percival,* The Roman Villa: An Historical Introduction *(London, 1976): Fig. 2.* REPRODUCED WITH KIND PERMISSION OF B.T. BATSFORD, PART OF PAVILION BOOKS COMPANY, LTD.

halls at right angles. The long gallery is a magazine for wine storage, and this and the *torcularium* are arranged around a *cortile*. The *torcularium* contained a central lever press flanked by two *lacus*. The *cella vinaria* contained some 200 dolia, representing a total capacity of some 2,500 hl (66,000 gal).[54] A Gallic industrial villa is represented by the huge complex at Vareilles (Saône-et-Loire) with multiple presses and treading vats and an L-shaped *cella vinaria* around two sides of the *cortile* which contained some 350 dolia constituting some 2,500-3,500 hectoliters (ca. 65,000-90,000 gals.) of wine.[55]

54 Thierry Odiot, "Le site du Molard à Donzère," in Brun, Poux and Tchernia (2009): 208-9; Catherine Petit-Aupert and Pierre Sillières, "L'installation viticole de Lestagnac," in Brun, Poux and Tchernia (2009): 210-11; Michel Hue and Abbaye de Flaran, *Lestagnac, un chai gallo-romain: La viticulture en Gascogne aintique* (Valence-sur-Baise, 2003).

55 Stéphane Mauné, "La villa de Vareilles," in Brun, Poux and Tchernia (2009): 212-5.

Cultural Procedures

Before we look in some detail at viticultural procedures it will be useful to say a bit about the vine's physiology and growth cycle.[56] The vine is composed of an upper and lower zone. The upper, above-ground zone is composed of a trunk that supports the canopy and conducts nutrients and water to it; arms branching from the trunk which function in the same way and are more than one year old; shoots, i.e., the succulent stems which arise from nodes on the arms and bear leaves; the leaves themselves; and flowers which ultimately give rise to berries. Canes are mature shoots which have lost their leaves and have lignified. Nodes occur at intervals of 15 to 20 cm (6-8″), and the areas between them are the internodes. Within the cane at nodal points is a diaphragm, a woody partition. From the nodes arise buds, and these in turn produce either leaf shoots which will bear the fruit clusters, or tendrils which will anchor the shoots. Each bud is actually composed of multiple buds, a primary one and two smaller ones which will develop only if the primary has been killed. Shoots are classified as primary shoots or summer laterals. Grape clusters on primary shoots are the primary crop, those on summer laterals a secondary crop. Leaf buds produce only leaves, whereas flower or fruit buds produce both leaves and the flower clusters which become fruit. Leaf blades are usually indented and have five lobes. Flowers are borne in clusters on a rachis, each individual flowerette on a pedicle. Flowers in the hermaphroditic (domesticated) vine are composed of a calyx; five fused sepals; the corolla with five green petals, fused at the top to form a calyptra; five stamens; and a pistil.

Roots are not divided into nodes and internodes. Branch roots instead arise from the interior of a root from meristematic cells. The grape root system is both spreading and descending. Under favorable conditions the roots spread over a wide area but the vine is then susceptible to severe stress when water is not available. When water is limited the root system will seek out water at depths of two to four meters, although it is not uncommon for them to reach depths of 7 meters and more.

Grapes are berries that, on the European vine, form in clusters. Clusters consist of a peduncle, a cap stem, a rachis, and the berries themselves. The berry is composed of an epidermis or skin (5-12% of the mature crop) which contains much of the color, flavor and aroma of the grape, a mesoderm or pulp which contains the majority of the sugars, and the seeds. The so-called 'bloom' is a

56 Robert J. Weaver, *Grape Growing* (New York, 1976): 13-34; Richard P. Vine, Ellen Harkness and Sally J. Linton, *Winemaking: From Grapegrowing to Marketplace* (New York, 2002, 2nd ed.): 29-36; Winkler et al, *General Viticulture* (Berkeley, 1974): 77-110.

thin, waxy layer on the epidermis which retards water loss. Since smaller berries have more skin, proportional to pulp, they generally have more of the components that make great wines than do large berries. The pulp is composed of solids and juice, the latter comprising 80-90% of crushed grapes. Most pulp is transparent with colorless juice. Seeds comprise up to 5% of crushed grape mass and may number from one to four per berry. Seeds are very high in tannin (5-8%) and oil (10-20%).

The vine is a deciduous plant whose growth cycle is determined by the seasons. Dormancy commences in autumn in temperate climates. During this period, leaves fall and starch in canes is converted to sugars to protect the plant against cold injury. This is a time when the vine is vigorously pruned. Toward the end of dormancy in early spring these cut canes may begin 'bleeding', when liquid flows from cut xylem tissues, often more than four liters per cane, rather alarmingly, but the phenomenon has no deleterious effect on the vine. When mean daily temperatures reach 10°C (50°F), buds begin to swell and shoots emerge, a process known as 'budbreak'. As temperatures continue to rise, shoots may grow as much as 30 cm (1′) per day. About 8 weeks after budbreak flowers bloom and shoot growth slows, and at the same time flower initiation for the following year's crop commences. As the flowers develop, the calyptras fall, a process called 'bloom'. As this happens pollen is released from the anthers of the stamens and falls onto the stigmas and germination occurs under favorable conditions. A pollen tube grows from the style to an embryo sac and delivers sperm to it. Several days after bloom, pistils and impotent berries 'shatter' from the cluster. This may involve more than half the cluster but is important because it provides space for the clusters to develop. Frost or rain during bloom may cause clusters to form small, seedless berries that fail to develop ('shot berries'), a phenomenon called *millerandage*. Berries that don't fall are said to have 'set', and the stage itself it 'fruitset'.

The growth of the berries occurs in three phases. In the first, the ovary and its contents, except for the embryo and endosperm, grow rapidly. Stage 2 shows little growth of the overall berry while the embryo and endosperm grow rapidly and the endocarps (seeds) become woody. Stage 3 sees rapid growth of the mesocarp (pulp), causing the final swelling of the berries to maturation. *Véraison* begins when berries start to color and soften, white berries turning from green to white or yellow, red berries from pink to darker reds. At the same time, sugars in the pulp intensify and acidity begins to slowly decrease. Left to its own devices the vine will produce 'overripe' berries, those with excess sugar and insufficient acidity to balance it.

Planting

Each cultivar matures at a slightly different rate, so a rational planting pattern is important. Today topographical maps and satellite imagery make vineyard design much easier, but the concerns are the same as ever: what blocks will there be, what roads and of what sizes to separate the blocks and give access to equipment, how will rows be oriented, placed, elongated. Irrigation design imposes additional constraints. Blocks are generally planted to individual cultivars, the size of the blocks determined by the tonnage desired of grapes of each cultivar. Row length is constrained by block size, but in general long rows are more efficient because they require less turning of farm machines. Row length today is also limited by the tensile strength of trellis wire, a limitation which the Romans did not face because they used strictly freestanding trellises. Furrow irrigation, the method used exclusively by the ancients, also limits row length, however, as it does also row orientation, relying as it does on gravity flow. Row orientation is also governed by sun exposure and canopy management, so some priorities, often conflicting, must be carefully weighed.[57]

Columella gives the most complete and rational exegesis of ancient planning. He recommends plots of 100 vines each, with regular footpaths (*semites*) between, or, as some prefer, plots of about one-half *iugerum* (RR 4.18). Footpaths are classified as *limites*, the broader ways which divide the field into sections (cf. Pliny NH 18.33-4), whereas *viae* are smaller paths which intersect these, much in the way of a Roman army camp.[58] Given Columella's recommendation for density of planting, each of these blocks will have been approximately 20 × 20 m (60 × 60′), that is, about the size of the Roman unit called the *clima* (Col. RR 5.1.5). Within each block Columella recommends five-foot spacing of vines to allow better sun exposure and circulation of air and to allow better inspection by the owner. The system also makes it relatively easy to compute labor costs in man-hours. Plus, Columella assures us, smaller plots have a good psychological effect on laborers, since they can complete work in smaller increments and see the results of their labors. Additionally, discrete plots allow the vigneron to judge the vigor and yield of the vines and adjust viticultural practices accordingly, and they allow easier access for the vintagers (*vendemiatores*) and for repair of trellises. Pliny (NH 17.169) says that blocks should be bisected by a main path (*decumanus*) 6 m (18′) wide, to allow the passage of carts going in opposite directions. These will be intersected by *limites* 3 m (10′) wide running through the middle of each *iugerum* or, in the case of plots larger

57 E. W. Boehm and B. G. Coombe, "Vineyard Establishment," in Coombe and P. R. Dry *Viticulture: Practices* (vol. 2) (Ashford, South Australia, 1988, 2nd ed.): 23-41.

58 Cf. F. H. Sandbach, "Vergil, *Gerorgics* II. 277," CR 42 (1928): 59-60.

than this (a *iugerum* is 2 × 4 *climae*), a transverse *cardo* for the *decumanus*, of the same width, but always divided up by lanes every fifth row, i.e., so that every 'square' (*pagina* is Pliny's term) is enclosed by every fifth picket. This will have created much smaller segments than Columella recommends, and Pliny's recommendation for the widths of the main paths is excessive even in this age of mechanical farm machinery; it seems that Pliny's academic and/or military bent has trumped the practical farmer.

Modern archaeology has shown that Columella's recommendations are in line with actual cultural practices. Re-excavation of several plots in the Vesuviana, especially in the large vineyard known incorrectly as the Forum Boarium, reveals the presence of pathways to allow access to vineyard workers and harvesters.[59] This large vineyard is roughly square, with two footpaths bisecting the north-south and east-west axes, as well as a footpath along the north wall, immediately parallel to the Via dell'Abondanza, permitting access to a wineshop along the street in the northeast corner of the vineyard and to a *torcularium* (pressroom) and *cella vinaria,* as well as to a *triclinium* located at the far western end of the east-west lane. The vineyard conforms closely to Columella's recommendations of half-iugerum quadrants (ca. .25 ha / .62 acres). Because of the location of the *triclinium*, the east-west *limes* is slightly more than halfway along the length of the north-south axis, so that the northern quadrants are slightly smaller than the southern. Large posts were set along the north-south *limes*, larger and more deeply embedded than the vinestakes, and these may be interpreted as posts for an arbor for the footpath. The posts were probably chestnut (*Castanea sativa*), also in line with Columella's and Pliny's recommendations (*RR* 4.30.2; *NH* 16.212, 17.147), and in accordance with the preference of modern Campanian vignerons as well.

Both Columella and Pliny want each block planted to a single cultivar, and for much the same reasons. Whereas in much of Italy smallholders planted different cultivars promiscuously to account for the vagaries of weather, the more rational approach allows the vigneron to harvest each variety at the height of maturity, to judge the vigor and viticultural requirements of each, to irrigate and fertilize discriminately, in short, to tailor viticultural practices to individual cultivars.

In either case, before planting can proceed the soil itself must be prepared and here the Roman viticulturist faced serious challenges. The Roman ard plow was incapable of tilling much deeper than about 60 cm (2′), and that in light, friable soils, nor did it have a true plowshare to turn over and mix the soil.

---

59 Wilhelmina Jashemski, "Large Vineyard Discovered in Ancient Pompeii," *Science* 180, No. 4088 (25 May 1973): 824-5.

FIGURE 14 *Design of the large urban vineyard, Pompeii.* PHOTO BY THE AUTHOR.

FIGURE 15 *Winery of the large urban vineyard.* PHOTO BY THE AUTHOR.

But hardpans, layers of dense clays which impede root growth and moisture infiltration, can be deadly to vines, and they most often lie beneath this depth. In that exigency the ancient farmer had recourse to *pastinatio*, the laborious process of digging the soil with the *bipalium*, a heavy shovel equipped with footrests.[60] Thus for a new plantation all the existing vegetation was grubbed out (Col. *RR* 3.13.6-13) and a level field (*campestris*) dug to 2 1/2 Roman feet, a sloped field (*acclivis regio*) to 3′ (hereafter measurements are in Roman feet) and a steep hill site (*praeruptior collis*) to 4′ so that soil stripped from the upper slope by erosion would still have sufficient depth. Since campestrine sites are really only suitable for cereals and rarely for vines, the laborious process of *pastinatio* is a real possibility. Obviously this labor-intensive work (we have evidence that this was let out to migrant contract labor) would be quite expensive, so vignerons had recourse to trenches (*sulci*) or simple holes (*fori / scrobes*) as well, so that a whole field need not be treated in this way (Cato, *Agr.* 43; Vergil *Geo.* 2.288-9; Col. *RR* 3.13, 4.1, 5.5; Pliny *NH* 17,.139-40 Florentinus in *Geopon.* 5.12; Palladius 2.10.[61] If the planting was to be in trenches they were dug to a uniform width and depth and were properly spaced, the overseer taking care that hired trenchers were not practicing deceit by throwing fresh dirt on untrenched ground to make it appear to be dug. Ideally the trenchers should use the 'stork' (*ciconia*), a clever device to ensure correct depth and profile. The stork is composed of a straight vertical bar with a transverse bar in the middle so that when the trench is the proper depth the horizontal bar lies on the surface of the soil on either side of it. Columella has improved the device by crossing these bars with an X-shape so that depth, width, and perpendicularity of sides of the trench can all be gauged simultaneously. Further, on one side he has attached a plumb line (*libella*) to ensure perfect straightness as well. Such an instrument quickly settles any dispute between the *dominus* (owner) and the *conductor* (contractor). Columella adds (*DRR* 3.13.1-5) that *pastinatio* is particularly valuable for provincials since they are accustomed to adopt planting schemes using holes. For planting in trenches, the soil is first harrowed in one of these ways and then planting trenches 3′ long and 2′ in depth are dug and rootings placed at opposite ends of the trench. Whatever the planting scheme, the space between rows should be 5′ to 7′, unless the slope allows tillage with oxen and plow, in which case 7′ to 10′ is proper (Cf. Pliny, *NH* 17-159-60). Several sites

60 Egypt: Ricci (1924): 39-40. Billiard [(1913): 158-60] says that in his day he saw Campanian vignerons using a spade identical to the *bipalium*.

61 Cf. Billiard (1913): 252-60. Billiard's reconstruction of Columella's improved *ciconia* seems far more logical to me than White's (K. D. White, *Farm Equipment of the Roman World* (Cambridge, 1975): Fig. 9). For Greek practice, Theophrastus *CP* 3.12.1-3.

excavated in Clermont (Hérault) in southern France seem to conform to Columella's prescription of short trenches with a vine bent to vertical at each end.[62] In a villa rustica partially excavated in 1993 during construction work on a new canal, part of the vineyard was exposed, including root holes as well as a *limes*. In this case the holes are 25-35 cm deep, but unfortunately no measurements of intervals are given in the report.[63] In another case, this a Republican-era farm near Lucera, Apulia, trenches show up as as dark, earth-filled streaks in the limestone subsoil.[64] And in the Apulian Tavoliere, aerial photography shows rows of continuous trenches interpreted as evidence of vine planting.[65] The average interval between rows was 2.15 m (7′) and the depth depended on the character of the topsoil.[66]

Obviously, regardless of planting scheme, vines must be obtained to plant since, as we have noted, all propagation of the vine is by cloning. Cuttings may be bought, but the most economical way to obtain them, in antiquity as today, is to propagate them oneself, and since vines must be pruned annually in any case, a ready source of cuttings is at hand. Furthermore, young vines with immature root systems require much more water than mature vines, and in almost all parts of Italy and much of the empire this requires irrigation. The careful vigneron will therefore wish to designate a separate plot as a nursery (*vitiarium, seminarium*: Col. *RR* 3.9.7-8; 3.13.1-5; Pliny *NH* 17.159-60). Soil of the nursery should be subjected to pastination at all costs. The plot should be in a sunny location, should be uniformly harrowed to a depth of 3′ and thrown back (*reici*) with the *bipalium* or *marra* (mattock) to form rows, then cleaned and leveled. Obviously, because of the necessity of irrigation it should be near a source of water.[67]

---

62 Hervé Pomarèdes, "Viticulture et équipements de production dans la cité de Lodève durant le haut Empire: apports des découvertes récentes autour de Clermont-L'Hérault," *Revue archéologique de Narbonnaise* 41 (2008): 1-4. Cf. Philippe Boissinot, "De la vigne au pressoir," in Brun, Poux and Tchernia (2009): 196-207.

63 Stefani in Casali (2000): 69-71.

64 A. H. McDonald, *Republican Rome* (London, 1966): 132 and Pl. 71.

65 Potter (1987): 101-02, citing G. D. B. Jones, "Il Tavoliere romano. L'agricoltura romana attraverso l'aerofotografia e lo scavo," *Archeologia Classica* 3.2 (1980): 85-100.

66 G. D. B. Jones (1980): 90.

67 Cf. Cato 46; Vergil *Geo.* 2.259-60; Varro 1.31. 3-4; Col. *RR* 3.5-6, 3.9.7-8, *DA* 1.5; Pliny *NH* 17.65-72, 17, 159-60; Palladius 3.10; Didymus in *Geopon* 5.3. For the modern nursery Billiard (1913): 264-71; P. R. Nicholas, A. P. Chapman and R. M. Cirami in Coombe and Dry (1988): 1-22; Boehm and Coombe (1988): 34-6. For comparative practices in Greece, Marie-Claire Amouretti, "Oléiculture et viticulture dans la Grèce antique," in Berit Wells, ed., *Agricul-*

The best way to maintain the genetic identity of varietals is to select and propagate cuttings from individual vines with desired characteristics, a process called mass selection. In mass selection, terroir provides the laboratory for testing vine performance. Today it is far more typical to buy vine cuttings from a commercial nursery, a procedure which is cheaper but is less precise and does not optimize terroir. Today cuttings are usually 'callused' before being set out in the nursery; a trench is dug, c. 50 cm (20″) in well drained sandy soil, and bundles of cuttings placed in this trench vertically. The trench is then backfilled with 5 cm (2″) of soil and watered periodically. Calluses form over cuts and root initials usually develop within 6 to 8 weeks, at which time the cuttings are planted in the nursery. This step was evidently omitted by the ancients. But the conditions for the nursery are quite similar: a well-drained site on reasonably level land with good sun exposure but protection from harsh winds and well supplied with water. The modern nursery soil is tilled to a depth of 30 cm (1′) and hoed to a fine tilth, the modern equivalent of pastination. For the ancient *seminarium* Columella recommends fertile soil, but not soil so different in fertility from that of the destined vineyard block that the cuttings will be traumatized when transplanted. The site should be cleared of stones and fenced to protect the nursery from livestock and poultry. Planting rows in the nursery should be 3 feet apart to allow for cultivation, 240′ long, and each planted with 600 slips, that is, 24,000 slips per *iugerum*. Slips should be selected from the most prolific canes from the most prolific vines, fertility to be judged by the number of clusters per shoot was well as the presence of fruiting shoots from every bud on the cane and even from old wood and secondary shoots (*nepotes*).[68]

Such vines are also selected for 'mallet-shoots' (*malleoli*),[69] that is, shoots shaped like a mallet, made by cutting a young shoot (*novellus palmes*) growing

---

ture in Ancient Greece: Proceedings of the Seventh International Symposium at the Swedish Institute at Athens, 16-17 May 1990 (Stockholm, 1992): 80-82.

68 Columella is mistaken. Winkler (1974): 198: "As long as the parent vine is healthy and reasonably hardy, it makes little difference whether the parent vine is a heavy producer or otherwise outstanding, since it has been demonstrated that climate, soil structure and cultural practices are responsble for almost all variations in crop size, berry size and intensity of color, and time of maturity."

69 Cf. Pliny *NH* 17.156; Isidore *Orig.* 17.5. The mallet shoot was still used in Europe in Billiard's time, though less and less for the same reason mentioned by Columella at *RR* 3.17, namely, a tendency to callus at the juncture of the shoot and cane and thus impede the growth of rootlings. Billiard also mentions a hybrid form in which one side of the cane segment is cut off to create an L-shaped cutting, so that it has the aspect of a horse's fetlock and hoof, the *zampa di cavallo* of the Italians, the French *bouture à talon.*

from a one-year cane (*innatus priori anni flagello*) along with a short segment of the cane itself to create the figure of an inverted mallet (an upside-down *T*). It is planted with three or four buds showing above ground. When the vigor of the shoot has been proved by several years of 'campaigning' (*annorum stipendiis*) it is deemed good for transplanting.

Cato's advice had earlier been somewhat different, so we can perhaps see an evolution in viticultural practices here. Cato (*Agr.* 46) would have us plant in rows 1 1/2″ apart, the slips to be planted 1 1/2′ apart as well, pressed into the ground with the foot or driven in with a mallet or maul. Here we are speaking of a simple cutting, not a mallet shoot, and the planting is rather crude as well, without the trenching Columella recommends. Cato further advises that cuttings should project 'one finger' above the ground and be smeared with cow dung. Presumably Cato mean the length of a finger, since the shoot will have little chance of survival unless at least a few nodes are exposed. Cato advises the attachment of identification tags and regular cultivation of the soil to promote rapid growth. Varro (1.31.3-4) apparently has Columella's longer shoots in mind when he advises that when the cutting first sprouts it should be cut back completely so that it will have a sturdier trunk and greater vigor when it sprouts the second time. A slender, weak trunk he calls a *flagellum* and is sterile, while the larger, fertile trunk is the *palma*.

New vines may also be propagated in the vineyard itself. The process, practiced far more in antiquity than today, is called 'layering' (Fr. *provignage*). Layering is the process of burying an uncut shoot of an existing vine and allowing it to root while still attached to the 'mother vine'. Today this is done when the cultivar is known to have cuttings which root with difficulty, or to replace a dead vine in an existing vineyard by layering from an adjacent vine. The cane is typically buried 20-50 cm (8-20″) deep along several inches of moist soil, with the further end out of the soil to expose one or two buds above ground. The shoot will root at each underground node and the following year these may be divided into separate rootings or one selected to remain and the others removed.[70] As noted earlier, nature herself may have taught early viticulturists the method; the vine naturally creeps along the ground if allowed and doubtless early on it was noticed that when covered by soil via erosion the shoots rooted and that, once rooted, they could be cut from the mother vine. Perhaps from there it was a short but momentous step to deliberate layering and thus to rooted cuttings and thereby clonal propagation.[71]

---

70 Nicholas, Chapman and Cirami (1988): 10; Winkler et al. (1974): 203-05.

71 Vernon L. Singleton, "An Enologist's Commentary on Ancient Wines," in McGovern, Fleming and Katz (2000): 74.

Layering in antiquity followed a similar regimen (Cato, *Agr.* 32.1, 52, 62; Vergil, *Geo.* 2.63-4; Columella *RR* 4.2.2, 5.5.16-17, 5.6.30; Pliny *NH* 17.96-7, 17.161, 17.205-06, 17.212; Palladius 3.10.6-7; Didymus in *Geopon.* 4.3). Cato (*Agr.* 32), whose training system is still the *arbustum*, that is, training to trees, advises layering in trenches, having pulled shoots from lower branches of trees; layers are to be left buried for two years after which they are detached. By Columella's time the *arbustum* system was moribund in many parts of the Empire, and he advises against it (*RR* 4.2.2); whereas Julius Atticus had recommended the method, Columella recounts the danger the Treveri and others have encountered with *provignage*, namely the entangling of roots which makes cultivation impossible and ultimately chokes the vine. But of course he is not speaking of layering to replace vines in rows or for transplantation but rather the indiscriminate *provignage* where vines are allowed to grow along the ground as a method of 'training' (if the term may even be used in this instance), the old French system of *provignage*. And this did indeed entail difficulties, as we will see when we discuss training systems in more detail. Columella also describes a form of *provignage* practiced with the Helvenaca cultivar (*RR* 5.5.16) in which sprawling canes are layered and the projecting part trained to canes of mother vines, but says the locals do this only so that fruiting canes can receive more nourishment. These 'divers' (*mergi*) or *candosocci*, as the Gauls call them, are cut off after the vintage as no longer useful. Columella says that they should be transplanted instead as quicksets.

Cato (*Agr* 62) and Pliny (*NH* 17.96-7) describe a variation on the practice in which layers are buried inside baskets or earthenware pots so that they can be transplanted quickly when they are detached from the mother vine. Palladius (3.10.6-7) advises this method in the *arbustum* but would have us hang the baskets in the tree to which the vine is trained. Interestingly, in the famous vineyard previously mentioned which was discovered by Jashemski (II. 5) were found two examples of layering, the shoots having been bent over and covered with dirt. Both layers had rooted, as revealed by plaster root casts, one having formed two branch roots, the other three. Elsewhere in the same vineyard Jashemski found examples of layering in pots as well, but in this case of fruit trees.[72] Finally, Pliny (*NH* 17.212) mentions a strange variation in which a whole vine is thrown down onto the earth, the wood of the canes split with wedges and these canes and their attached shoots led to several furrows and buried, the projecting parts of shoots then being tied to rods (*hastiles*).

72 Jashemski (1973): 824.

Transplanting

Whether from cuttings or layers, the propagated plant is transplanted from the nursery to the vineyard. But it is well to note one standard part of the modern procedure that was not practiced by the ancients at this point, namely grafting.[73] Today propagation is most often by grafting the scion of a *Vitis vinifera vinifera* cultivar to the rootstock of a native American grape such as one of the *Labrusca* or *Muscadinia* species. This is necessary because of the ravages of the Phylloxera root louse which was accidentally introduced into Europe from the New World in the mid-nineteenth century by way of cuttings being used to hybridize vines. Native American species are resistant to the parasite, but have an unpleasant taste when vinified into dry table wines, whereas the noble European cultivars which give us all the world's most popular wines had no resistance. The upshot was that the parasitic insects devastated European vines in a relentless tsunami of destruction and threatened literally to end European viniculture as we know it until the secret of grating *Vinifera* scions to American rootstocks was discovered and accepted.[74] The ancients were experts on grafting, as we will see, but used it exclusively as a way to renew the vigor of old vines or to change the variety of a vineyard plot.

As to transplantation (Theophrastus *CP* 5.3.1; Columella *RR* 3.9.7-8; *DA* 7.1-6; Pliny *NH* 17.67, 17.123; Palladius 3.9; Florentinus in *Geopon.* 4.11),[75] first the vigneron must establish the timing for the operation. Today quicksets, rooted slips grown in the nursery, are usually lifted midwinter, their roots cut back prior to lifting with a special digger attached to a tractor. Tops are also frequently cut back. Good quicksets have several major roots spreading in various directions as well as several shoots 4-5 mm (1/4″) thick and 20 cm (8″) long. Quicksets are bundled and labeled, and if they cannot be transplanted immediately, are kept for several days stacked in the shade under a tarpaulin which is periodically sprinkled with water. Ancient practice was similar. Theophrastus had advised planting both slips (unrooted cuttings) and quicksets in rainy and cold climates shortly before the fall equinox, but in mild or hot climates after the winter solstice. Cato suggests from the time of the winter sowing (*Agr.* 27), which Columella (*RR* 2.8.2) elsewhere fixes on October 2, to the winters

---

73 Vines et al. (2002): 104-07; Winkler et al. (1974): 197-237.

74 For an engaging popular account of the battle against the depradations of Phylloxera (*Daktulosphaira vastatrix*) see Christy Campbell, *Phylloxera: How Wine Was Saved for the World* (London 2004), replublished in the U.S. as *The Botanist and the Vintner: How Wine Was Saved for the World* (Chapel Hill, NC, 2006).

75 For the modern practice, Nicholas, Chapman and Cirami, "Grapevine Porpagation," in Coombe and Dry (1988): 1-22.

solstice. Varro (*RR* 1.34) agrees. Palladius (2.9.5-10) gives the more general advice to transplant during calm, warm days of February, taking care that the slips and quicksets are not parched by sun or wind. But, he adds, in colder regions or rainy ones the whole of Spring may be used. Though the vigneron may plant either slips (*palmes*) or quicksets (*viviradices*) or both, quicksets are universally preferred where feasible. Both come from the previous year's fruiting canes, since roots arise from meristematic tissues near the surface of the cane, usually near the buds at nodes. As the roots develop, branch roots extend the system and these, in turn, branch into fine roots called 'feeder roots'. The root zone includes an area of active growth at the tip, a zone of absorption of water and nutrient minerals behind this, and a zone of conduction behind this which delivers the water and nutrients to the stem.[76]

The ancients were at pains to choose a good section of cane. Columella (*RR* 3.17-18) says that some think slips may be taken from the whole cane as long as five or six eyes are buried, but he himself agrees with those who prefer the part of the cane next to the old wood and completely rejects the use of the 'arrow' (*saggita*), i.e., the extreme end of the cane, since the wood is only fruitful to the fifth or sixth bud. Mallet shoots may be rooted in the nursery, as we have seen, but if planted as slips they are inserted using a special 'trench fork' (*pastinum*), a two-pronged iron dibble. Some actually loop the shoot of the mallet shoot to place more underground, a practice Columella condemns. The length of a slip will depend on the length of the internodes between the 5-6 buds recommended, but they are generally 23-30 cm (9-12″), so that some 100-150 cm (40-60″) projects above the surface when the soil is backfilled. Billiard,[77] the practicing vigneron, says that the ancients' misgivings about certain sections of canes being used as slips are entirely correct, but I have been unable to confirm this from any contemporary source. By Pliny's time (*NH* 17.156) mallet shoots were no longer cut so as to retain the section of cane which gave them their name, so they were essentially simply slips, and Pliny avers that no other kinds of slips root better. But Pliny refutes Columella and says that 'arrows' root even better, especially when they are twisted at the end to impede bleeding.[78] He concedes that they are easily broken, however, at the place where they are twisted. He also mentions 'three-bud shoots' (*trigemmes*) which are not twisted before planting. To be avoided are leafshoots (*pampinaria*) as well as shoots that have not yet fruited; both reservations perfectly correct. Pliny agrees with Columella that slips should be at least 1′ long and contain 5-6 buds and

76 Weaver (1976): 10-13.

77 Billiard (1913): 266-8.

78 Billiard says that twisting the ends of slips still had adherents in his own day.

internodes. Shoots which have been allowed to dry may be soaked in water several days to rehydrate (*reviscere*) them before transplanting, a disastrous recommendation.

Much to be preferred, again, are quicksets which have been proven in the nursery.[79] Cato (*Agr.* 28) says that quicksets are dug up carefully to maintain as much of the rootball as possible and are carried in bundles in a basket (*corbula*) or a wooden tray (*linter*) to the vineyard, but not on a windy or rainy day to avoid desiccation or root rot. Quicksets are planted in trenches, topsoil spread around the roots and thoroughly tamped with rammers and bars (*festucibus vectibusque*). Correct time for transplantation is from October 24 till the winter solstice (cf. Varro 1.34). Columella (*RR* 3.15.3) says they must be as freshly dug as possible and dug carefully to avoid damage to roots, shorn of most shoots as in fall pruning so that the vine is reduced to one strong cane which will become the trunk, and shorn as well of calluses (*nodi*) and scars (*cicatrices*) and any damaged roots. If a two-year-old mallet shoot is to be used, it should be lopped off to 1′ to increase its vigor (*RR* 4.7.3). Pliny (*NH* 17.159-60) says that shoots are cut back at the end of the second year to the bottom node, allowed to sprout again, and then transplanted in the thirty-sixth month. He also mentions (17.161) a 'luxury method' of transplanting in which four quicksets are bound together at the bottom and the bundle threaded through the hollow shank bone of an ox or through a terracotta pipe and then buried with two buds projecting. The shoots are said eventually to merge. The pipe is eventually broken or the bone removed and the vine then left to bear on all its shoots. Another strange method (cf. Columella *DA* 9) employs a mallet shoot split down the middle, the pith carefully removed so as not to harm the buds, the shoot bound up and then rooted in a mixture of soil and dung and pruned back when it throws out shoots. Columella guarantees that such a vine will bear seedless grapes, though Pliny, for once more skeptical than the practical Spaniard, quite sensibly expresses amazement that a shoot deprived of its pith will survive.

As previously noted, the ancients had two basic planting schemes, in trenches (*sulci*) or in holes (*fori, scrobes*).[80] For the former, Columella (*RR*

---

79 Cato, *Agr.* 27-8; Col. *RR* 3.14-16, 4.7.3; Pliny *NH* 17.173; Palladius 3.9.5. Cf. Weaver (1976): 186-7. Today one-year-old rootings are used, kept moist in burlap sacks or under canvas. Before planting the tops are trimmed to a single spur of 1-2 buds and roots shortened to 8-10 cm (3-4″) and all those within 20 cm (8″) of the stem removed.

80 In Calabria in the area of Locride, hand trenching is still practiced; the ground is first dug with mattocks to a depth of a meter and then, in the case of plantings, the soil is left to weather for a year. Cf. Orlando Sculli, *I vitigni autoctoni della Locride* (Cosenza, 2004): 114-31.

3.13.1-5, 3.15-16) says trenches are dug 3′ long and 2′ deep and quicksets placed at both ends of the trench, their rootballs toward the center of the trench and the shoots bent vertically along the far walls of the trench so that two buds are above the surface level. Several stones are placed on the plant to keep it positioned and also to protect it against extremes of temperature and moisture and then the trench is backfilled. Additionally, Mago had advised an application of grape pomace mixed with brown manure and, if the soil was extremely lean, an admixture of richer soil from elsewhere as well. Trenched ground, says Columella, may be planted when moist or dry but not when excessively moist. Unrooted mallet shoots are planted between trenches so that they may be used to replace quicksets which die. The gaps between trenches should be 3′ so that the vines will be evenly spaced. If the expense of *pastinatio* is prohibitive, trenches 6′ long are dug, 6′ are skipped and then another such trench dug, and so on. These trenches are to be 3′ deep. Quicksets are placed along the trench at 3′ intervals, and backfilled in the same way. Even more economical (but presumably less effective) is to trench 2 3/4′ deep and 5′ long, then skip 15′ and cut again, then put in quicksets and unrooted cuttings (*palmites*), putting in a great number of the latter so that those which happen to root can be used to fill in gaps in spacing. Pliny (*NH* 17.167-8) describes a cheaper variation on the trench, the furrow (*pastinatum*), as well as the trench and hole. His trenches are to be three spades' width and 3′ deep, the transplanted vine at least 3′ long with at least two buds projecting above the surface. Holes should be backfilled with a bit of dung at the bottom, and on sloping ground (*clivosa*) should be deeper than normal with the top of the downhill side banked up. He adds that sometimes holes are elongated so that vines can be planted at both ends and are then called 'troughs' (*alvei*), but this sounds exactly like Columella's 3′ trench, so perhaps a distinction has been added. Young vines so planted should be trained to a marsh-reed (*calamus*).

Modern archaeology has shown that the agronomists' recommendations were in line with actual cultural practice. To judge strictly by this evidence, plantings in trenches are more common than in holes, but trenches are obviously easier to detect in archaeological contexts. We find attestation in various parts of Italy as well as in eastern and western provinces. For example, in Greece and in the Black Sea region[81] we find *sulci* at Thermi, close to Thessalonike, where trenches are 20 cm (8″) deep, 65-80 cm (26-32″) wide, and 1.4-2.7 m (4′ 6″- 8′ 8″) apart. At Pella, the inter-row spacing was 1.9-2.9 m (6′ 5″-9′ 6″), the width of trenches 45-70 cm (18-28″), but depth was not possible to

81 Eudokia Skurlatidou, "Techniques and Methods of Viticulture in Ancient Times: The Examples Found in Greece and the Black Sea," *Eirene* 42 (2006): 73-88.

measure. At Megara interrow spacing was 1.4-1.7 m (4′ 6″ - 5′ 7″), trench width and depth both 15 cm (6″). In the foothills of Mt. Olympos, another plot has intervals between rows of 1.5 m (5′ 11″) and the trenches are 70-100 cm (28-39″) wide. Almost all these plots, incidentally follow the north-south orientation recommended by Columella (*DA* 1.6, 3.12.5-6, 4.2). Other trenched plots occur near Kato Achaia in Patrasso, at Nemea, on Salonico, near Pieria in Macedonia, and on the Erakliotike Peninsula in Tauride, modern Crimea. Dates range from the fourth to first centuries BCE.[82] In the west, trenches are attested in southern Gaul.[83] At Saint-Jean du Désert, near Marseilles, for example, trenches are at regular intervals and at a depth of more than a meter. Both *sulci* and *scrobes* are attested here, often with the addition of organic material as compost. Because the plantings were quickly refilled, we can actually see traces of the marks of digging tools. Widths of Gallic trenches vary between 20 and 40 cm (8-16″), and spacing between rows varies from .55 to 2.75 m (22″-8′). Depth is difficult to estimate because the surface has been disturbed so often. At Saint-Jean du Désert, trenches were at least 40 cm (16″) deep. Trenches are also found at La Quintarié, Clermont (Hérault). And because trenches were frequently backfilled with rich soil, manure, and compost, as Columella recommends, we see trenches by way of aerial photography at any number of sites; for example, in the Tavoliere of northern Apulia.[84]

Planting in holes (Col. *RR* 3.15.5, 5.4.2-3; Palladius 3.9.14) is indicated for extreme economy or for the trellising systems of *vitis capitata* or *vitis arbustum,* to be discussed below. Columella recommends a planting hole 2′ × 3′ × 3′ in which the slip or quickset is to be placed obliquely, the projecting part of the slip to be tied to four stakes at the corners of the hole. Mago had said that *scrobes* should be backfilled only halfway to encourage deep root growth, but Columella says this is impractical in a wetter terroir since rainwater puddles and causes root rot, though he qualifies the practice for dry Africa (*RR* 5.5.4-6).[85] The same treatment of grape pomace, manure and/or 'fat' soil is recommended here as for trenches (*RR* 3.15.5, Pliny *NH* 17.173). Holes should be backfilled and the soil tamped carefully around the roots (Cato *Agr.* 28; Pliny, *NH* 17-159-60). Cato recommends for the operation the rammers and bars (*festuces* and *vectes*) we have seen before, Pliny a dibble (*paxillus*). Columella says

82 Yanis Pikoulas, "Oinou istoro (Narro il vino). La ricerca in Grecia (1990-2006)," in Ciacci, Rendini and Zifferero (2007): 42-7.

83 Philippe Boissinot, "L'archéologie des vignobles antiques en France méredionale," in Ciacci, Rendini and Zifferero (2007): 35-41.

84 Jones (1980): 85-100; Potter (1987): 101-3.

85 Cf. Greene in McGovern, Fleming and Katz (1996): 311-22.

that in dry, windy conditions a 'garment' of mulch may be used to cover the projecting vine and prevent desiccation, exactly as is recommended today.[86]

An interesting side issue is whether the vines are to be interplanted, especially with legumes which nitrogenate the soil, as is so common in modern Italy. Jashemski [87] found in the small vineyard attached to the House of the Ship Europa (I.15.3) broadbeans (favas) and horsebeans, suggesting that this practice has an ancient pedigree as well.

Another issue, obviously, is the overall planting pattern of a vineyard plot (Theophrastus *CP* 2.5.6; Vergil *Geo.* 2.273-87; Varro *RR* 1.8.5; Columella 3.14-15, 5.3).[88] Two considerations seem to have exercised the ancient viticulturists, terroir and cultivation practice. Thus Vergil says that planting should be closely spaced in rich, campestrine fields but less so in lean and hilly ground. Columella seems to prefer quite the opposite: 5′ intervals between rows in lean soils, 6′ in medium and 7′ in fat ones. The conflict is only apparent, since Vergil is speaking of the *arbustum* and Columella a trellised system. Elsewhere (*RR* 5.3) he gives precedence to 5′ × 5′ spacing and gives a simple formula for calculating the number of vines required for a plot using this spacing, but also gives calculations for intervine spacings of 3′, 4′, 6′, and 7′ and implies that 5′ should be the minimum interrow interval. Pliny (*NH* 17.171-2), says that 5′ is a typical interrow interval in medium-density soil, in rich soil at least 4′, and in thin soil up to 8′. In Umbria and Marsia, he adds, where they interplant cereals, they leave 20′ to create a 'ridged field' (*porculetum*). Rainy and misty climates demand wider spacing, drier ones the opposite. Apparently Pliny is concerned about the transmission of mildew. The other consideration is mentioned only by Columella (*RR* 3.13.3), who says that if the plot will be cultivated only by the spade-man (*fossor*), spacing between rows is 5′-7′, but if it will be tilled with oxen and plow, the spacing will be 7′-10′. Obviously the determining factor here is the conformation of the land, since plowing is impractical on hilly slopes. Others, Columella adds, set vines at 10′ intervals in the *quincunx* pattern (basically like the five pips on a die) to allow both cross plowing and diaginal plowing, but this is not profitable except in very rich soils which will support luxuriant growth.

Again Columella returns to the issue of varietals, this time (*RR* 3.20-21.5; cf. Palladius 9.11-13) to urge that they not be planted promiscuously but in

86 For modern practice, Weaver (1976): 187. Today vines are carefully tamped, the hole filled completely and, under dry or windy conditions, the vine shoot completely covered with loose, dry soil to prevent desiccation.

87 Jashemski (1979): 242.

88 For modern practice, Boehm and Coombe in Coombe and Dry (1988): 23-34.

separate *crús*. It is essential that several cultivars be planted, since the vagaries of weather which harm vines are innumerable and each cultivar is more sensitive to or tolerant of any given set of circumstances. Columella suggests four cultivars, planted in proportion to their caliber. This is difficult, since vines in an existing vineyard or bought elsewhere are difficult to distinguish visually. And so, yet another advantage of the nursery is that by ingrafting the vineyard with scions from select vines from the nursery only, within a few years it is possible to reliably separate cultivars. Besides the advantages of separate plantings already mentioned, varietals infloresce and ripen at different times, making the harvest of the mixed vineyard at the ideal time impossible. Furthermore, some cultivars have better aging potential than others, so the vigneron can age his noble wines separately and thus reap greater profit. Pruning is also more effective with separate *crús*, and aspect can be selected discriminately. Finally, separate *crús* provide a more esthetically pleasing vineyard for the owner.

Obviously, the density of planting was a key concern for the ancient vigneron. For comparison, in the eastern U.S. where availability of water is not a major concern, 2 × 3 m (6′ × 10′, i.e., 6′ between vines and 10′ between rows), is recommended, whereas in the Napa Valley 2.4 × 2.4 m (8′ × 8′) spacing is common. This represents 650 vines per acre (1,600/ha) and 600 vines per acre (1480/ha) respectively. By contrast it is typical to see 3,000 vines per acre (7,400/ha) in Champagne. But there is a tendency toward greater density in California, on the European model, on the theory that vines water stressed in this way produce less fruit per acre but fruit of a higher quality, and the greater density of vines means that overall yield is the same or more.[89] Planting densities throughout the modern world vary enormously, from as low as 500 vines per hectare (235/acre) to as high as 50,000/ha (23,500/acre). Today an additional constraint on vineyard design is the use of vineyard machinery which reduces labor costs but requires wide avenues between rows and considerable headspace at the ends of rows for turning radius.[90]

89 Jeff Cox, *From Vines to Wines: The Complete Guide to Growing Grapes and Making Your Own Wine* (1999): 28-30.

90 Boehm and Coombe (1988): 27-9. James Halliday and Hugh Johnson, [*The Art and Science of Wine,* 2nd ed. (Buffalo, NY, 2007):17] refer to recommendations in Roman times of 125,000 vines per acre (50,000/ha), but this is surely a misinterpretation. The highest density I can find recommeded is 9,000 per *iugerum*, congruent with 8,000/acre or 20,000/ha, a figure typical of mid-nineteenth century Europe. For a 4th c. BCE Greek vineyard of 21.4 ha (8 acres) in the Tauric Chersonessus planted to 5,000 vines/ha (2,000/acre) cf. Vandermersch (1994): 119.

Archaeology is beginning to reveal a bit about actual vineyard density, as well as overall size of plots.[91] Excavation and aerial photography in southern Gaul show vineyards delimited by paths, ditches, hedges and walls, and in a few cases we can actually determine the size of the vineyard; e.g., more than 20 ha (50 acres) at Girardes, Hérault, two plots measuring 330 × 75.45 m (1,083 × 248′) and 330 × 96 m (1,083 × 315′); at La Quintarié, Hérault, one measures 137.1 × 89.3 m (450 × 293′). In general Gallic vineyards seem to be elongated rectangles. Planting patterns in the environs of Rome, especially south of the city where growing conditions were ideal, are so regular that they are regarded as a marker for vineyards during excavation. Width of trenches is 80-90 cm (32-35″), but distance between parallel rows of trenches varies widely, from 3 to 10 m (10-33′). The wider spaces seem to imply intercultivation with cereals, vegetables or legumes. Unfortunately it is not possible to determine inter-vine spacing with trench plantings. There is evidence from the second and first centuries BCE at several sites of replanting with different orientations and at smaller intervals between rows, perhaps an indication of a trend toward pure grape culture as opposed to interplantings, probably to supply the increasing demands of the burgeoning city. For comparison, Columella's minimum inter-vine spacing is 1 m (3 Roman feet) and minimum inter-row spacing recommended is 1.5 m (5′) while his corresponding maximums are 1.5 m and 3 m (5′ and 10′). This yields densities ranging from 900 to 3000 per acre (2200-7400/ha). Unfortunately we do not yet have enough data to make even rudimentary estimates of how this is confirmed or not in the archaeological record.

The procedure for marking out the plantings has not changed appreciably over the millennia; today row positions are surveyed and pegged and planting sites on rows are marked by using a planting line with fixed 'buttons' at planting intervals.[92] In the same vein, Columella (*RR* 3.15; cf. Palladius 1.9.10) suggests taking a cord with brightly colored rags at the predetermined intervals. The line is stretched along the ground in the correct orientation and a reed thrust into the ground at each rag.

During the course of the first few years it will be necessary to replant vines which die or fail to thrive (Cato *Agr.* 33.2-3, 49; Columella *RR* 4.15-16). Cato advises the vigneron to cut furrows in longer gaps and plant these with quicksets. Whole vines may also be transplanted provided the trunk is no thicker

91 I. Figueiral et al., "Archaeology, vine growing and wine production in Roman Southern France: The site of Gasquinoy (Béziers, Hérault)," *Journal of Archaeological Science* 37 (2010): 139-49; Boissinot (2007): 37-8; Santangeli Valenzani and Volpe in Ciacci, Rendini and Zifferero (2007): 48-54.

92 Boehm and Coombe (1988): 32-4.

than the human arm. The vine is pruned back so that no more than two buds are left on each arm. Then roots are dug out carefully and the vine planted in a hole or trench, backfilled and tamped, trained and tied and cultivated often. Columella recommends that replanting be done in late autumn after the vintage using the mallet shoots propagated in the trenches between the quicksets or by layering from an adjacent vine. If layering is the system chosen, after one year the part of the layer next to the mother vine is cut, but only to the pith so that it is gradually 'weaned' from the mother. The next year it is completely detached, the extraneous part of the vine cut back to the roots and a hole dug beneath the roots and backfilled to encourage deeper rooting. If mallet shoots are to be used they are dug up sometime after October 15, except under adverse weather conditions, in which case the vigneron should wait until early spring. Unused mallet shoots are dug up—recall that Columella has advised planting 20,000 mallet shoots per *iugerum*—to reduce competition for nutrients, water and sunlight, though slips may be kept in the vineyard for three or four years if closely pruned.

Culture of the Vine

Training

Now begins the serious work of raising the vines. There are a number of viticultural practices that apply to all vines, regardless of age—cultivation of the soil, irrigation, fertilizing—though young vines require particular attention because their immature root systems make them especially sensitive to environmental stresses. But early on, the vigneron must make some serious choices as to the method of training he will use, since this will determine how he prunes and trains immature vines.[93] The term 'training' refers to the design and development of a vine framework. 'Trellising' is a subset that refers to the training of a vine to a structure which supports this framework, and 'pruning' is, in part, cutting away portions of annual vegetative growth to maintain the desired number and spacing of nodes per vine on these structures. Training is done to maximize production, to facilitate cultural operations such as tilling and harvesting, to improve canopy microclimate, and to support the fruit load. A trellis supports the vine to optimize production by regulating capacity, i.e.,

93 Boehm and Coombe (1988): 36-40; B. M. Freeman, E. Tassie and M. D. Rebbechi, "Training and Trellising," in Coombe and Dry (1988): 42-3; Jamie Goode, *The Science of Wine: From Vine to Glass* (Berkeley, 2005): 84-91; Billiard (1913): 356-68; Vine et al (2002): 59-61; Winkler et al (1974): 272-86. Training in Egypt: Ricci (1924): 33-9; Greece: Amouretti (1993): 83-7.

the productivity of the whole vine, and vigor, i.e., growth of individual shoots. Many systems of training are still used worldwide today, but, contrary to the pattern of even fifty years ago,[94] most vines today are trellised, the most common system being trellising to two or more wires strung from posts.[95]

There are several general rules for modern training. First, there is generally no pruning or training done the first growing season. The first dormant season, all top growth is trimmed except the strongest, well placed cane, which is usually trimmed to 3-4 buds to become the trunk of the vine. During the second growing season the strongest shoot from this cane is tied to a stake and all others trimmed off except for one reserve shoot. The second dormant season the trunk is trimmed back, if necessary, to the correct height of the head (i.e., the part from which the arms will diverge). The third growing season, all shoots are broken off from the base of the trunk, those at the head are allowed to grow and as growth progresses weaker ones are trimmed away to encourage growth of the stronger, that it, those which are destined to be the arms of the vine. In the third dormant season all canes are pruned away except 3-6 spurs, each 2-4 buds long. These will become the fruiting canes. During the fourth and subsequent seasons the vine is pruned to match the designated training system.[96]

Now compare Columella (*RR* 4.16.2-17.1): When the new vineyard is 30 months old it receives permanent props. Stakes are set 1′ away from the trunk (*truncus*) so as not to injure the roots and to allow space for cultivation. A stake is placed to the north of the vine to receive the brunt of winter cold and wind. If placed between vines (as for the *vitis pedata* system to be discusses momentarily) it must be sunk in an especially deep hole to support more weight. Pliny (*NH* 17.178-81) adds that when the trunk reaches the height of the crossbar used in trellising, some vignerons 'blind' it by removing eyes to make it grow longer, but this, he rightly insists, is harmful. Better to trim off leaf shoots to the point where the trunk is supposed to lignify. Pliny gives a rough-and-ready rule for the time of head training, namely, when the vigneron judges that the trunk is strong enough to bear a fruit load, and then arms are trained from the head, but tied under crossbars in trellised systems to provide a more sheltered position. In general, the trunk should be as thick as a thumb before head training begins, but soil fertility also dictates timing. Arms should be pruned to 2-3 buds to produce fruiting canes (*materia*) and thereafter should be bound more tightly to restrict the flow of sap and thereby encourage growth behind this

94 White [(1972) 232] says that in his day unsupported, head trained vines (*en gobelet*) were the universal practice in Europe except in Italy and Germany.

95 Freeman, Tassie and Rebbechi (1988): 42-3.

96 Weaver (1976): 36-40.

ligature, but the end of the arm (*cacumen*) should not be bound in this way. Also left under the crossbar is a reserve shoot called a 'guardian' (*custos*), no more than 3 buds long, to provide a fruiting cane the next year if the selected shoot should fail, and even a third shoot, the 'petty thief' (*furunculus*) should both shoots fail.

Vines propagated from slips in the vineyard are trained to a T-shaped support called a 'horse' (*canterius*) in spring before budbreak, but in this case the stake is only 1′ tall with a correspondingly short crossbar for tendrils to grasp so that the tender shoot does not have to be tied (Colulmella *RR* 4.14.1). Obviously the rest of the training system will have to be adjusted for slips as well, since their root systems are so much less developed than those of quicksets. In the case of quicksets, Cato (*Agr.* 33; cf. Columella *RR* 4.21.3) is at pains to ensure that the future trunk is straight by training it to the stake with a number of ties. A practice approved by modern viticulturists, incidentally.[97]

There were essentially six training systems used in antiquity (the *locus classicus* is Pliny *NH* 17.164-6), along with their variations.[98] Two of these are based on the vine's natural habits, namely, creeping along the ground and climbing up trees. All other systems supply some permanent, artificial supports for the vine to cling to. Although admittedly speculative, it is perhaps useful to suggest a possible evolution in ancient training systems. The unsupported system which allows the vine to creep, Latin *vitis prostrata* (Fr. *provignage*) has severe disabilities in all but the driest climates and generally produces fruit of lower quality, and it appears likely that the so-called *arbustum* system (Latin *vitis arbustiva*), in which vines are trained to trees, was the original supported system since it simply encourages the vine's natural habit yet still provides cultural advantages and superior fruit. A sort of paradomestication, as it were. We have seen that the system was almost universal in Etruria from a very early time. At some point it was natural to plant more densely and shorten the vine to a level a worker could handle without using ladders. Thus, the *vitis capitata,* the head-trained vine (Fr. *en gobelet*). The next natural step would be to support the vine on a stake, a sort of artificial 'tree'. Hence the Latin *vitis pedamenta,* the staked vine (Fr. *en échalasse*). At some point a cross piece was added at the top of this stake to allow arms to extend themselves and thus increase sun exposure, as well as to support heavier fruit loads, and thus the Roman *vitis iugata/ vitis canteriata* (Fr. *vigne a jong simple*), the 'yoked vine'. We know that such 'yokes' (*iuga*) were oriented along rows and sometimes extended so that they almost touched adjoining ones. Thus the next logical step would be to run a

97 Boehm and Coombe (1988): 38.

98 Billiard (1913): 355-73; White (1970): 231-6.

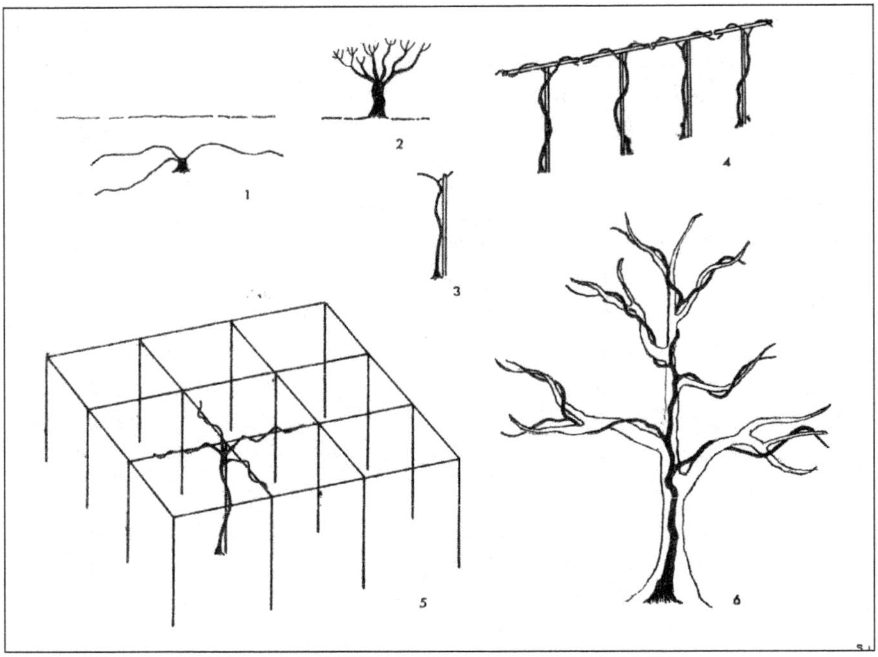

FIGURE 16 *Training schemes: 1)* vitis prostrata, *2)* vitis capitata, *3)* vitis pedamenta, *4)* vitis iugata, *5)* vitis pergulata, *6)* arbustum. *From Jashemski (1977-1993): Fig. 309.*

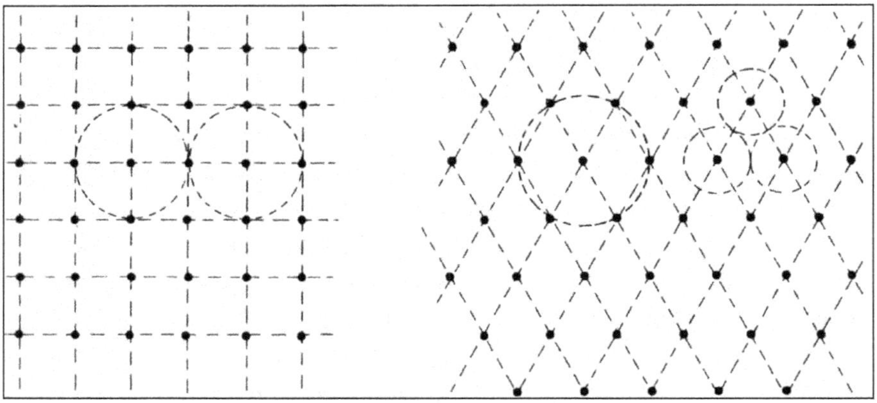

FIGURE 17 *Vitis pergolata vs. quincunx, showing the increased cultural area of the latter. From Jeffs,* Sherry *(London, 1992): 139.*

continuous cross piece along the tops of stakes to provide greater support and simplify the building process. Given this continuous frame, it is logical to run crosspieces at right angles so that the vine can support four arms instead of two, and thus we have the arbor (*L.pergula*, Fr. *vigne en berceau*). This system relies on stakes that are equidistant from the cardinal points, but in this case adjacent vines are competing for moisture and soil nutrients with three other vines. But by staggering the rows we obtain an advantage for spreading root zones. For example, if we posit a 5 × 5 planting pattern, then it is obvious that the root systems of each vine will intersect those of its four neighbors at 2 1/2′, but if we stagger the rows so that cross pieces define a checkerboard pattern, root zones and canopies of adjacent vines have 20% more space to spread. This is the Latin *Vitis compluviata*, since the intersection of the crosspieces suggested to Romans the sloping valleys of their impluviate roofs. Alternately, it is known as the *quincunx* pattern. Both *pergolata* and *quincunx* patterns are still used in the sherry vineyard of Spain, where they are known as *marco real* and *tresbolillo* repectively. The *tresbolillo* accomodates 15% more vines, while *marco real* is more easily cultivated.[99]

Be that as it may, of the dozens of trellising systems attested over the centuries, all fall into four categories: stakes, vertical trellises, wide topped trellises, and arbors. The choice of type is more dependent on the availability of affordable materials and congruity with cultural and harvesting practices than with productivity, since all such systems have been thoroughly proven over the centuries.[100]

Of the two unsupported systems, *Vitis prostrata* or *provignage* was not highly regarded by the Roman agronomists, although they recognized that it promoted the most rapid fruition (Cato *Agr.* 52; Varro *RR* 1.8.8; Columella *RR* 4.15, 5.4, *DA* 4; Pliny *NH* 17.185; Palladius 3.11, 3.16; Didymus in *Geopon.* 4.3; Florentinus in *Geopon.* 5.2). Varro mentions the use of the practice in Spain, and the dry climate there makes it not unlikely, but he says that the fruit is still scavenged by foxes and mice unless they are trapped out, as they are on the island of Pandateria. The determining factor for Varro is the degree of soil surface moisture, since the vine is highly susceptible to molds and mildews and, when producing fruit, does not need moisture so much as sun, the reason that vines climb up trees in the first place. Pliny says that the system was used in the greater part of the world, being customary in Africa, Egypt, Syria, and the whole of Asia (Turkey), as well as many parts of Europe.[101] Archaeological evidence attests the practice in southern Gaul, as, for example, at Les Girardes,

99 Julian Jeffs, *Sherry* (London, 1992, 4th ed.): 138-9.

100 Winkler et al. (1974): 254-71.

101 Billiard [(1913) 271] says that the system was still used in his day in Burgundy. Piccoli [(2004): 35] says it is used only in northern Italy today, specifically in Liguria, the so-called *scriscianti*, "creeping' vine, although it is still widely practiced in Greece as well. Cf. M. Fregoni, *Origini della vite e della vitcoltura. Contributo dei populi antichi* (Rome, 1991): 87.

Lapalud.[102] The greatest problem with the system in the long term is that roots become so entangled that cultivating the soil is almost impossible, and this is absolutely essential to good vine health. The famous example occurred in the ancient Moselle, as previously mentioned.

The second unsupported system, *Vitis capitata* (also *Vitis subrecta per se*), the 'headed vine', had antecedents as far back as Assyria.[103] For ancient Israel we have evidence from the Nineveh relief of the capture of Lachish in 701 BCE on which Judahite captives are accurately depicted. In the background are vines, depicted as freestanding and low to the ground but headed. The practice was possible here at such a low height (almost *provignage*) because of the total absence of summer rains, but small animals will still have been a problem.[104] Interestingly, Columella (*RR* 5.4) calls it "the most highly approved method," in which a single sturdy trunk supports several branches. The next best is *Vitis iugata,* next *Vitis characata,* a 'palisaded' trellis,[105] and last in esteem is the procumbent vine, which he calls *Vitis stratura*. He further explains (*RR* 5.5.8) that there are actually two similar methods, true *Vitis capitata* in which a trunk is raised to several feet and fruiting canes (in fact these will have erupted from spurs at the head) are left in a circular arrangement at the top and trimmed back to 1′ lengths, and *Vitis bracchiata* in which four arms are left and more numerous fruiting canes are allowed to produce. In either case, when the canes are trimmed back they are called 'horns' (*cornua*). Both methods have the advantage that they do not interfere with tillage by oxen and plow. Some people actually tie up the canes in a dense 'head' so that they will not be injured by oxen or a careless plowman. Thinning is done as for trellised vines, heavily in damp and cold regions to promote air flow and sun exposure, lightly in dry, hot climates to provide more shade for fruit

The other Roman training systems employ the vine's natural climbing habit. The vine may be trained to a living support, such as a tree, or to a non-living support made from various materials, especially wood. In the former case, we have *Vitis arbustiva* or simply the *arbustum*, the training of vines to trees (Columella *RR* 5.6-7; Pliny *NH* 14.10, 17.89, 17.141, 17-199-203; Palladius 3.10; Florentinus in *Geopon.* 4.1). The method has a long and relatively well attested

102 Boissinot (2007): 37-8.

103 Jean-Pierre Brun, *Archéologie du vin et de l'huile dans l'Empìre romain* (Paris, 2004): 34-6; Brun and Fanette Laubenheimer, "Le vignoble gaulois," in Brun, Poux and Tchernia (2004): 144.

104 Walsh (2000): 110-19. Piccoli [(2004): 35-6] says the modern Italian practice, *alberello*, 'little tree', is practiced along the coast of Liguria, in part of Val d'Aosta, and in Piemonte at Oltrepo Pavese and Piacentino.

105 Cf. Sereni (1965): 136-8.

history, as we might expect. For example, though there is no evidence of training vines to trees in lower Mesopotamia, the method was clearly known in upper Mesopotamia from the first millennium BCE, as witnessed by Assyrian relief sculptures.[106] Hittite laws discuss compensation in areas where grape vines have been burnt down in conjunction with fruit trees, suggesting the possibility that the Hittites were using the *arbustum*. But another text suggests that even this early (ca.1600-1400 BCE), vines were being trained to supports as well, so the law texts may simply be a case of interplanting vines among fruit trees.[107] An Attic black-figure amphora from c. 540-30 BCE, now in the Louvre (Inv. AM 1008) shows satyrs at the harvest, and in their midst is a vine trained to a tree. In this case we even see the props recommended by Columella (to be discussed below) to reinforce branches overladen with the vine's fruit load. The method has been traced back in Etruria to the roots of viticulture there and was so tenacious that it was regarded as typical of the area in historical times.[108] Forni follows others in seeing training stakes and yokes as typical of southern Italy and perhaps even a tradition of the Oenotrians (Gr. *oinotron* = vinestake, according to one conjecture) or the Greeks of Magna Graecia, while long pruning and training on trees was the tradition of the Etruscans and their areas of influence in northern Latium and the Po Valley.[109] Certainly the method is well attested in Cisalpine Gaul, which, as we have seen, derived its viticultural practices from the Etruscan motherland.[110]

Thus the Romans probably learned the methods from northern Latium. Columella (*RR* 5.6) does his usual thorough job describing the *arbustum*, including the species of trees preferred for the job and their culture, the spacing of trees and vines, planting, pruning, etc. Briefly, preferred species of trees are Gallic and Atimian elms, the poplar and ash. Spacing depends on the intended method of cultivation and whether cereals will be interplanted. Vines and trees are 'married', and if there is an age disparity between the 'partners' one will not thrive. Planting of vines in the *arbustum* may be in either holes or trenches. As usual, quicksets are preferred to slips. Plantings should be at least

106 Sereni (1964): 143-8.

107 Ronald L. Gorny, "Viticulture and Ancient Anatolia," in McGovern, Fleming and Katz (1996): 157.

108 Myriam Giannace, "I comprensori indagati nel 2005. l'alta Valle dell'Ombrone (SI), l'alta Valdese (SI), la Val d'Orcia (SI), la bassa Val di Cornia (LI), l'alta Valle dell'Albegna," in Ciacci, Rendini and Zifferero (2007): 210.

109 Forni in Ciacci, Rendini and Zifferero (2007): 74. Paolo Braconi ["*Vinea nostra*. La via romana alla vitcoltura," in Di Pasquale (2010): 152-59] makes a case that the Latin term *vinea* originally specified the *arbustum*.

110 Piccoli (2004): 37-43.

1 1/2′ from the tree trunk so that the roots of each will not compete. *Pastinatio*, where indicated for the system, should be done in the autumn, the planting itself to follow in the spring. Trenches will contain two vines 1′ apart. Both vines and their 'husbands' should be planted in the same orientation as they had in the nursery. In warm, sunny, moderately dry climates, planting may also proceed in autumn. The transplanted vine should be 'caged' to prevent damage by livestock. In a hot climate the vine should be planted to the north of the tree, the opposite in a cold climate, and on the east or west in a moderate one. Celsus, says Columella, had recommended no pruning after the first growing season, but Columella disagrees. Those who want productivity and ease of culture and harvest train the vine to the tree's lower branches, but the best quality clusters come from upper branches. During fall pruning old fruiting canes are pruned away, new shoots trimmed of laterals and allowed to hang down. Long cordons are sometimes conveyed to nearby trees and trained there to give better exposure. Also, as in other training systems, ligature of fruiting canes may be used to encourage lignification, but Columella recommends only to the third bud. Cross-cordons should only be conveyed to adjacent trees after the vine is mature enough to support the additional fruit load and after another two years these should be pruned off and others trained across (*RR* 5.6.29).

A special kind of arbustum called the *rumpotinum* (presumably a Gallic word) is also described by Columella in some detail.[111] Found in Gaul, it is a system in which vines are trained to dwarf trees such as viburnum, cornel cherry, hornbeam, mountain ash, or willow. But willows grow only in watery places and this, says Columella, spoils the flavor of the grape (and, we might add, will quickly kill the vine), and so are not recommended. The elm can be adapted to the system by being pruned severely when young so as not to exceed 15′ in height. If one does not intend to interplant cereals, there should be 20′ between trees, but if so, 40′ between trees and 20′ between rows of trees. Culture is as for the Italian *arbustum*. If a cross-cordon (*tradux*) will not reach from one tree to another it may be trained to a rod (*virga*) suspended between the trees. Dwarf trees which become overladen with the vine fruit may be propped (*adminiculi*). One of the famous cupid frescoes from the House of the Vettii at Pompeii shows cupids as vintagers, and to the left and right of a lever-and-capstan press are vines trained to dwarf trees. Between the trees can be seen such cross-cordons (*traduces*). One is so substantial that a cupid actually perches on it to gather clusters while another steadies a ladder leaning against it, from which his fellow vintager has evidently clambered onto the cordon. It would seem that these cross-cordons have been trained to substantial rods of

111 Cf. Sereni (1964): 182-97: linguistic evidence for *rumpus* in paleo-Ligurian.

the sort recommended by Columella, unless this is artistic license. Here we may see the origin of two systems still widely used in various parts of Italy and first discussed by Sereni. In the first, *alberata*, vines are planted around the trunk of the tree as described by Columella for the *arbustum*. In the second, the so-called *piantata*, trees, often dwarf trees, are planted in files and supports, now generally wires but not uncommonly wooden rods, are attached to adjacent tree and the vines planted along the file thus created and trained to the supports instead of the trees.[112]

Pliny (*NH* 17.199-203) echoes Columella's directives, right down to the spacing, but adds that the white poplar is also used for the 'husband', especially in Campania (*NH* 14.10), and elsewhere that the fig tree, because it produces little shade, may be planted among vines (*NH* 17.89). Does this mean as a trellis? It would seem not. In Jashemski's excavation of the large vineyard at Pompeii[113] root cavities were found for 58 trees, usually in the center space among four vines but occasionally in the rows between two vines. Small root cavities next to larger ones were evidently seedlings allowed to grow until they could be culled for vigor. Rows of widely spaced trees were also located around the perimeter of the vineyard as well as between the first and second rows of vines on each side of footpaths. The density of the trees is more than Columella recommends for interplanting (*RR* 3.21.11), but perfectly consonant with modern Campanian practice where fruit and nut trees are generously interplanted among vines. Some of the trees in the vineyard could be identified as species recommended by Columella for the *arbustum*, but the standard training method for this vineyard was otherwise quite obviously the trellis and so it is probable that we have here a case of simple interplanting. The reason for the generous number of trees is simply the intensity of the Campanian sun.

The other methods require materials for stakes, trellises, bindings, etc., and the agronomists are at pains to instruct the careful vigneron on the best materials for each function. Materials are needed for stakes, for crossmembers where indicated and for binding vines to each. Wood is preferred for stakes for obvious reasons. The reader will recall that access to woodland, reed beds and/or osier thickets was a *desideratum* for the ideal villa farm. Cato (*Agr.* 17.1) tells us that wood for stakes is ready for cutting at the winter solstice. Chestnut is preferred because of its ease of working, "obstinate durability" and regenera-

112 Emilio Sereni, *History of the Italian Agricultural Landscape* (Princeton, NJ, 1997. Tr. by R. Burr Litchfield): 295-303. Cf. Boris Basile, Gaetano Di Pasquale, Antonella Monaco and Mauro Vella, "Sopravvivenze dell'antico paesaggio della vite in Campania," in Di Pasquale (2010): 97-105.

113 Jashemski (1973): 825-6; *idem* (1979): 209-10.

tive power when cut (Pliny *NH* 17.147). Vergil (*Geo.* 2.354-61) would have his vigneron prepare such materials during the dormant season and recommends canes (*calami*), shafts of peeled rods (*rasae hostilia virgae*) and ash stakes (*fraxineae sudes*) along with strong forks (*furcae valentes*). Willow rods of great length were used for both props and crossmembers and strips of willow bark were also used for bindings (Pliny *NH* 16.174). Other woods used for props were holmoak, olive, juniper, cypress, laburnum, and elder, in order of preference. Stakes of all other kinds must be replaced every year, but these are more durable. Vergil (*Geo.* 2.368) is our sole witness that stakes were typically decorticated, and Billiard[114] points out that this is good vineyard hygiene; the ancients knew that all kinds of noxious pests lodge under tree bark.

Woody plants of smaller size are also used as crossmembers, and here again the preferred types are branches of chestnut and willow, but reed is also frequently used for the purpose.[115] The reed or cane (*Arundo pliniana*) grows widely in the Mediterranean, and when cut and dried provides a lightweight woody support. Reeds were often bundled to provide more strength (Varro 1.8; Pliny *NH* 17.174). Pliny assures us that, bound in this way, reeds will last for five years, and Columella (*RR* 4.17.1) says that when they are bundled they are oriented in opposite directions alternately so that the thickness of the bundle remains consistent. Pliny also describes a peculiar system in which shorter shoots of the vine itself are bundled to living vine canes and the resulting crossmembers known as 'cables' (*funeta*). The Romans, it seems, were even using parts of the vine itself as parts of trellises.

Binding of shoots and canes to props and crossmembers (*palmare*: Col. *RR* 11.2; *alligare*: Cato *Agr.* 6 and 33; Columella *RR* 11.2; *ligare, religare, nectere*: Columella *RR* 4.13) was typically effected with willow withies, strips of willow bark, or broom straw (Col. *DA* 29.1) and Columella recommends that a willow copse and a patch of broom (L. *genista, Genisteae sp.*) be planted near the vineyard specifically to provide such bindings. Cato (*Agr.* 33.5) tells us that willow withies are cut and dried to be used later and that willow bark is tied in bundles and dried in the same way. When needed for the vineyard they are soaked in water to make them pliable again. Columella (*RR* 4.13-14) says choosing ties for young vines is especially important since willow and elm withies can cut the tender vine, and so advises the application broom, marsh rush (*Juncus gerardii*) or sedge (L. *ulva*, perhaps *Cladium mariscus*), or even dried reed leaves. Umbrian Ameria was famous for willows with slender brown withies (Col. *RR* 4.30.4), useful for *Amerina retinacula*, 'Amerine ties" as well as for

114 Billiard (1913): 360.

115 For the use of reed in Egyptian viticulture, see Ricci (1924): 32-4.

baskets, part of the equipment of the typical vineyard (Vergil *Geo.* 1.265; Cato *Agr* 11).[116]

The simplest form of trellis is simply a prop or stake (*ridica*) up which the vine can climb, and this training regimen is called *vitis pedamenta* (also *cum amminiculo sine iugo*: Pliny *NH* 17.164). The practice among the Romans almost surely derives from Magna Graecia, either directly from the Greeks or as mediated by the Oenotrians. It is mentioned by Homer (*Il.* 18.561) and is well attested from Attic vases, from the tablets of Heraclea, as well as from Pharaonic Egypt.[117] Props for immature vines are typically reeds dried for two years (younger ones will take root) or briar canes (Col. *RR* 4.12). The prop may be the *palus*, a length of unsplit wood; *teres*, round branches of such evergreens as juniper, laurel, cypress, forest pines (*Pinus sylvestris*) and elder (*Sambucus nigra*). Stakes are supports split from strong wood such as holm oak (*Quercus ilex*) or cork oak (*Quercus suber*); poles (*perticae*) are of the same woods but are unsplit branches, not trunks. Thus we have stakes, poles and roundels. All such frame elements are inspected in the fall after pruning and if still largely sound are reversed and the rotting part formerly in the ground is removed. Those beyond reuse or too short from being lopped in this way are replaced and vines rebound to new props (Col. *RR* 4.24.14; 4.26).

Columella describes a special form of *vitis pedamenta* invented by Julius Atticus, the famous viticultural writer, in which a single vine is trained to two stakes on opposite sides of it (*RR* 4.2, 4.16) but he rejects the method. He is also our only witness that a special dibble, *paxillus*, was used to drive the hole into which the prop was to be placed. White[118] points out that modern dibbles used for the same purpose are made of metal and have a crossbar at the top to give better purchase in driving the hole.

A subset of the staked vine is *vitis characata.* The name derives from the Greek *chárax*, 'vine stake', and consists of a vine with cordons bound to palings arranged in a circle. It is widely attested in modern times in Liguria, Piemonte, Emiglia-Romagna, and Lombardy and was only completely abandoned there in the twentieth century. In Piacentino (Emilia-Romagna) it was called *crasse*, but variants of the term show up in dialects of many northern Italian areas.[119]

116 In modern Calabria vineyardists still use willow strips as binders, as well as rope grass (*Ampelodesmos mauritanicus*): Orlando Sculli, *I vitigni autoctoni della Locride* (Cosenza, 2004): 105-31.

117 Brun (2003): 37.

118 Cf. White (1975): 23.

119 Piccoli (2004): 36-7.

Billiard[120] also attested the use of the system in Saint-Émilion (Aquitaine) in the early twentieth century.

Evidently the most common system in Italy in the first century BCE was *vitis iugata* (Varro 1.8), in which a trunk was trained up a prop and then cordons were trained across a transverse 'yoke' (*iuga*) (Col. *RR* 4.13, 4.19-20, 4.22.1-8). Yokes could be poles (*perticae*), reeds (*harundines*), cords (*restes*) or bundled vines (*vites*), the first prominent in the Ager Falernus, the second among the Arpi, the third in Brindisi and the last in Milan. The practice had precedent in Egypt.[121] Columella (*RR*. 4.19) says the props should be at least 4′ tall but no taller than 7′. The determining factors are soil and atmospheric moisture, the moister the terroir, the higher the frame. This is sound viticultural practice; higher frames discourage fungal infections by promoting sun exposure and ventilation. Ideally the frame should be 5′ tall to facilitate culture and harvest, but higher frames yield better wine quality. At the head of the vine the cordons or arms (*bracchiae*) are splayed in opposite directions and the fruiting canes (*palmae*) are tied to the yoke so that they recurve and hang down below the frame to shed rain. But when *véraison* (*mitescere*) arrives these are tied to the yoke along their length to support fruit load and to resist fungus and the pillaging of pests (Col. *RR* 4.13). If a vineyard has been taken over which has been otherwise trained, it may be profitably retrained to the yoked system. But it is apparent that in continuous iugate trellising (i.e., when crossmembers continue from one prop to the next without gaps), some have been in the habit of not replacing dead vines but simply letting cordons grow on them, sometimes to the length of a fourth or fifth prop. This involves more expense for the new owner, since new vines will have to be propagated by layering to renew vigor and productivity. If the trunk is old and eroded, layers are brought from canes on cordons. If the trunk is still sound a wound is created in it with the point of the vinedresser's knife, about 3 1/2′ from the ground, which stimulates budding. If the bud sprouts a vigorous shoot this is layered, if a shorter one, it is pruned back to a 'thumb' (*pollex*) and allowed to prove its vigor by resprouting. If the sprout is too short in the first instance it is pruned back completely and allowed to form a calculus (*furunculus*). Some actually lop off the trunk above the spot where it has been wounded to stimulate budding, but Columella condemns the practice as leading to disease and death. If the vine must be cut down completely it is cut off below the ground and the hole backfilled with soil

---

120 Billiard (1913): 360.

121 T. G. H. James, "The Earliest History of Wine and Its Importance in Ancient Egypt," in McGovern, Fleming and Katz (1996): 205-06, a horizontal bar on vertical forked props.

to completely cover the stump. But the stump may need to be rooted out and the hole replanted (*RR* 4.22.1-8).

The yoked system, combined with regular planting pattern, led almost inevitably to the last two training methods, generically called *vitis pergulana,* in which crossmembers were carried across transversely as well as along the lines of rows. If the frame is raised high enough so that culture and harvest are not otherwise impeded—and we have seen that this was often done for the health of the vine— it is only logical in even moderately fertile terroirs to double sun exposure and fruit load by adding transverse members to create a regular grid pattern, Pliny's *vitis quadripartita/vitis compluviata* (*NH* 17.165), "which produces more wine since the crossbars are doubled to make four." If vine rows are staggered the system creates a reticulate pattern, the 'compluviate' system.

Jashemski[122] thought this the most likely configuration for the large vineyard at Pompeii. The vines were definitely staked, as evidenced by the stake holes found next to each root cavity. Jashemski correctly adduces that a pergolate system is most productive, most resistant to wind damage, and best for hot, dry climates with rich soils. Further, it is still a standard pattern in the Vesuviana today. Nails were found in the backfill removed from root cavities (the whole vineyard had been indiscriminately backfilled after the original excavation), so some sort of frame was apparently used, though a simple yoked frame is possible. But Jashemski thinks the pergolas were along the footpaths only. None of the stakes were found *in situ,* but a graffito found in the Villa del Fondo Prisco (*CIL* IV.6886)[123] reads, "840 sharpened stakes, 460 unsharpened, a total of 1300." Jashemski surmises that the unsharpened stakes may have been used for the crossmemebrs of frames. Another graffito, this one from the villa rustica at Giuliano, near Pompeii (*CIL* IV.6887) reads, "1,023 stakes in a big pile." We also find archaeological evidence of pergolate training in southern Gaul[124] and Trentino Alto Adige.[125]

122 Jashemski (1973): 826-7; (1979): 212-14.

123 Cf. Grete Stefani, "La Villa del Fondo Prisco in località Civita, " in Pietro Giovanni Guzzo, ed., *Casali di ieri, casali di oggi* (Naples, 2000): 45-7.

124 Boissinot (2007): 37.

125 Piccoli (2004): 36.

Grafting

As previously noted, grafting is almost universal in the modern vineyard.[126] But the ancients did not graft for the most common reason of their modern counterparts, but rather used their nurseries to prove the fruitfulness and viability of vines and then grafted them to established rootstocks to bring them into production more quickly. Columella, for example, mentions a two-iugera vineyard which he replanted in only two years by using scions from one early ripening vine from a friend's vineyard and grafting them to established rootstocks (*RR* 3.9.6). The process was much the same in antiquity, if the motives were not.[127] The three parts of a graft are the stock, the scion and the union. The stock is composed of the root ball and part of the trunk. The scion is a segment of cane cut from another vine. The union is the juncture of the two. Grafts are effected to provide a rootstock which is more resistant to disease and insect pests or certain soil conditions such as high lime content, to cull mixed cultivars or to change cultivars in an established vineyard, or to propagate a new cultivar quickly. All cultivars of the *Vinifera* vine are readily grafted. The callus that forms at the union and protects it forms best at high moisture levels and between 24-30° C (75-85° F). Today the graft is generally covered with a moist, porous material such as soil, sand, moss or sawdust. Since the cambium layers of stock and scion are the first to effect a union, it is important that they are tightly joined and coincide as exactly as possible. Then the union must be held firmly in place so that it is not broken. The younger the tissue grafted, the greater the chance of an effective graft but the more difficult it is to stabilize, so in general woody but young graftstock is preferred. Dormant tissues are harder to graft than growing ones and thus summer or at least late spring is the best time for grafting. Stocks may affect the nutrition of parts that grow from the scion, but will not affect its varietal character.[128]

Today there are two basic variations on the process: bench grafting, where the rootstock and scion are joined in the workroom and then transplanted together to the nursery or vineyard; and field grafting, which, as the name implies, takes place in the vineyard and utilizes a rooted stock. The most common configuration for the union is the so-called cleft graft, in which the stock is sawed straight through to expose the full interior of its diameter, then a sharp instrument is used to split a cleft across the diameter, about 2.5-4 cm (1-1 1/2″) deep, then one or two scions with one or two buds each are sharpened to form

126 Nicholas, Chapman and Cirami (1988): 10-20; Winkler et al. (1974): 197-237.

127 Billiard (1913): 277-89; Sereni (1964): 138-40.

128 Winkler et al. (1974): 205-19.

a sharp wedge and these placed at the sides of the cleft in the stock so that bark and cambia are aligned.

Columella (*RR* 3.10) as usual gives us our most precise description of the process as practiced in Roman vineyards. Scions are chosen carefully. The most fruitful cuttings are not the extreme part of the shoots, the *saggita* previously mentioned (Col. *RR* 3.17.2; Isidore *Orig.* 17.5.6-7), since the vine's sap is naturally directed from the trunk to the extremities and thus the seeming fruitfulness of these sections of shoot is a function of position. The shoots closest to the trunk are vigorous but barren. The vigneron should look for intermediate shoots which are exceptionally fruitful despite their position. Country people call this position on the vine the *umeri*, 'shoulders'. Others think any shoot will do except a leaf-shoot (*pampinarium*; also *spadones*, 'eunuchs'). Columella scolds vineyardists who stint on the cost of a skillful *vinitor* (vine dresser) or even assign this critical task to those incapable of more challenging work. According to Cato (*Agr* 40-41), grafting should be done in the dark of the moon, after midday when the south wind is not blowing, preferably in spring after florescence. Some years later Varro (*RR* 1.40.3.) says that the preferred season for grafting some species (his dialogist is speaking of grafting trees, but this includes vines) has changed, but he gives no specifics. He asserts that moisture promotes decay and therefore suggests grafting during the hot, dry dog days, clearly a mistake. Columella (*RR* 4.29.1-12) says that Julius Atticus had suggested grafting from November 1 to June 1 and to a stock cut just below the 'head', both of which practices he deplores. The proper season is early spring, just at budbreak. Autumn is acceptable, though not as good. Pliny (*NH* 17.108) simply says that grafting should be done in the waxing moon. Obviously there was no consensus among our agronomists as to the proper time of grafting.

The general term for Roman grafting is *insertio* or *insitio*, and the Romans practiced three basic forms: the cleft grafting (*insitio* again), patch or bud grafting (*inoculatio*, *emplastratio*), and bore grafting (*terebratio*) (Cato *Agr.* 41; Varro *RR* 1.40.5-41; Col. *DA* 8, *RR* 4.29; Pliny *NH* 17.103-10). The most common form as today was cleft grafting, and the method was much the same: the stock (*truncus*) was cut across through the pith (*medulla*), split in the center, and into this a scion (*surculus*) with a sharpened end ('after the fashion of a reed pen," says Columella, that is, on the oblique, rather than as a chevron) was inserted so that the piths aligned. Columella suggests that the stock be cut slightly above or below the ground so that the union can be protected with earth. In the latter case the hole dug is simply backfilled with earth, in the former the union is daubed with kneaded clay and then overlaid with moss to retain moisture. In cutting the scion, Columella suggests a clever expedient: use the sharpened portion of the cane cut from the scion to hold open the cleft in the stock while inserting the scion. Obviously we are speaking of field grafting here; given the Romans' motives for grafting, there was no need for bench grafting. The cleft, he says, should be above a node to keep the split from running further down the trunk than desired and also to provide purchase for the binding. Thus the union is bound and tapes brought around this node to anchor the tie. Tapes are of willow withies, inner bark (liber, bast), or rush, the last-named best because withies shrink as they dry and cut into the vine. Softer tapes can be drawn tight if they slacken by driving a small wedge between the union and the tape. Afterwards the scion is stripped of all superfluous growth.

Cato and Varro had earlier mentioned the peculiar practice of 'bleeding' the scion of 'excess' sap. The scion is cut three days before grafting to allow it to bleed, but even before this a cut is made in the cane below the site of the intended cut to allow sap to exude. Billiard says that bleeding the vine in this way was still practiced in his day in France and Italy (where it was called *il salasso*, 'the bleeding') and deplores the practice, as he should. Pliny adds that scions should be two years old, swelling with buds, not thinner than a finger, and cut at the smoothest part of the internode. The grafter should take pains that the pith of the scion is in contact with the cambium layer of the stock. The oblique tongue of the scion should be no longer than 3″ and a clean cut, more easily achieved by dipping the cane in water before cutting it. The vinedresser should also take care not to strip bark from the pith of either section when cutting them. The scion is then inserted to the point where its bark begins. Scions which must be taken some distance to be grafted may be kept temporarily in a dry ditch under straw with earth piled on top and the leafy ends of the canes protruding.

Cato is the first to describe a second grafting method, rather strange from our perspective, a sort of hybrid between grafting and layering. If the vines are adjacent, the ends of canes from the two vines are both cut on the oblique but in opposite orientations and the matching ends joined, pith to pith, and tied with willow bark. Cato does not say so, but presumably after the union is callused the cane from the desired vine is cut loose from its 'mother' and all other canes on the less desirable vine are eliminated so that the cane becomes the *de facto* stock of the new vine. A variation on the method is 'terebration' (Cato *Agr* 41; Col *RR* 4.29.13-17, *DA* 8.3-4). An awl is used to drill a hole through the trunk of a well rooted vine, then two canes of the desired cultivar are cut obliquely and inserted through this hole from opposite sides in such a way that they are joined pith to pith. If the union is effected the vineyardist now has two cordons for his new vine. Cato recounts what can only be described as a bizarre variation: The canes, which are 2′ long, are bent toward the ground and back up toward the stock and fastened in a loop. The bottom of this loop is fastened to the ground with forked sticks and covered with dirt and the union then smeared with clay. Columella more sensibly says that one 'whip' is brought over from a mother vine and inserted through the hole. Sometimes the union is effected the next spring, sometimes even in the second growing season, at which point the whip is cut loose from the mother vine and the upper part of the stock vine cut away. If a whip is too far to be brought over from a mother vine it is simply cut as a scion in the usual manner, the bark of the internode stripped away for a short distance to expose the pith, and the scion then inserted into the hole in the stock so that piths are in contact and the union daubed in the usual way. The ancestors, so says Columella, had used a gimlet (*terebra*) which creates sawdust and burns the tissue in the hole so that the union seldom took, so he has himself invented a new tool called the 'Gallic augur' (*terebra Gallica*), a sort of ream which makes shavings instead of sawdust. Shavings are easily removed and the hole thus created not burned so that the union is more often successful.

The ancients also practiced bud grafting (*inoculatio*) much as is done today. In this method a bud (*oculus*, 'eye') along with a section of bark is cut from the desired cultivar and then inserted under a slit made in the bark of a stock vine. Similar is the *emplastratio*, patch graft, in which a bud is cut so that it is surrounded by an escutcheon of bark, an escutcheon of the same size and shape incised on the stock vine, and then the bud patch is inserted into the stock and tied securely there. Today a special punch is used to ensure that the patches are exactly congruent, but we do not hear of such a tool from our agronomists, so it would seem that this form of grafting involved some very meticulous work on the part of the vinedresser. The agronomists do not mention the method specifically for the vineyard, which may or may not be significant, but Billiard assumes that the ancients left this delicate form of grafting to the gardeners.[129]

Finally, Columella (*DA* 8.3) describes a sort of 'poultice' to daub the union composed of pulverized clay and brown manure, mixed with water and kneaded to the right consistency and then applied to the union. Manzi[130] says that in his day (late nineteenth century) Columella's poultice was still being used on vines and was called 'Saint Fiacre' (Saint Fiacre is patron saint of gardeners).

---

129 Billiard (1913): 284-5.

130 Luigi Manzi, *La vitcoltura e l'enologia presso i Romani* (Rome, 1883): 96, n.1.

Irrigation

Grape vines derive most moisture from soils, though absorption from dews and fogs on leaves is not negligible. Soil moisture is today categorized as gravitational water, typically found in the upper soil zones, which drains quickly; capillary water, found in the intermediate zone, which drains slowly and only when soil is saturated; and hygroscopic moisture in the lowest soil zone, which essentially never drains. Today there are a number of analytical tools available for determining a soil's water retention and drainage potential as well as its instantaneous moisture level, but one simple empirical method still used, one available to to Roman vigneron, is wilting point. In spring the rate of shoot development is also an excellent empirical indicator of water availability. As moisture levels approach the wilting point, shoot growth decreases and internodes near shoot tips become shorter. At the same time the typical yellow-green color of shoots becomes dark green. As a deficit increases, tendrils become flaccid and droop, older leaves turn yellow, their margins dry out and curl, and the leaf dies and drops.

In summer, a sudden reduction of available moisture causes wilting of leaves and shoots and dropping of basal leaves. This is especially common in shallow soils where temperatures rise quickly. Under drought conditions young clusters at the pre-bloom or bloom stage desiccate quickly, though more mature clusters are more tolerant. Eventually berries become wilted and shriveled. Oddly, there is some evidence that moisture levels only slightly above the wilting point have advantages, promoting deeper root growth and deepening the color of berries, among others.[131]

In many parts of Europe, today as in antiquity, the answer to moisture deficits is irrigation.[132] In the nursery, vines must be watered regularly for the first month, since droughty soil is the number-one cause of death in rootings. But beyond what is essential, excess moisture simply discourages new root growth. For the mature vine, the most critical time for irrigation is winter, when our empirical methods are not at hand. Fortunately for the Romans, the Mediterranean climate, in which the majority of rainfall occurs between the months of October and March, makes winter irrigation largely superfluous. Otherwise the next most critical time is spring during the early growth cycle, but only when the vine shows stress. Specifically the two most critical times are during

131 Weaver (1976): 141-59.

132 Cox (1989): 52-3; Winkler et al. (1974): 385-409; Goode (2005): 78-83; Halliday and Johnson (2007): 63-5; M. G. McCarthy, L. D. Jones and G. Due, "Irrigation—Principles and Practices," in Coombe and Dry (1988): 104-28. Egypt: Ricci (1924): 45-6.

florescence, when water stress causes poor fruit set and/or 'shatter' (*coulure*), a condition in which undeveloped berries fall off; and during *véraison*, when water stress causes reduced berry size. But again, within reason, the latter may actually improve quality of fruit by increasing the proportion of skin mass to juice. A secondary purpose of irrigation is in flooding vineyards for short periods to kill parasitic pests, and Billiard[133] points out that the ancients had this aim as well (Theognis of Megara in *Anthologia lyrica* 879-84; Diodorus Siculus, *Bibl.* 1.36; Pliny *NH* 17.249).

There is widespread evidence for irrigation in the Near East and Egypt, as we might expect,[134] not to speak of Greece. Roman agronomists have a great deal to say about irrigation[135] but mostly in connection with meadows and gardens, and very little to say about irrigation of vineyards and olive orchards. The reason is obvious; these are two plants which are ideally suited to withstand the hot, dry Mediterranean climates. Pliny (*NH* 17.249-50) says, very sensibly, that the time and extent of irrigation depends on the terroir as well as the cultivar, but in general irrigation is good during summer but bad in winter. In Spain, he assures us, they even irrigate during the vintage. Irrigation in Italy is most beneficial at the rising of the Dog Star (April 25), but in moderation. This will be during florescence, so Pliny is entirely correct. Vines that habitually produce harsher wines require that their vines be irrigated more, as for example, in the Fabian district of the territory of Sulmo. In this district they even irrigate in midwinter to 'warm' the vine. Archaeology now supplements the rather exiguous testimonia. Water, of course, is fundamental to the establishment of the villa, not only for domestic use and luxurious amenities, but for irrigation. Water was drawn from deep wells, from cisterns, from nearby springs, from rivers, and from aqueducts, both public and private, though most of the private ultimately connected to the public. We are reminded that the Aqua Crabra permitted water to be withdrawn for villa use during certain hours (*CIL* XIV.3676 and VI.1261) and that the aqueduct at Tusculum which serviced Cicero's famous villa there was reserved for villa use.[136] Likewise, as part of the complete restructuring of the Campanian villa at San Rocco, Francolise around 30 BCE, a vast system of cisterns was created which, added to those from an earlier phase, was capable of storing some 1,100,000 liters (291,000 gals) of water for both domestic and agricultural use. The system was necessitated by the geology of the site on a limestone spur of Monte Maggiore, a conformation covered

133 Billiard (1913): 343-4.

134 Brun (2004a): 144.

135 White's [(1970): 151-72] is still the best general discussion of the literary sources.

136 Marzano (2007): 165-71.

by argillaceous soil which permits very little water to percolate. The new system comprised five cisterns in two interconnected barrel-vaulted structures. Water was supplied by an aqueduct bringing water from a spring or watershed on the western flank of the hill as well as from eaves drip from the displuviate roofs of the buildings.[137]

Cultivation

Cultivation[138] is defined as loosening, turning or stirring the soil by mechanical means, either around or between established plants. Purposes are to destroy weeds, to facilitate irrigation, to prepare soil between rows for planting of cover crops, to incorporate cover crops or fertilizer into the soil, to help control pests, and to promote the absorption and retention of moisture in compacted soils. Weeds are eliminated because they compete with the vine for soil moisture and nutrients. Cultivation of wet soil, however, promotes formation of 'plow sole', a semi-impervious, compacted layer of soil that develops immediately below the level of cultivation. Today weeds and cover crops are often allowed to grow in the 'middles', the areas between rows, but not in the rows themselves, because limited competition for water and nutrients retards vine vigor and thus allows wood and fruit to mature. Weeds are turned under in the fall before they can set seed, and annual grasses are allowed to die. Vineyards generally overwinter with bare top soils, allowing frosts to penetrate deeply into the soil to prevent early budbreak and consequent damage from late frosts. Cultivation is especially important during the first several years of establishment of vines until root development is adequate to compete with weeds for soil moisture and nutrients, and the canopy for sunlight.

Roman agronomists seem almost obsessive about cultivation[139] until we consider two limitations. First, as previously noted, while modern plows are capable of deep-tilling soils to 2 m (7′), the Roman ard plow was capable of very little depth and essentially no turning, and therefore intermediate and lower root zones of the vines could be severely stressed. The second and perhaps more important consideration was the limitation on moisture absorption and retention in the Mediterranean climate where rainfall is so relentlessly

137 Cotton and Métraux (1985): 35-42. On the use of an irrigation pump in Lucera, Puglia, of the double-action type such as that at Silchester, though not specifically in a vineyard, see Jones (1980): 95-6.

138 Winkler et al. (1974): 371-84; Cox (1989): 54-5; Vine et al (2002): 63-4.

139 Billiard (1913): 318-44; White (1974): 237-9.

seasonal. Further, Billiard points out that the ancients thought of plant and animal life as analogous, and thus all parts of plants must 'breathe', including the roots. We now know that, in addition to promoting moisture retention, aeration of soil by cultivation makes nitrogen in the soil available to the vine and thus it does 'vivify' the roots in a sense.

Again, the Romans were heirs to a long tradition in regard to cultivation.[140] For example, in the fourth century BCE, the lease of prebend lands of Zeus Temenites at Amorgos (Cyclades) requires that the vineyard be deep-cultivated twice a year, once in Anthesterion (February-March) and a second time in Taurion (October). In the same period Theophrastus (*CP* 3.10.1-3 and 11) recommends several cultivations a year with the hoe. In general the Romans adhered to the idea that there was no limitation on cultivation except the expense in man-hours (Cato *Agr.* 33.2; Vergil *Geo.* 2.354-61, 2.398; Col. *RR* 4.5, 2.2.14; Pliny *NH* 14.50, 18.240). A good compromise, says the practical Columella (*RR* 4.5) is every thirty days from March 1 to October 1 (eight times), rooting out all weeds and grasses. But the consensus seems to be that three cultivations are the minimum, once in early spring, once in late summer, and once in the autumn after the vintage.

Cato (*Agr.* 11) in his detailed list of equipment for the 100 iugera vineyard lists as cultivation equipment 2 carts, 2 plows, 2 yokes, 10 iron plowshares, 6 iron spades (*ferrae palae*), four shovels (*rutra*), 2 four-toothed hoes (*rastri quadridentes*).[141] Elsewhere (*Agr.* 54) Cato recommends a sort of muzzle (*fiscella*) which prevents oxen from eating the tender vine shoots during spring plowing.[142] From other authors we hear of cultivation with the two-toothed hoe (*bidens/ sarculus bicornis:* Palladius 1.43) and foot-rest spade (*vanga/bipalium*: Palladius 1.43.3)[143]

---

140 Brun (2003): 42. Egypt: Ricci (1924): 39-40; Greece: Amouretti (1993): 83-7.

141 Cf. Billiard (1913): 320-26; K. D. White, *Agricultural Implements of the Roman World* (Cambridge, 1967): 123-45 and s.v. *palae*: 17-20; 180 *ligo*: 37-40; *bidens*: 47-52; *bipalium*: 20-24; *marra*: 40-43; *rastrum*: 52-6; *rutrum*: 28-31; *sarculum*: 43-47; *capreolus*: 58-9; *dolabra/dolabella*: 59-66; *ascia/rastrum*: 66-8; *falx vinatoria*: 93-7; *pastinum*: 109-10; *acus*: 110; *crates/ irpex:* 146-51.

142 Cf. Pliny NH 18. 177; Billiard (1913): 321-2.

143 Cf. Billiard (1913): 323; White (1967) ss. vv. The *bidens* and *vanga* remained unchanged well into Medieval Italy. Cf. Claudio Gulli, "L'iconografia della vite nella produzione figurativa sienese dalle origini al XIX secolo," in Andrea Ciacci and Myriam Gianacce, *Senarum vinea: il paesaggio urbano di Siena: forme di recupero e valorizzazione dei vitigni storici* (Siena, 2012): Fig. 4, p. 139. In a fresco discovered only in 2000 and painted c. 1341-44 in the Complesso musicale Santa Maria della Scala, Siena, we see three peasants, one of whom

Varro (1.31.1) says that the spring cultivation occurs between the rising of the Pleiades (at the spring equinox) and the summer solstice, the second (1.32) between the summer solstice and the rising of the Dog Star. As also Columella (*RR* 2.2.24; cf. Palladius 2.3.2-3): the vineyard should be deeply tilled to encourage deep root growth, and wet campestrine soils should be tilled first after April 13, second after July 13 and the third time roughly September 1. More amenable soils give the vigneron more flexibility. Soil should never be worked when muddy or half-soaked, conditions the farmers call *varia* and *cariosa*. In fact, muddy ground which is nevertheless tilled becomes unfit for cultivation for a whole year, and 'carious' soil when tilled is sterile for three years running. Columella recommends what we call 'contour' plowing, but not as a preventative of erosion, but rather to reduce strain on plow oxen and plowmen (*RR* 2.4.10). Pliny, who reverses Columella's sequence of cultivations, says that the third is from mid-Februaary to mid-March (*NH* 18.240) but acknowledges disagreement about many issues connected with cultivation. Still, he asserts, there is general agreement that it should never be done before the hottest part of the day (thus drying the soil), and never in wet, sticky soil (*NH* 17.188-9).

As we have seen already, the method of cultivation depends entirely on the conformation of the land, the training method used, and the financial resources of the vigneron. For example, some land is so steeply sloped that plowing is impossible (unless the land is terraced) and so must be hand-cultivated (Pliny *NH* 18.178). Pliny also notes that cultivation is expensive in terms of contract labor or the man-hours of slaves.

For spring cultivation the vigneron will plow (*arare*) or hand-cultivate (*fodere*) and then harrow (*occare*), this last the process in which clods are broken up. Columella specifies the two-toothed hoe for the process (*RR* 4.14.2).

Late summer cultivation is called 'pulveration' (*pulveratio*) and has a secondary purpose which the modern vigneron will approve. This tillage occurs after *véraison* has begun, during the Mediterranean dry season when moisture levels are not so important. When the topsoil is perfectly dry the soil is so thoroughly broken up that it becomes dust (*pulvus*). Clouds of dust raised in the process settle on leaves and fruit and protect them against intense sun and fogs (Col. *RR* 4.28, 11.2.60, *DA* 12.1; Pliny *NH* 17.49). In colder districts this generally occurs in mid to late August (Col. *RR* 11.2.60; cf. Theophrastus *CP* 3.16.3-4; Aristotle *HP* 2.7.5).

The timing of the third cultivation is not so critical since the vine is dormant, and so we hear recommendations from October 15 (Col. *RR* 4.8) to

---

is planting the vine while another hoes with the *bidens* and a third digs next to a pergola with the *vanga*.

FIGURE 18 *Viticultural implements: 1) rastrum, 2) dolabra, 3) bipalium, 4) bidens, 5) marra, 6) sarculum.* DRAWING BY ASUMI HASAN.

January (Palladius 2.1). This tillage is the *ablaqueatio* and has as its primary purpose the loosening of soil around the upper root zone to admit air and moisture (Cato *Agr.* 33; Col. *RR* 4.8; Pliny *NH* 17.49; Palladius 2.1, 2.11).[144] Simultaneously summer roots (surface roots) must be cut off with a knife to encourage intermediate and deep root growth. Workers must be sure to cut off all such roots one finger-breadth from the 'mother root' so as not to injure it. Afterwards soil is pulled away to leave a depression (*lacus*) and the stock is left exposed in mild climates and otherwise is backfilled before the middle of

144 Cf. Billiard (1913): 327-8; White (1975): 237. Egypt: Ricci (1924): 41-2. In modern Calabria in the area of Reggio *ablaqueatio* is still practiced and for the same reasons: Sculli (2004): 118.

December. This is also an opportune time to fertilize where indicated. Ablaqueation should be done every year for the first five years, then at least every three years. Billiard[145] discerns another benefit of the practice, the killing of pestiferous insects which overwinter in the upper root zone.

Fertilizers

The necessity of fertilizing vines was as highly debated by the Roman agronomists as it is today.[146] Grapevines are adaptable to a wide range of fertility, require less soil nutrient than most other horticultural crops, and will survive under conditions of poor fertility which will stifle other crops. The fruit itself represents a very small fraction of the total mass of the plant and so continuous cropping is possible with little loss of soil fertility, particularly when vegetative and unneeded woody parts are plowed back into the soil after pruning. There are only four elements which have been shown to cause deficiency disease—nitrogen, potassium, zinc and boron—though minute traces of manganese and iron are also used. All these elements are used in such small quantities that excess may be a bigger problem than deficiency since it promotes vegetative vigor at the expense of fruit quantity and quality, and may even be toxic. Today various bulk fertilizers are often used for convenience, but not from necessity; nitrogen is readily supplied by organic material such as decayed plant material, including grape pomace, and by animal manure, as well as by the enzymatic activity of certain legumes sown as cover crops. Potassium is readily supplied in sufficient quantities by almost all soils, but may be added in the form of potash, i.e., potassium sulfate. Zinc deficiency is somewhat more common in heavily sandy soils but is a trace element of which vines need only about .23 kg (1/2 lb) per acre, and is supplied artificially as zinc sulfate. Boron deficiency is most common in acidic soils subject to high leaching of soil salts, two conditions which rarely obtain in a Mediterranean climate. Boron is supplied artificially as boric oxide in very small concentrations. The main symptoms of mineral deficiency in the vine are leaf discoloration, especially chlorosis, i.e., yellowing of leaves, and black leaf, but may also include dead leaves (leaf scorch), and small clusters of small berries which mature poorly. In general, a good organic compost supplies sufficient amounts of all the trace minerals the vine needs except nitrogen, which can readily be supplied by brown manures and legumes.Today there is some debate, especially in

145 Billiard (1913): 328.

146 Winkler et al. (1974): 160-71; Cox (1999): 94-6; Billiard (1913): 328-38; White (1975): 125-45.

Bordeaux, about the necessity of *any* fertilizer, since it is thought that the modification of the soil increases productivity at the expense of quality of fruit. Billiard[147] astutely compared this with Columella's account (*RR* 2.15) of his uncle, who absolutely rejected manuring because it ruined the bouquet of his wine.

Be that as it may, it is evident that many Roman farmers, including villa farmers, aimed at increased productivity over fruit quality, inasmuch as the geoponics give specific directions for four types of fertilizer in the vineyard, namely green manure, brown manure, leguminous crops and mineral fertilizers. The timing recommendation is during rainy winter weather when other vineyard activities are precluded (Palladius 3.8.1). Presumably this does not include cases where the manure must be tilled into the soil. For the Romans, brown manure was frequently mixed with compost in specially designed compost pits (Varro 1.13.4; Col. *RR* 1.6.21-2, 2.14.1-9). Varro describes the best manure pit (*stercolina*) as one close to the farmstead for convenience of composting, either two separate pits or one divided into two compartments, one for fresh manure and the other for aged manure ready for use as fertilizer. These pits should be sited so that the sun never dries out completely the manure. Both recommendations are entirely correct, by the way; new manure is practically useless in dry, arid climates and may in fact scorch young plants, and the organisms which compost both brown and green manures need ample moisture in order to thrive.[148] Some actually place the privies of slaves over the former, though Varro gives no details as to how this arrangement may have been designed. Palladius explains that the aging of compost for a year also kills the seeds of noxious weeds. He suggests that aged urine (ammonium nitrate) be kept separate and applied to vines in February mixed with *amurca*, the bitter glycoside from olive mash which is separated from the oil. Columella (*RR* 2.14) suggests that the compost heap be thoroughly stirred in summer to promote rotting.[149] This is more good advice, of course, since the action of composting organisms is aerobic and aeration of composts is essential.

For brown manure, Roman farmers used the excreta of poultry, cows, sheep, goats, donkeys, horses, pigs, and humans.[150] As early as Varro (1.38), brown manure of various sources was being rated for efficiency as fertilizer. Best brown manure is guano, except that of waterfowl, and the best guano is that of pigeons. Guano is broadcast like seed. All are rated high today for their

147 Billiard (1913): 329.

148 White (1975): 133.

149 White [(1975): 133] apparently misses this passage.

150 Cf. White (1975): 126-29 for a good discussion of the relative merits.

nitrogen content. Next come human excreta, next dung of sheep, donkeys and horses. Columella (*RR* 2.14) generally agrees but specifies that guano of chickens and other fowl (again, excepting waterfowl) is almost as good as pigeon guano. Human excrement is especially good if mixed with compost, human urine better suited for young vines if aged for six months. Next is dung of livestock, ass dung being best, followed by that of sheep, goats, and draft animals. Pig dung is not reccommended, and White points out that pigs fed on mast have dung which is very low in nitrogen.[151] Animal dungs may be combined on farms which produce only cereal crops but should be separated for mixed farming. Pliny (*NH* 17.53) adds that some advise the addition of the table sweepings of banquets as well as hair soaked in wine lees or wine lees alone mixed with water.

Most brown manures will have been administered during the winter dormancy, but Columella (*RR* 2.5) adds that especially lean land may also be manured before the late-summer cultivation (*pulveratio*). He also specifies some general proportions: on level terrain, piles of five *modii* (a *modius* is ca. 8.8 l/ 2.3 gal) about 8′ apart on level land, 6′ apart on hilly. Heavy manuring is 24 loads (= 80 modii; cf. *RR* 11.2 86 and Palladius 10.2.1) per *iugerum* Once this is spread it is plowed in immediately, and so no more should be brought to the vineyard at a time than can be incorporated in a day. Cato (*Agr*. 17.4) prescribes a special cart for the purpose, the *scirpea stercoraria*, on which was placed a huge basket of wicker or plaited withies, in which the manure was delivered. Such carts were still common in Campania until the mid-twentieth century. From the cart it was heaped in piles, then carried to the vines on litters (*crates stercorariae*), as seen on a mosaic found at Saint-Romain-en-Gaul and now in the Louvre.[152]

Roman green manures (Cato, *Agr*. 33, 37: Col. *RR* 2.13; Palladius 1.33, 5.1) included lupine, fava, vetch, straw, bean pods, chaff, leaves of holmoak and oak, ocinum (perhaps clover, but the identification is disputed), lentil, chickpea, pea, lucerne, and seaweed washed in water to leach out salts, as well as grape pomace and fern bracken. Many of these, it will have been noted, are leguminous plants designed to supply the nitrogen in which Mediterranean soils are so often deficient. Typical time of application is early spring, before florescence, and Billiard points out that this timing is excellent since this is the period during which nitrogen fixation of *Rhizobium leguminosarum* slows down its metabolism. We should also note that many of these legumes will have been sown as cover crops or as interplantings in the vineyard itself, and

151 White (1975): 129-30.

152 Cf. Billiard (1913): 36-8 and Fig. 107.

the logical time to 'till in' leguminous crops such as chickpea, pea, lentil, and fava is after the beans have been harvested in early spring. Columella says that the cover crop lupine, on the other hand, requires least labor and provides the greatest benefit to exhausted vineyard soils. In colder climates lupine is not plowed in until mid- to late-August, before the pulveration, which will 'harrow' the lupine into the soil.

Mineral fertilizers seem to have been used more commonly in the provinces. In this category we find smelting slag (Vergil *Geo.* 1.84; Col. *RR* 2.14; Palladius 4.7), river sand (though this is really to improve tilth), cinders (Palladius 12.9; *Geopon.* 5.40), gravel (Pliny *NH* 17.49. Since a type is not specified, this may also be for tilth), marl, and chalk (Pliny *NH* 17.49). Application of ash was common in the Po valley, where farmers preferred it even to cattle dung, to the extent that they burned the latter to obtain the ash from it. Also used were ashes from lime kilns (Pliny *NH* 17.49, 17.53). Application of marl and other stones was practiced in Britannia and Gaul. The process, known as lithic mulching, improves tilth, adds key minerals to the soil, helps retain soil moisture, and radiates heat back into the leaf zone at night when temperatures dip.[153] Marl (*marga*) comes as white, red, 'dove-colored', argillaceous, tufa marl, and sandy marl, and in two consistencies, rough and grainy. The Aedui of Gallia Lugdonensis and the Pictones of Gallia Aquitania have made their lands very fertile by application of chalk (*calx*), especially beneficial for vines. Marl is thrown over the top of soil immediately after plowing so that its 'medicinal qualities' (*medicamentum*) will begin to be absorbed immediately (presumably by the leaching action of rain). Pliny adds that marl should be mixed with stable dung (*fimum*) (*NH* 14.42-9). White points out that stable dung is especially good because it mixes excrement with stable litter and the latter absorbs large amounts of nitrogen as well as decomposed organic material.[154]

Diseases and Injuries

As he tended his growing vines, our ancient vigneron will have been constantly watching for signs of trouble. Vines are hardy and incredibly adaptable, but they are nonetheless susceptible to a number of diseases and environmental stresses.[155] Pliny writes a long disquisition on diseases of trees (*NH* 17.216-40),

153 Dale R. Lightfoot and Frank W. Eddy, "The Agricultural Utility of Lithic-Mulch Gardens: Past and Present," *GeoJournal* 34.4 (1994): 425-37.

154 White (1975): 125-6.

155 Winkler et al. (1975): 439-555; Hornsey (2007): 386-436.

among which, as we have noted, he includes vines. Common to all trees are damage by various insect and animal pests, 'star-blight' (*sideratio*), 'pain in the limbs' (*dolor membrorum*), hunger (*famis*) and 'indigestion'(*cruditas*) and sometimes even 'obesity' (*obesitas*). All, obviously, by analogy with human and animal diseases. On the other hand, the ancients' lists are most notable to a modern viticulturist for the diseases *not* mentioned that are terrible menaces today. We have already mentioned the phylloxera root louse, but we can add the devastating New World fungal diseases of oidium and gray rot as well. Viticulturists have paid and will continue to pay a high price indeed for intensive monoculture.

Sideratio is the ancients' generic term for weather-related injuries to plants.[156] One of the most dreaded of meteorological problems for any viticulturist, ancient or modern, is an untimely frost (*pruina*) (Pliny *NH* 17.222). Frost is particularly harmful after the plant has budded in spring because it blights the swelling eyes of the buds (*lactescentes oculos*), the part which is called the *carbunculus* in the flower. Frost immediately before the vintage sometimes blights the grapes or injures the flavor of the wine (Col. *RR* 3.2). Severe frosts in winter (i.e., killing freezes) may even kill vines, especially in northern or mountainous areas. Frosts are especially bad on calm, windless nights. Against these the ancients prescribe measures which are still common, namely smokescreens and dustings with powder to absorb dews before they freeze (Col. *DA* 13; Pliny *NH* 18.70; Palladius 1.35; Deophanes in *Geopon.* 5.31). Vignerons in Billiard's day used plaster and chalk dust toward the same end. The reader will recall that dusting the clusters was one of the aims of the late-summer cultivation, the *pulveratio*. But a modern vigneron will doubtless be alarmed by advice to burn three crabs among the vines, or to roast the flesh of the sheatfish (*silurus*) in a slow fire upwind of the vineyard, or Varro's advice to place a picture of a bunch of grapes among the vines at the setting of the Dog Star, as a votive offering to the frosts. But he will doubtless sympathize with such desperate measures. And he will also understand the ancients' concern to plant cultivars not subject to premature budbreak (Theophrastus *CP* 5.1.12).

Parching wind (*vapor siccatus*) is another weather pattern that can be disastrous. Such winds are especially common at the rising of the Dog Star (April 13) when they kill newly grafted vines and saplings of figs as well as mature vines (Pliny *NH* 17.222). Pliny specifically mentions seasonal winds such as the sirocco (*Atabulus*) in Apulia, and the Olympias (the west-by-northwest wind) in Euboea (*NH* 17.230-3).

156 The best discussion is still Billiard (1913): 375-92.

Dew disease (*roratio*) (Pliny *NH* 17.226) is the ancients' misplaced etiology for *coulure*, the shattering of clusters during florescence so that the grapes do not form, and of shotberry (Fr. *millerandage*). Coulure may be caused by environmental stresses as well as microbial attacks. But the ancients clearly understood that some cultivars are more susceptible or resistant (Col. *RR* 3.2). Another errant etiology for coulure was excessive vigor or 'growing pains' (*articulatio*, *silvescere*: Col. *RR* 4.11.2, 5.6.23). Finally, coulure was said to be caused by the "worst of all injuries", a heavy rain while it is shedding its blossoms (setting fruit) (*deflorescens*) since the fruit is washed off at the same time (Pliny *NH* 17.228). This is precisely right. Another meteorological disaster is hail (*grando*), still a dreaded event, especially in fall when fruit is large, soft and vulnerable. The only palliatives the ancients have are to avoid areas known to be subject to hails and and to plant hard-skinned varieties.

Finally, there is a sort of generic term, *carbunculus* (the disease, not the calyx of the flower or the soil type) which seems to cover a variety of vine defects for which the ancients could find no specific cause (Col. *RR* 3.2.4; Pliny *NH* 14.33, 17.251, 18.272-7). After the rising of the Dog Star the grapes begin to be 'charred' (hence the term) as though scorched by a hot coal, and whole territories may be affected. Pliny recommends a desperate measure, 'scarification', in which the vinedresser makes a long incision along the trunk to 'bleed' the plant. The term is associated with scabies and rubigo, fungal diseases, as well as with sun or cold after rains, humidity, lowlands, strong dews, and stagnant air. Remedies were smoke and sacrifice, equally useless against fungal diseases, and planting on slopes, excellent advice. One cultivar, Helvenacca, is noted as especially resistant. A second manifestation, probably imaginary unless it is root rot, is excess moisture in the roots (*radices lacrimosae*). The initial cure is to pour salt water on the roots as a desiccant, but if the fruit then drops, one must knead ashes with vinegar and smear this on the vines themselves, or ashes and realgar (sandaraca, arsenic sulfide) if the clusters rot. If the vine does not bear, smear it with the ash and vinegar mixture; if it bears but does not ripen its fruit, the same but let it dry and then prune the vine back to the roots and drench the stump with strong vinegar and stale urine and cover it with the mud thus created. Then cultivate around the roots often (Pliny *NH* 17.261-2). Some of the issues cited by Pliny may be associated with excess nitrogen, as attested by Theophrastus (*CP* 4.14.6) and the *Geoponica* (5.40) and Brun[157] thinks this condition is very logically combated with the use of ashes containing potassium (Cf Palladius 12.5). But by mixing ash with strong urine Pliny has defeated himself.

---

157 (2003): 43.

Pliny correctly attributes susceptibility to hot and cold to both injury and disease. The former is largely avoidable since it is caused by carelessness such as excessively tight binding of the vine to the trellis, clumsy hand cultivation which cuts roots and scales off (*desquamavit*) bark, or pruning with a dull knife (*NH* 17.226). Incorrect pruning also causes *morbus articulatio*, 'joint disease', when cuts on canes face upward and admit rain and cold, leading to rot. 'Joint disease' is also associated with improper cultivation and harsh weather (*NH* 17.226-7).

Insect damage is even more of a threat to modern vineyards than it was to the ancients,[158] though we now have a greater chemical arsenal to combat the threat. Today insect damage is caused by various moths, beetles, especially Japanese beetles, leafhoppers, and microscopic and submicroscopic lice. Many of these pests occur as larvae (worms and caterpillars) as well as in adult forms as beetles and moths, a fact which leads the ancients to two general categories of pestiferous insects, the larval (*vermiculatio*) and the flying forms (*volucrae*). Larval pests are divided into worms (*vermi*: Pliny *NH* 17.220-1, 17.229, 17.264-7) and caterpillars (*convolvuli/involvuli/urucae*: Cato *Agr* 95; Pliny *NH* 17.264, 17.229). Various efforts have been made to identify the pests mentioned. Thus Billiard[159] thinks the worms mentioned may be the moth *Sparganothis pilleriana,* the leafroller Tortrix, or a cochylis moth, and that the caterpillar is a leafroller of some sort. Manzi[160] had earlier compared *convolvulus* to the *ponteruola* of his day, perhaps the same leafroller, and Brun[161] agrees that this is indeed the *Spraganothis pilleriana* posited by Billiard. It would seem, therefore, that pests identified by the ancients are probably larval and adult phases of the same species. Especially telling is Pliny's comment (*NH* 17.229) that 'cobweb' (*araneum*) is peculiar to olives and vines, when fruit is wrapped up in a sort of webbing (*telae*) and that caterpillars (*urucae*) are bred from them by heavy rains and damp, sticky heat (*NH* 17.229). As a preventative Cato (*Agr*. 95) recommends that we apply an insecticide of strained *amurca* which has been heated until it has the consistency of honey, to which are added 1/3 pint of bitumen, 1/4 pint of sulfur, to be applied when cooled to the trunk and under branches. Pliny adds (*NH* 17.264-5) that some are content to boil Cato's mixture upwind of the vineyard and let the smoke coat the vines. This is excellent

158 Vine et al. (2002): 65-72; Weaver (1976): 265-80; G. A. Buchanan and T. G. Amos, "Grape Pests," in Coombe and Dry (1988): 209-31. The problem is monoculture again.

159 Billiard (1913): 415-22.

160 Manzi (1883): 101-02.

161 Brun (2003): 43.

practice; sulfur is still a standard insecticide and amurca has now been experimentally proven to be an effective insecticide as well.[162]

Pliny says that gnawing caterpillars may also be killed by sprinkling the leaves with cow dung mixed with water, to be applied before a shower so that the rain distributes the insecticide over the vine (NH 17.267). But his strongest cure (NH 17.266) is to have a woman who is just beginning her menses to walk around the vineyard with bare feet and her belt undone! The flying phase of the leafroller (Pliny NH 17.265) also gnaws at young vines, and Pliny says that some wipe the blade of a sharpened pruning knife on a beaver skin (for the odor?) before pruning as a preventative of the moths, while others smear the vine with bear's blood after the pruning for the same purpose. Other insects attack woody parts of the vine, in which case Columella advises chopping away dead wood and applying *amurca* mixed with lupine. Other pestiferous insects mentioned are locusts (various species of the genus *Acridium*); the *biurus* (Pliny NH 30.146), unidentified but perhaps a weevil or mole cricket; the root louse, though in this case not the infamous *Daktylosphaira vastatrix;* and the *spondyl*, some sort of beetle (Pliny NH 27.143). We can certainly empathize with some of the more desperate measures of the ancients against insect pests, nor were such measures restricted to the ancients. In 1566, for example, the people of Beaune petitioned the *curé* of Autun to excommunicate a vine louse and in 1603 one prominent vigneron, tired of fighting a particularly noxious coleoptera, called for its excommunication as well.[163]

Larger animals are also a threat.[164] Birds frequently eat immature fruit, the most common culprits being starlings, linnets, jays, crows, magpies, and thrushes (*Digest* 19.2.15.2). Deer eat both vegetation and fruit, as do rabbits. The ancients also mention bears (Pliny NH 10.199) foxes, rats, mice, and shrews (Varro 1.8, Col. DA 15).

Unquestionably the most curious 'pests' mentioned by Pliny are plants and other materials which are acrid, bitter, or fetid (NH 17.239). The vine has a sensitive 'nose', as it were, and does not like strong-smelling odors nearby. Because of their acrid smells and juice, radishes and laurel are both baneful to vines, which also detect cabbage and other acrid garden vegetables, hazel (*corylus*), bean pods, and bitter vetch. In the latter case Pliny directly contradicts himself since he has recommended the legumes named here as insecticides and green manure. Furthermore, one wonders how the vine's sensitive nose will have

162 Brun (2003): 43.

163 Jeffs (1992): 149.

164 Weaver (1976)): 281-3; Winkler et al (1974): 552-3.

reacted to *amurca*, the common vineyard insecticide, which one modern author has described as "horrid, blackish, greasy, foul-smelling stuff."[165]

Fungal diseases[166] are a major threat to vines, affecting roots, the base of the trunk, canes, leaves, and fruit. Soil fungi in particular attack roots and trunk bases. Most common soil fungi today are Armillaria Root Rot and Collar Rot, the latter killing the cambium and bark of the trunk at the base and thereby girdling it so that the vine is starved. Trunk fungi such as canker rot affect mature canes and are especially common in older vines. These include Eutypa Dieback, Dying Arm Disease (*Cytosporina*) and Black Measles (*Phaeoacremonium*) which rots the heartwood of the trunk. Fungal disease may also originate on or spread to clusters and cause 'bunch rot'. Some of these, especially Powdery Mildew or Oidium (*Uncinula necator*) are New World diseases and will not have been known to the ancients, but the most common fruit fungus, *Botrytis cinerea*, also known as slipskin disease or grey mold, was most definitely a hazard in the ancient vineyard. Grey mold forms a blue-grey film on grape skins and may rot the grapes quickly, a disaster, or, under certain climatic conditions, very slowly, in the process desiccating the berries and concentrating the sugars, acids, and flavor components. Grey mold develops quickly where conditions are consistently humid but slows when nights are cool and dry and middays warm and humid. Additionally, the 'noble' version is more common in dry, chalky soils with high nitrogen concentrations. The destructive phase relies more on sugar, especially fructose metabolism, whereas the noble phase relies more on acid metabolism, thus reducing acidity of musts.[167] Billiard[168] wonders if this may be Pliny's 'cobweb' (previously mentioned as the larval stage of the leafroller) since the fungus produces a network of fine filaments which envelope the fruit and consume it. Ironically, Botrytis is the active agent of some of the modern world's most expensive sweet wines such as Sauternes, since these filaments pierce the berry's epidermis, desiccating the pulp and intensifying sugars and flavor compounds. Today chemicals are the cure of choice for fungal diseases; ancient remedies mentioned are all cultural and all excellent: plant resistant varieties, plant on slopes in airy, dry microclimates, thin foliage to promote airflow, light fires in the vineyard in dank weather to create air currents, 'dust' the clusters (*pulveratio*) to absorb heavy dews, and amend soil with sand and stone to promote better

165 Terrenato (2012): 86.

166 Patrick Forbes, *Champagne: The Wine, the Land, and the People* (London, 1967): 270-74; Weaver (1976): 254-62.

167 Hornsey (2007): 367-82.

168 Billiard (1913): 384-92.

drainage and thus prevent soil fungi. In other words, the best cure is prevention. One cultural practice not mentioned by the agronomists in this regard but clearly effective is trellising on high frames (5-7′) to promote air flow and remove the canopy as far as practicable from the source of most fungal spores, the dank soil. We are reminded that Pliny lists generic fungal diseases (*roratio*) as a defect of weather, as indeed they are; fungal diseases are universally spawned in damp, warm environments.

Doubtless included among Pliny's climatic diseases as well are viral diseases[169] and for obvious reasons; without the aid of electron microscopy, only the symptoms of viral diseases are observable. For example, Grape Leafroll, also called White Emperor disease, is so called because leaves roll downward and turn red toward the shoot tips. Could this red color be our *carbunculus*? The result is decreased vine vigor and productivity and delayed fruit maturation, all symptoms mentioned by Pliny. Other viral diseases implicated in poor fruitset or maturation are Yellow Vein, caused by the grape strain of the tomato ringspot virus, and Corky Bark, i.e., Grapevine Virus B.

Finally, there is a condition in which excessive production of wood causes the vine to flower poorly or not at all. For the ancients the corrective procedure was the ablation of a narrow band of bark on the shoot above the site of the clusters (*circumrasio / circumcisio corticis*: Pliny *NH* 17.246; Palladius 2.15). The temporary wound impedes the flow of sap beyond the cluster and allows the vine time to complete its reproduction. The procedure was known by the Greeks (Theophrastus *CP* 1.5.5, 17.10) and could also be effected at the base of the trunk (Palladius 2.15, 3.30, 12.10; Quintilius in *Geopon*. 5.35, 9.8, 10.61).

Thinning

About eight weeks after florescence the process called *véraison*, change of color, begins the ripening process. From now on the vineyardist's main goal is to tend the clusters so that they mature completely and at the right time. Dead leaves in the vicinity of clusters are plucked and destroyed (they may harbor pests or disease) and green vegetation is repositioned so that clusters are exposed to proper sunlight, what modern vignerons call 'dappled shade', and to air circulation. But even exposure to direct sunlight is now proven to be beneficial at this stage.[170] As the maturation process continues the vineyardist may need to remove green vegetation and even clusters to ensure proper health and

169 Weaver (1976): 251-2; Hornsey (2007) 431-3.

170 Cox (1989): 967.

ripening of the fruit that remains. Today the process is 'thinning', in Roman antiquity *pampinatio* (Cato *Agr* 33.4; Varro *RR* 1.31.2; Col. *RR* 4.6-10, 4.27-8, 5.6.31, 11.2.61). As usual the Romans were heirs to a rich tradition; thinning was practiced in Greece (Theophrastus *CP* 3.16) and Egypt.[171] Recommendations for timing vary, as we might expect for a procedure where timing is not so critical, and the procedure may be repeated several times. Cato suggests thinning when the vine begins to leaf, a date which seem quite early. Varro advises the period between the rising of the Pleiades and the summer solstice. Pliny recommends the time around the spring equinox or when the shoots have grown 4′, but suggests a second thinning from mid May to the summer solstice. The ancients do not explicitly say so, but the very early recommendations must surely refer to cluster thinning of flower clusters before they bloom to prevent the vine from overcropping.[172] Today this is practiced in conjunction with light spring pruning, which allows the vine to retain vigor, another practice that the Romans knew well and approved, as we will see. Columella also recommends early thinning (*RR* 4.27-8) and frequent thinning thereafter (*RR* 4.6-7, 4.10.2). He adds (*RR* 11.2.61) that *pampinatio* may even proceed from mid to late August where rainy conditions obtain.

Thinning involves the removal of shoots and tendrils (Fr. *épamprage*), leaves (Fr. *effeuillage*) and pinching off shoots (Fr. *pincement*) to suppress vigor. Though the Romans used the generic term *pampinatio* for all three, they were keenly aware that there was a qualitative difference in the skill level required for each.[173] Columella, for example, says that shoot-thinning should be done strictly by a trained *pampinator*, since the procedure, in his opinion, is even more important to productivity than pruning itself and is less injurious to the plant. In the spring some buds generate as many as three shoots, two of which must be removed. At the same time the *pampinator* will evaluate whether the vine has produced more clusters (again, presumably flower clusters) than it can mature and so will remove excess clusters as well. No special tool is needed for the procedure since spring growth is so tender it can be snapped off by hand. The other forms of thinning can be done by unskilled labor, even children. This includes pinching off tips of shoots to check rank growth (*pincement*), and removing water-shoots, suckers, leaf-shoots and tendrils. Two productive shoots may be allowed to grow at first so as to have a reserve shoot, but one should be later removed (Col. *RR* 4.6-7; Varro *RR* 1.31.2). Columella mentions a clever system used by his uncle in Baetica because of the hot, dry climate and

171 Ricci (1924): 43-5; Brun (2003): 42.

172 Cf. Cox (1989): 82-5.

173 Billiard (1913): 368-72.

intense sun, a sort of reverse *pampinatio*. His uncle actually shaded the plants by placing palm mats over the frame of the trellis after the rising of the Dog Star (June 22) when the sirocco begins to scorch the area. Here only minimal thinning was practiced, since fungal disease was not an issue and the grapes needed all the shade they could have; it is possible for grapes to become 'sunburned.'

Pruning

One of the most critical viticultural procedures takes place during the vine's dormancy. Indeed, viniculture as we know it would be impossible without pruning, defined as the cutting away of lignified tissues. It is in fact one of the three bases of viticulture, along with selection for hermaphroditic vines and clonal propagation. That is because the vine's natural habit is to produce luxuriant vegetation under conducive conditions, with a great number of fruit clusters. Unfortunately, none of these clusters will ripen properly, at least by human standards (birds and wild animals which disperse the seeds are content with much less sugar in the berry), and the vine will react to overproduction by producing a very poor fruit load the following year, or in some case the year after this. Fortunately, in all areas except its northern limits the vine is incredibly resilient to even severe pruning (for a novice, watching an expert lop off knurls, dead wood and canes with seemingly reckless abandon can be a shocking experience), and will react by producing a balanced crop which reaches optimum maturity and does so reliably year after year. The origins of pruning were codified by the Greeks in the form of a myth which explained the etiology of goat sacrifice to Dionysus, god of the vine. According to the myth, goats are hateful to the god because they are so fond of eating the vine's tender spring vegetation—as indeed they are, along with other cervids. As 'punishment' for their brazen affront the wine god took them as his sacrificial animal. Behind the myth is a less poetic but far more profound reality. Almost surely early man observed that vines which had been selectively nibbled by ravaging animals produced the next year less fruit of a far superior quality. We could just as easily see the god's choice of our little caprine hero as a signal honor, for teaching man one of the basic tenets of good vine management.[174]

Pruning, especially in modern trellising systems, can be quite a complex science, most of which we needn't even consider. But it proceeds from several

174 Billiard (1913): 469-70.

basic tenets which any Roman viticulturist would have recognized.[175] The basic purpose of pruning, of course, is to produce good fruit, either of the highest quality but of lesser quantity or the reverse. Specifically it is done to select superior fruiting canes, to maintain vine shape so as to optimize sun exposure and air circulation, and to regulate the number of buds retained per vine to maintain long-term vigor. Pruning can be done anytime during the dormant season, but many modern viticulturists wait until late winter since pruned vines are somewhat more susceptible to winter cold injury and since the pruner can then judge better which canes and buds have better survived the winter. An average grape vine is capable of developing two to three hundred buds during a growing season, far more than the vine can properly nourish, and so various formulae have been developed for 'balanced pruning'. Pruning techniques and number of buds maintained will be modified according to the age of the vine and the designated training system. Fruitful buds arise from one-year-old woody canes, which arise in turn from two-year-old wood and older, so it is easy to distinguish them. In addition, the vine will produce any number of shoots which are unproductive and sap strength from fruitful parts, not to speak of knots, knurls, and other woody protuberances which interfere with cultural practices. All are typically pruned away.[176] The two main types of pruning today are also the ones practiced by the ancients. Spur pruning—a spur is the basal portion of a cane, one to four buds long, which the ancients called a 'thumb' (*pollex*) or a 'horn' (*cornu*)—is practiced on varieties which have clusters and fruitful buds all along the base of the cane. Cane pruning (canes have eight to fifteen buds) is practiced on varieties with small clusters or less vigorous basal buds.[177]

As noted, the ancients were keenly aware of all these issues.[178] Columella, for example scolds those who equate luxuriant vegetation (vigor) with productivity (*RR* 5.6.36); he is well aware that different cultivars must be pruned differently to maintain both long-term vigor and fruitfulness (*RR* 3.21.7). He identifies first-year fruiting canes (*palmites*) as non-fruit-bearing the first year (*pampinaria*), but fruit-bearing the second (*fructuaria*) (*RR* 5.6.29). He recognizes that other shoots must be rigorously pruned. These include (presumably) water-shoots and suckers, as well as 'throat-shoots' (*focanei*), that is, water shoots

175 Winkler et al. (1974): 287-337; Weaver (1976): 175-84; Vine et al. (2002): 64-5; Halliday and Johnson (2007): 50-56.

176 John R. McGrew, "Basic Guide to Pruning," in McGrew et al., *Growing Wine Grapes* (Ann Arbor, MI, 1993): 30-32.

177 Weaver (1976): 175-84; McGrew (1993): 30-32.

178 For Egypt: Ricci (1924): 42-3; Greece: Amouretti (1993): 83-4; Judaea: Walsh (2000): 119-22.

which arise from the 'crotch' of the cordons, what the modern vineyardist calls the 'saddle'. Throat shoots may be maintained, however, if a cordon is old or unhealthy, to replace it. Reserve spurs (*custodes*) are pruned back so as not to grow unless needed. Cuts in reserve spurs are made obliquely to face downward so as not to promote rot. Next year's fruiting canes are cleared of secondary shoots and tendrils and the number of buds to be retained is evaluated based on the vine's vigor and the previous growing season's yield (Col. *RR* 4.24; Pliny *NH* 17.181), i.e., 'balanced pruning'. Columella adds some general principles which seem rather arcane to the modern pruner, given the vine's resiliency. For example, cordons are to be pruned according to their orientation: northerly ones with minimal pruning and only one fruiting cane; southerly ones with vigorous pruning and two fruiting canes, though these will be thinned less to shade fruit; easterly and westerly cordons according to soil fertility and the cordon's demonstrated vigor (Col. *RR* 4.24). But we must remember that modern trellising systems generally have only two orientations, whereas the ancient compluviate or quincunx systems had four.

The ancients noted that the timing of pruning was a matter of judgment. It could be done anytime during the dormant season, from leaf fall in autumn to early spring bud swell. If vines are hard-pruned in late fall or early winter, the ancients noted, they are subject to winter cold injury, but spring pruning is more time consuming because care must be taken to avoid injury to swelling buds. A compromise for the ancients as for modern viticulturists is a rough pruning in the fall or early winter, leaving two or three times the wood thought necessary, then final pruning in the spring.[179] In France in Billiard's day[180] spring pruning was also used to delay budbreak and thus reduce risk of injury from late frosts in this northerly climate, a technique that Columella (*RR* 4.11.2) obviously knew, perhaps from Theohprastus (*CP* 3.13.1).

Roman thinking on pruning becomes more refined through time. Cato (*Agr.* 32) offers only the cryptic remark that vines should be pruned 'early' (*mature*) and should be 'well knotted' (*bene nodentur*), i.e., should have several nodes left for budding and grafting. Vergil (*Geo* 397-407) recommends fall pruning after leaf fall. Varro (*RR* 1.34.2) advises fall pruning as well, immediately after the vintage, which in his area falls in late September, but also recognizes the efficacy of spring pruning in areas of severe winter freezes. He specifically forbids the thirty days surrounding the winter solstice, the coldest part of winter. Columella (*RR* 4.9-11) recommends fall pruning following *ablaqueatio*, i.e., from October 15, but also recommends dual prunings in cold climates or where

179 McGrew (1993): 29 for modern theory.

180 Billiard (1913): 345-6.

there are signs of the approach of a severe winter (*RR* 4.23). Dual pruning is practical in a small vineyard, but a large vineyard may require fall pruning exclusively because of labor considerations. In general, he adds, early pruning promotes vigor, later pruning productivity. He follows Varro in expressly forbidding midwinter pruning (*RR* 2.8.2). Pliny (*NH* 17.191, 18.240-41) seems to follow Columella but specifies that spring pruning proceeds from mid-February to mid-March and adds the humorous note (18.249) that country people make fun of their derelict neighbors who have not finished their spring pruning before the equinox by imitating the sound of the cuckoo, a bird which appears at this time.

Today pruning is accomplished with a variety of shears and loppers, all of which were encompassed by one pruning tool in Rome, a masterpiece of rational design, the *falx vinitoria/vineatica/putatoria* (Col. *RR* 4.23; Varro *RR* 1.22.5; Palladius 1.43).[181] *Falx* is the generic term for a billhook knife used for pruning in arboriculture, but this billhook is modified in a number of ways to adapt it for viticulture. Columella gives an exact description of its parts and uses, and comparison with many dozens of examples from the archaeological record allows us to precisely visualize its use. The straight part of the blade above the haft is called the *culter* because of its resemblance to a knife. The inner curve of the hooked part is composed of a *sinus*, the lower part, and the upper *scalprum*. The point of the hook is the *rostrum*, 'beak'. All these are standard parts of the billhook. But on the outer perimeter of the hook we find a chevron-shaped point (*mucro*) directly opposite the handle, and on the part of the hook opposite the *rostrum* a rectangular or lunate *securis*, 'axe'. When the pruner uses a thrusting or sawing motion he uses the inner part of the well honed *culter*; to make precise cuts he uses a quick drawing motion of the *sinus*. To smooth rough wood of knurls or stumps of shoots he uses the *scalprum*, to hollow out wood the *rostrum*, to cut away thicker wood with a hacking motion the *securis*, and to winkle out something from a confined area or to bore holes in wood the *mucro*. The most effective part for pruning per se is the *scalprum*, since the vinedresser can grasp a cane or shoot with his left hand and, with a deft tug, make a clean cut which is less susceptible to infection or rot. Examples of these versatile little tools are so standard in the Roman vineyard that their discovery at archaeological sites is regarded as reliable evidence of viticulture there.[182] Further, the tomb relief of L. Cornelius Atimetus (Vatican 1009418) shows a cutler and his wife Epaphra proudly displaying their wares, nicely

181 E. de St. Denis, "*Falx vinitoria*," *Revue Archéologique* 41 (1953): 163-76; Christian Vernou, "À la recherche des outils antiques des vignerons," in Brun, Poux and Tchernia (2009): 260.

182 E.g., Boissinot (2007): 35-41.

displayed in and above an amoire designed for the purpose. The top row of wares is an assortment of billhooks, among which we recognize several examples of our *falx vinitoria.*[183] Exact and modified forms of the handy tool have been used continuously in many parts of western Europe right down to the present,[184] and although Billiard[185] says it had been abandoned in France in his own day for the *secateur* (shears), the same tool is still used in Italy where it is sold as the *pennato*. One variation of the latter offered by purveyors of agricultural tools is in fact identical to the tool described by Columella.

The pruning procedure itself varies depending on the age of the vine and the training system adopted. For example, both Columella (*RR* 4.9.11) and Pliny (*NH* 17.35; cf. Theophrastus *CP* 3.13.1-3) recommend pruning away most of the young vine during the first and second dormancies to encourage better root growth, just as is done today. Even slips and mallet shoots should be pruned. Columella specifies that the second year, spurs should be trimmed of all shoots, leaving two buds at the saddle, *pace* Cato (*Agr.* 3.3.2) and Vergil (*Geo.* 2.362-70). The third year the vinedresser really begins to shape the vine since it will reach the height of the trellis this year and begin bearing fruit (Col. *RR* 4.17.2). By now the trunk is strong enough to support cordons with fruit, so more canes are left and splayed along crosspieces and tied in such a away that the following season's growth will run along the crosspiece and not droop to the ground. Canes will be limited this year as in succeeding years, based on the vine's vigor. In fertile soil three or four cordons will be left for the pergulate trellis, which Columella prefers because it makes regulating exposure easier, not to speak of cultivation and inspection of fruit.

Pruning of mature vines follows principles adduced before. Concerning the procedure itself, Cato is positively cryptic again (*Agr.* 32.1, 33.1): "Begin pruning early. See that the canes are well knotted and be careful not to bend them downward or tie them too tightly. See that the trees are well 'wedded' [Cato uses the *arbustum*] ... Leave fruiting shoots and reserve buds at proper intervals. Train the vines as high as possible and tie them firmly, but without choking them." Vergil is even terser (*Geo* 2.362-70): "Prune only after vines are firmly wedded to trees. " By the time of Columella we can see how the science has progressed. Columella (*RR* 4.21) has us prune cordons, sometimes called *duramenta*, 'hardened canes', and tie them about 1" below the four crossmembers. These have been left with only one fruiting cane (*palmes*) until now, but we may now choose based on the fertility of the terroir, remembering that vigor

183 Marcone (1979): 46-8.

184 De St. Denis (1953): 168-73.

185 Billiard (1913): 352.

without productivity simply leads to *coulure* and improperly ripened fruit, and that a weak vine overburdened with too much fruit load will quickly lose long-term vigor. We should leave two fruiting canes per cordon in rich soil, unless the soil is extremely productive. Cordons are cut back to fit the frame. Some laterals which show special vigor may be trained to the frame as well, but if they sprout in a position where next year's growth will be, they should be pruned back to a reserve spur (*pollex / custos / resex / praesidiarium*) of two to three buds, and once a fruiting cane has sprouted from it the next spring the old wood beyond it will be pruned away and the spur then shoot-trained as the new cordon.

Columella advises fall pruning in conjunction with *ablaqueatio* so that all excess wood and vegetation can be evaluated at the same time (*RR* 4.24). The shank of the trunk (*ima vitis*) is exposed with the small mattock (*dolabella*) and suckers removed. 'Legs', that is, main branching roots, and trunk, are pruned of shoots and warty knobs (*furunculi*). If the trunk is diseased, the mattock is used to chop away dead wood and the *falx* is then used to trim down to live wood so that a callus will form, and the wound is daubed with earth mixed with *amurca* to keep out the wood borer (*teredo*, presumably the fruit-tree borer, *Crytophasa melanostigma*, but possibly the termite)[186] and ants. Dry, shaggy bark is removed from the upper trunk.

Pruning for the 'prostrate' vine (Col. *RR* 5.5.17) is as for trellised vines except that woody canes are more severely pruned and spurs pruned back to resemble knurls. Then vegetative growth is pulled into line with rows so that the interspace can be cultivated and then pulled in the transverse direction so that the intervine spaces can be cultivated as well. In the *arbustum* (Col. *RR* 5.6.31; Pliny *NH* 17.207-11) tendril shoots are eliminated from immature vines unless required for placement, but those on old vines are useful and so are pruned to the third bud to produce wood the next year. Shoots are tied to hardwood and are called 'droopers' (*praecipites*). These are heavy bearers but overburden the vine and so should not be allowed to droop unless the cordon has reached the end of the tree's branches and its top. If the vine is ready to support droopers the vinedresser should twist the shoot and then bend it down so that the interruption of sap will cause the part before the twist to liginfy. Droopers should be allowed to trail for only one year. Another kind of shoot which is allowed to depend is called 'firmwood' (*materia*) and is allowed to grow because it produces good crops of fruit and new shoots. It is unclear to me, however, what distinction the ancients are making here between regular cordons and 'firmwood'. Throat-shoots must be eliminated. Many people think that luxuriant

186 G. A. Buchanan and T. G. Amos, "Grape Pests," in Coombe and Dry (1988): 224.

growth in the *arbustum* is more productive, but they are mistaken here as elsewhere; not only does excess vigor repress long-term productivity, but luxuriant vegetation holds fogs and dews too long and encourages *coulure*. The *arbustum* vine should be pruned diligently, just as other systems.

During or after pruning (Col. *RR* 4.26.1-3) the vinedresser is responsible as well for repairing frames and binding the vines to trellises. The vineyard is purged of all vine trimmings, a natural vector for pests and diseases, and this should be done in dry weather so that the soil will not be compacted by trampling.

Properly pruned to one of these systems, the vine is capable of overwintering without serious injury and beginning its next growth cycle in a healthy, moderately vigorous but productive state. But of course we have neglected (temporarily) the climax of that growth cycle, at least from the human perspective, namely the harvest of fruit. To that we now turn.

CHAPTER 4

Vinification

Harvest

The climax of the viticultural year, the vintage[1] (*vindemmia*) also marks the beginning of the winemaking process per se. Obviously, successful fermentation depends upon adequate sugar concentration in berries which depends in turn upon proper culture and harvest of the berries at peak ripeness. But good wine also contains significant amounts of acid to counterbalance the residual and perceived sweetness, and as the grape matures after *véraison*, sugar concentration increases dramatically while acidity gradually decreases. In warm climates maturity is often reached in as little as twelve days, during which the total acids can drop from 17 g/l to 8 g/l and sugars can increase from 103 g/l to 137 g/l. Additionally, sugars and acids may be balanced but, when pips are still green, the grape has not achieved what is called 'physiological ripeness', in which case the wine will have a 'green' or underdeveloped taste. Today there are a number of analytical tests and instruments to determine the best balance among these elements, but it is interesting that none has entirely replaced empirical methods available to the Romans. Modern winemakers still taste grapes, often compulsively, from several areas in the vineyard, over the course of many days, examining pips in the process to make sure they are brown. When sugars and acids are balanced, pips have matured, and tannins are not too astringent, then picking can commence.[2]

But before the ancient harvest can begin, preparations must be made (Cato, *Agr.* 23). Treading vats, settling vats and fermentation vats are cleaned and repitched if necessary; harvest and transport baskets are mended and repitched, and inferior, unripe grapes (*uvae miscellae*) may be gathered in advance to make a 'sharp wine' (*vinum praeliganeum*) for the farm workers to drink during the year. As the ripening grapes announce the approach of the vintage, there is increasingly intense guarding of the crop by the farm slaves, sometimes

1 For comparative practices in Egypt, Ricci (1924): 47-79; Leonard H. Lesko, "Egyptian Wine Production during the New Kingdom," in McGovern, Fleming and Katz (1996): 215-17; Brun (2004a): 144-6 (papyrological and literary evidence); Greece: Amouretti (1993): 85; Modern Vesuviana: Sannino (2009): 205-47. My discussion owes much to Robert I. Curtis, *Ancient Food Technology* (Leiden, 2001): 372-80. Cf. Thurmond (2006): 311-64.

2 David Bird, *Understanding Wine Technology* (Warwickshire, 2010, 3rd ed.): 29-30.

 | DOI 10.1163/9789004334595_005

ensconced in temporary huts (*casae repentinae*) out in the vineyard (Varro *RR* 2.10).[3] The good will of the gods is also invoked: The Vinalia rustica or Vindemiales feriae, to be discussed in more detail later, were held on August 19-20 in honor of Jupiter, Minerva and Venus to invoke a good harvest (Varro *RR* 1.1; Ovid *Fast* 4.297-8). In the city of Rome itself, the *flamen dialis* leads a grand procession from the city, and beyond the *pomerium*, immolates a lamb to Jupiter and then, holding a grape cluster, expresses the juice into a cup and offers this to Jupiter as well (Varro *LL* 6.16.20, Festus 269M).

In an area as geographically diverse as Italy, not to speak of the provinces, harvest might very well take place any time from mid August until well into November, and we even hear of harvesting after the first frosts have partially desiccated the grapes, a sort of prototype *eiswein*. But in general in central Italy the vintage occurred in late August and September, in the warmer regions, such as Spanish Baetica and along the coast of Africa in mid to late August (Col. *RR* 11.2.60). Billiard[4] laments the fact that so many ancient authorities such as Cato (*Agr*. 25) advise harvest when grapes are dead ripe, a "deplorable practice" by modern standards since at this point much of the acid content of the grape so necessary for a balanced product is gone, a fact which necessitated the rather alarming degree of modification suggested by ancient geoponics as compensation. But in the absence of pure cultured strains of wine yeast, all organoleptic considerations give way to the singular necessity of promoting a rapid colonization of wine yeasts to the exclusion of all others and the need to achieve alcohol level sufficient to inhibit secondary fermentations. To put it bluntly, sulfur dioxide has given us moderns the luxury of criticizing ancient vincultural practices.[5] On the other hand, Pliny (*NH* 18.315-16) indicates that the vintage had been moved up in his day: "In the old days they thought the wine crop was never ripe for picking before the equinox (September 21), but these days I see they are gathered indiscriminately," Tchernia[6] posits (correctly, in my opinion) a gradual shift in the Empire from taste for sweet wines to drier ones, reflected in a tendency to advance the date of harvesting.

But of course the individual vigneron will have made an exact determination as to the specific day that the vintage will commence, just as today. Perfect

3 Billiard (1913): 292-3.

4 Billiard (1913): 428-30.

5 On the difficulties of preserving wines vinified using ancient techniques at an experimental *cella vinaria*, cf. André Tchernia, "La vinification au début de notre ère et le goût des vins romains," in Tchernia and Jean-Pierre Brun, *Le vin romain antique* (Grenoble, 1999): 118-9.

6 André Tchernia, "Le vignoble italien du 1er siècle avant notre ére au IIIe siècle de notre ère: répartition et évolution," in Marie-Claire Amouretti and Jean-Pierre Brun, edd., *La production du vin et de l'huile en Mediterranée* (Paris, 1993): 286.

ripeness today is judged to be a combination of sugar, acid and pH, the ideal being 22 Brix, .75 acidity, and 3.4 pH respectively. Overcropped vines are typically deficient in acidity because the fruit must hang on the vine so long to achieve sufficient sugars that acid levels will have fallen precipitously, yet another way that really fine wines are made in the vineyard, not the pressroom. In addition to numerous 'rules of thumb' to gauge the ripeness of the berries, Pliny (*NH* 18.74-75) indicates the empirical nature of the process: harvest "when the grape shoot droops down to the stem" or "when, after removal of a grape from the cluster, it leaves a gap." Columella, Palladius and Deiophanes (Columella, *RR* 11.2; Palladius 10.11; Deiophanes in *Geop.* 5.45) all give the sensible advice to taste the grapes for sweetness. Others recommend judging by color of the berry. Today grapes are sampled from the middle of clusters (grape clusters ripen from top to bottom) taken from a number of locations in the vineyard. Varieties are sampled separately, and the same variety grown in different fields likewise.[7] To all of these factors should doubtless be added the exigencies of weather, availability of contract labor, presence or absence of molds, etc.[8] For example, rain just before harvest swells and cracks berries, dilutes flavors and exposes musts to spoilage organisms, and therefore a rain forecast at the time of harvest means grapes are gathered beforehand, even if they are not at optimum ripeness.[9] Hail during the growing season strips leaves from the vines and damages grapes, but hail near harvest can completely destroy a crop, and for unknown reasons there are microclimates that are more prone to hailstorms than others in the same viticultural area. The vigneron in such a microclimate may well advance the vintage to avoid such a disaster. And in coastal areas so-called 'salt dews' from salt-laden sea air can settle on vines and grapes with sometimes disastrous results, especially if it comes during harvest. In modern Bordeaux, salt dews are attested 80 km (50 mi) from the Bay of Biscay. Again, these dews are predictable and may advance the harvest.[10] In addition, ripe olives can remain on trees for several weeks and remain good, whereas grapes ripen and peak in a very short time and, if not harvested, will wither, drop, and /or be eaten by insects, birds and mammals, for which reason

7 Winkler et al. (1974): 575-92; Cox (1999): 97-8.

8 For comparative modern methods, cf. M. A. Amerine et al., *The Technology of Winemaking* (Westport, CT, 1980, 4th ed.): 77-91.

9 Cox (1999): 106.

10 H. J. de Blij, *Wine: A Geographic Appreciation* (Totowa, NY, 1983): 83. The harvest in Sauternes was completely wiped out by hail three times in moderns times within a thirty year span (1951, 1952 and 1973).

the grape harvest is coordinated with the olive harvest in many Mediterranean areas where both are grown.[11]

Modern vintners know that berries are best gathered on a dry day because they readily absorb water which dilutes their sugar concentration and flavors.[12] The ancients also avoided rainy weather, and the ancient geoponics advised that the harvest should not be conducted too early in the morning, to avoid dew on the fruit. But in this Mediterranean context, our vintner is advised to avoid midday hours as well, when the grapes are too hot (Cato, *Agr.* 25; Pliny, *NH* 18.74). Billiard[13] notes that this would be a serious problem because excessive heat will have impeded the spontaneous fermentation our ancient winemaker relied upon. Equally likely is the fact that elevated but not excessive temperatures would favor what modern vintners know as an explosive fermentation, one that proceeds so quickly that it strips the wine of flavor and bouquet.

Today grapes are harvested by hand with knives or shears (Fr. *secateurs*) or (increasingly) with mechanical harvesters. In manual harvesting, shears are preferable because the harvester is less likely to pull the clusters and thus break the skin of the grapes, an action which may lead to oxidation of the must. In antiquity contract laborers using our *falx vineatica/vinatoria* cut the grape clusters and others carried them to the processing plant. Cato says that 40 such billhooks were suitable for the vineyard of 100 iugera (Cato, *Agr.* 11; cf. Columella, *RR* 12.18.) Columella recommends that pickers use the *sinus* of the billhook instead of the *scalprum*, since a drawing motion makes a cleaner, more precise cut, whereas the hacking motion of the *scalprum* is less accurate and tends to damage the stock. We should imagine our *vinator* grasping the cluster with one hand and, with the other, cutting the peduncle of the cluster with a quick, drawing motion.

Clusters were then placed in harvest baskets (*corbulae, fiscinae, fiscellae, quali*, all apparently interchangeably) finely woven of withies, vine cuttings, or small tree branches, and made in-house during winter days or bought, the best of the latter reputedly coming from Nola, Sinuessa, Casinum and Rome. They were pitched on the inside to make them leak-proof and the standard size held 3 *modii* (3 bushels, 26 l). Campanian pickers still use such baskets, identical in shape and still pitched in the same way. Pickers attached them with cords or

11 E.g., ancient Israel: Walsh (2000): 39. For a modern vintage determined by availablility of labor, A. T. Calabresi, "*Vin Santo* and Wine in a Tuscan Farmhouse," in M. Douglas, ed., *Constructive Drinking: Perspectives on Drink from Anthropology* (Cambridge, 1987): 122-34.

12 Rodney Boothroyd, *Home Winemaking: Techniques and Recipes* (New York, 1986): 33.

13 Billiard (1913): 430-33.

FIGURE 19 *A vintage scene from a Christian sarcophagus in the Lateran Museum. After Billiard (1913): Fig. 137.*

leather straps around their necks to leave hands free, especially important when vines were trained to trees and pickers were obliged to clamber up tall ladders. When harvest baskets were full, the contents were transferred to larger baskets of 10-modii (10 bushels/ 86 l.) capacity, the so-called *decemmodiae fiscellae, corbes, cophini* or *corbulae amerinae,* and these were carried by porters to donkeys supplied with panniers (*clitellae*) or to large trays fashioned from single blocks of wood (*lintres*) which were mounted on carts.[14]

14 Billiard (1913): 432-3. Egyptian pickers simply plucked clusters rather than cutting or shearing, witness the iconography and the etymology of the Egyptian word for picking: James in McGovern, Fleming and Katz (1996): 206. Egyptian contracts specify that women will be used to carry grapes from the vineyard to the winery: Ricci (1924): 53. In ancient Israel the billhook was used, though inferior clusters were left on the ground as gleanings,

Harvest from the *arbustum* necessitated the use of ladders, and we have already mentioned the fresco in the House of the Vettii depicting Erotes who have placed such a ladder against the *tradux* of a vine trained between two trees. Additionally, during the excavation of the peristyle of the Pompeiian House of Polybius (IX.xiii, 1-3) in 1973, the imprint of a ladder was found when the original soil surface was uncovered. It was exceptionally long and narrow, 8 m. (26′) long and only 50 cm (19′) wide at the base and 30 cm (12′) wide at the top. This narrow profile was apparently designed to fit into the branches of fruit trees grown right in the peristyle garden of the house.[15] The ladder would also have been ideal for harvesting grapes from vines trained to the upper branches of trees.

As the grapes were brought to the winery, they were carefully sorted (Fr. *triage*). If defective clusters had not previously been culled for use in making *vinum praeliganeum,* as Cato prescribes, they were doubtless culled at this time. At the same time the finest clusters might be sorted out for use as table grapes, lesser clusters left for wine (Varro *RR* 1.54), according to quality and variety (Palladius 3.9, 10.17).[16] Today during triage grapes will typically be destemmed to reduce tannins, particularly for white wines, but this is done exclusively by machine, the only cost-effective way to effect the procedure.[17] Some of these machines are quite simple, well within the technical capability of the Romans, but we have no evidence of their use. And, as we shall see, white wines were apparently filtered of gross lees immediately after treading, and this will have eliminated stems very quickly.

The Winery

From Cato's elaborate descriptions as well as from archaeological evidence we have an excellent idea what the winery, the *torcularium vinarium* (Cato *Agr.* 12.1, 18.1, 66.1, 112.3; Vitruvius 10.1.3, 10.1.5) *or cella vinaria* (Plautus *Mil.* 857) or *cella torcularia* (Columella *RR* 1.6.9) must have looked like on a typical villa farm. It is important to remember, however, that much simpler methods were used and may even have prevailed on subsistence farms. Rossiter, for

---

in accordance with the Biblical principle. Clusters were placed in baskets and carried to the winery by workers or on animals, especially the donkey: Walsh (2000): 167-86.

15 Jashemski (1979): 28-9 and Figs. 41-2.

16 Billiard (1913): 433-4.

17 Yair Margalit, *Concepts in Wine Technology* (San Francisco, 2004): 21-49.

example,[18] points out that wine making on a small scale was carried out in a portable treading tub and seems to have involved no pressing of the grapes. In the early Mideast where vinification was developed, vineyards were modest and rain infrequent so that a building dedicated to this end was unnecessary. In the Aegean, spouted tubs, so small that their use as treading vats has been questioned, are found from Bronze Age Crete at Arkhanes, Gournia, Mallia and Kato Zakro. Black-figure vases of satyrs treading grapes in such tubs are unambiguous: such tubs are large enough for one person to tread a small batch of grapes.[19] A Greek red-figure crater in the National Museum in Athens even shows a satyr treading in a wicker basket, in this case placed over a stone or terra cotta spouted basin to catch and channel the must.[20] And, of course, there are our *palmenti*. Alternately, a flat area could simply be enclosed with low stone or brick walls, the gaps sealed with clay, and clusters thrown in and trodden. Must ran into a channel in the floor and was ladled from it into various vessels to be taken to the cellar to ferment. Such simple, open-air facilities figure on a Greek bas-relief in the Naples Museum.[21]

But such open-air installations are less effective for larger vineyards, particularly in cooler and wetter climates, and thus the development of the winery.[22] As will be imagined, here too there is great variation in the size and complexity of such facilities. Perhaps an indication of the 'evolutionary' nature of Roman wineries is the confusion in the terminology; there is simply no specific word in Latin for the winery. Strictly speaking, the term *torcularium* refers to the press or the press room, while the term *cella vinaria* refers specifically to the room, or more typically the courtyard, where wine was fermented and aged. The size and arrangement of the winery were naturally proportional to the size of the vineyard. For the *torcularium* per se, Vitruvius (*de Arch* 6.6.9) poses this rule: "If the press is not a screw press but a beam press, it will be necessary to

18 J. J. Rossiter, "Wine and Oil Processing at Roman Farms in Italy," *Phoenix* 35 (1981): 348-9. For an excellent review of the archaeology in the rest of the Empire, especially the western Mediterranean basin, cf. Jean-Pierre Brun, "Pressoirs et chais," in André Tchernia and Jean-Pierre Brun, *Le vin romain antique* (Grenoble, 1999): 77-90.

19 Albert Leonard, Jr., "'Canaanite Jars' and Late Bronze Age Wine Trade," in McGovern, Fleming and Katz (1995): 233-4.

20 Brun in Tchernia and Brun (1999): Fig. 4.

21 Billiard (1913): Fig. 143.

22 Billiard (1913): 436-9; Brun in Tchernia and Brun (1999): 68-9. For comparative evidence, Egypt: Brun (2004a): 148-51; Magna Graecia: J. C. Carter, *Discovering the Greek Countryside at Metaponto* (Ann Arbor, MI, 2006): 143-5. For an experimental reconstruction of a Roman winery, including a Catonian beam-and-capstan press, Brun in Tchernia and Brun (1999): 91-105.

make the room at least 40′ (12 m) long and 16′ (4.7 m) wide, to permit easy operation. If two presses are required, the width must be enlarged to 24′ (7.1 m)." Cato, to provide for a *torcularium* with 4 beam presses, the standard type in his day, recommends (*Agr* 18) a room 66′ × 36′ (19. 5 × 10.6 m). And regardless of size, the room should be between the kitchen and the *cella vinaria* to allow boiled musts to be made in the kitchen. Moreover, the *torcularium* should be light, whitewashed, and roofed to disperse heat and isolate it from odors, and should be airy to ventilate dangerous gases during primary fermentation (Florentinus in *Geopon.* 6.1). Cato also mentions an arrangement with 4 beam presses in a 2 × 2 configuration.

Archaeology has added enormously to that rather exiguous testimony. For example, the large Pompeiian vineyard north of the amphitheater (II.v) discovered by Jashemski[23] had a small but fully equipped winery in the northeast corner. One small room had fittings for a beam-and-capstan press. Adjacent to this *torcularium* was an open shed which held 10 dolia. This *cella vinaria* has large open windows facing northeast, just as Pliny (*NH* 14.133) recommends. Each dolium has a capacity of 2 *cullei* or 40 amphorae (10.4 hl/275 gal.). Thus the 10 dolia had a total capacity of some 20 *cullei*, which Jashemski thinks may approximate the yield of this vineyard. In the case of the small vineyard just inside the Porta di Nocera (I.xx.1)[24] there is a small, one-room *torcularium* in the northeast corner of the site, with a treading vat but no press. The *cella* in this case is a large, underground, barrel-vaulted room accessed by stairs near the east wall. A large vat on the western end Jashemski interprets as a *lacus* for fermentation. Directly above this vat in the ceiling is a large hole in which is embedded a dolium whose bottom has been removed, ostensibly so that must from the treading vat above could be poured directly into the vat. Another bottomless dolium embedded in the ceiling in the center of the room provided air and light. But if Jashemski's interpretation is correct, this *cella* is a disaster in the making. Carbon dioxide is so much heavier than air that there is no way the meager ventilation provided by the two embedded dolia and the stairway can have vented the huge amounts of gas produced in primary-phase fermentation. More likely this phase was permitted to proceed in the treading vat above, along with the settling of gross lees and the maceration of solids, and then the partially clarified must was poured into this lacus where further clarification and the beginning of secondary-phase fermentation proceeded for several

23 Jashemski (1979): 226-7.

24 Jashemski (1979): 227-8.

FIGURE 20 *Underground wine cellar at Pompeii. From Jashemski (1979-1993): Fig. 334.*

days before the wine was strained and apportioned into dolia for the remainder of fermentation and aging.[25]

The *cella vinaria*, meanwhile, is typically part of the larger storage magazine along with oil vats and containers for boiled musts, the *cella olearia* and *defrutarium* (Col. *RR* 1.6). It should be adjacent to the *torcularium* for convenience, and should face north and have northern windows if enclosed and preferably be in the northern or eastern quadrant of the *pars rustica* (Vitr. *de Arch.* 6.6; Col. *RR* 1.6; Pallad. 1.18). It should be far from baths, bakehouses, smokehouses, and all sources of strong and especially fetid odors, far from cisterns where stagnant waters may taint the wine. Fine wine is delicate and highly susceptible to off odors. Columella even advises the burning of incense (*RR* 12.18.28). As to size, our authors advise a cellar big enough to handle more than one vintage, to allow for extended aging of vintage wines and, more importantly, since carefully timed sale during periods of peak demand can even double the sale price (Varro *RR* 1.13). Varro also advises a pavement inclined toward a reservoir so that leaking wine can be recovered, though this does not seem to have been the norm in central Italy, where most cellars are open air courtyards with sand floors in which the dolia are embedded. Cato (*Agr.* 11) prescribes a number of dolia sufficient to receive the product of five vintages, a total capacity of some 800 cullei (416,000 l / 110,000 gal), an extraordinary amount, unless this is a manuscript error. In any case, very few of the extant examples of cellae approach anything like this huge capacity. On the other hand, it is astonishing how often the archaeology reveals the efficacy of the ancients' prescriptions in

25 Billiard (1913): 463-5.

regard to orientation, as even my cursory survey of the villas in Marzano's large survey[26] reveals; in the vast majority of cases the *pars fructuaria* is on the north or east side of the complex, often at the northeast corner, while bath complexes in the majority of cases are on the south and west. The Romans clearly understood the basic principles of passive solar heating, and *cellae* were habitually located on the cooler sides of the complex which did not receive afternoon sun, and often at a lower level in terraced villas. For example, at the famous Settefinestre villa complex, excavators locate the *cella vinaria* in one part of a *cryptoporticus* and along one side of a ramp which led to it.

Until the time of Palladius (1.18), our ancient geoponics say nothing of placing the *torcularium* in a superior position relative to the *cella* so that gravity flow can be used to move must from one to the other, but this is such a commonsense arrangement that it was doubtless used early on. Obviously this is one of the refinements of the *palmento*. The winery of the Villa of the Mysteries provides an example of the arrangement; in the *torcularium* wine was transferred from a combination pressbed and treading vat (*forus*) to a settling vat (*lacus*) and was conducted thence by means of a gutter to and along the wall of the *torcularium* to the *cella* outside and below. Palladius specifically recommends that the treading vat / press bed be elevated at one end of the *torcularium* "in the manner of a tribunal of a basilica," and that two *lacus* be located below and in front of this platform to receive the overflow and strained wine from above. There will be a stairway of 3 or 4 steps to provide access to the 'tribunal'. From the *lacus*, canals are constructed or terra cotta pipes are arranged around the perimeter of the walls to the *dolia defossa* in a *cella* at a yet lower level. If there is an oversupply of must, wooden casks are placed in the middle of the *cella*, on wooden platforms above the dolia so that no additional room is needed and walkways are clear. Just such an arrangement is now attested in the archaeology, right down to the particulars, in the Moselle.[27]

The Roman writers seem far more concerned with the siting of the *cella vinaria* (Varro *RR* 1.13.6; Colum. *RR* 1.6.11, 12.18; *Geopon.* 6.2.2-3). Palladius provides an excellent resume: "We ought to have the wine cellar positioned toward the north, cool and almost dark, far from baths, stables, ovens, compost bins, cisterns, waters, and other sources of unpleasant odors, of a size commensurate with requirements such that it will not be overwhelmed by the crop." But Palladius adds that if the cellar is designed specifically for fermentation and aging in *cupae*, wooden casks, the room should be paved in the manner of a

---

26 Marzano (2007).

27 A. Neyses, "Drei neuentdeckte gallo-römische Weinkelterhäuser im Moselgebiet," *Antike Welt* 10.2 (1979): 56-8.

FIGURE 21A *Tribunal of the* torcularium *at the Villa of the Mysteries, Pompeii.* PHOTO BY SANDY THURMOND.

FIGURE 21B *Gutter for gravity-flow delivery of must from the* torcularium *to the* cella vinaria *at the Villa of the Mysteries.* PHOTO BY SANDY THURMOND.

treading vat with low perimeter walls and a tile pavement, so that if casks leak, the wine can be recovered. The floors of traditional *cellae*, of course, had been unpaved so that dolia could be buried in sand or soil.

Equipment

Cato (*Agr.* 13.1) gives us a detailed description of the equipment needed for the winery: 1 pitcher (*urceus*), a bronze 5-quadrantal vessel (*aheneum quod capiat Q. V:* = 5 amphorae/113 l. / 30 gal.), 3 iron hooks (*unci ferrei*), 1 bronze disk (*orbis aheneus*), ? mills (*molae*), 1 cribble (*cribrum*), 1 sieve (*incerniculum*), 1 ax (*securis*), 1 bench (*scamnum*), 1 half-dolium vessel (*seria*), 1 pressroom key (*clavis torcularius*), 1 bed (*lectus*) for 2 freeborn guards (a third, a slave, will sleep with the contract workers), news press frails (*fiscinae novae*) as well as old ones, 1 net-cord (*epidromum*), 1 bed cushion (*pulvinum*), lamps (*lucernae*), 1 hide (*corium*), 2 gridirons (*graticulae*), 1 meat rack (*carnarium*), 1 ladder (*scala*). Obviously, much of this equipment pertains to the feeding and housing of contract and household workers, a reminder that the vintage continues for many hours a day and for several days on a large estate. Elsewhere, in his description of the pressroom of the olive vineyard (*Agr.* 12), Cato mentions equipment that will have pertained to the winepress as well: capstans (*suculae*), ropes (*funes*), hoisting ropes (*funes subductarii*) cables (*meliponti*), pulleys (*trocliae*), crowbars (*vectes*), heavy wooden pins (*fibulae*) to shim the press's anchor posts (*arbores*), and wedges (*cunei*). Reference to iron hooks, lifting ropes and pulleys reminds us that the dead weight of large press beam was so great that it required equipment to raise it. Forbes in his pioneering work on ancient technology[28] estimates that press beams as long as 50 Roman feet would have weighed over 500 kg (1500 lb), and Brun and Tchernia[29] found that the dead weight of their reconstructed Catonian press at the experimental winery at Beaucaire, Provence, was roughly 2.5 metric tons. Thus a pulley attached by iron hooks to the ceiling joists of the torcularium allowed hoisting ropes to lift the press beam when it was not in operation. It could be kept in this position by belaying the rope or by placing a sort of crutch underneath it; a bas relief from the Palazzo Rondanini shows such a wooden post planted in the ground and forked at the top to receive the press beam (*prelum*).[30] Brehaut[31] correctly interprets the *vectes* as levers used to torque the capstan which brought down this same beam.

28 R.J. Forbes, *Studies in Ancient Technology, vol 3* (Leiden, 1965, 2nd ed.): 144.

29 Brun and Tchernia (1999): 99.

30 Brun in Tchernia and Brun (1999): 74 and Fig. 89.

31 E. Brehaut, *Cato the Censor On Farming* (New York, 1933): 9, n. 7.

Treading the Grapes

The one essential element of the winery was the treading vat and, to reiterate, treading could be as simple as a single person treading a small batch of grapes in a simple, portable vessel, or as elaborate as a large, purpose-built, tiled or plastered vat. Treading is simply the breaking of grape skins and pulp and the maceration of the internal components with human feet or some other element.

The grape skin is made up of three layers, an outer cuticle which is covered with a waxy coating, the pruina, an under layer, the epidermis, and the innermost hypodermis, which contains the color cells.[32] There is no doubt that, even today, treading by foot is the gentlest, most effective way to produce the best wines, though it is relatively slow and expensive. But the modern bladder press, if gauged to extract no more than 85% of potential must, is now a close second.[33] The advantage of human feet is that they gently break up and mix skins and internal components of grapes without crushing seeds, stems, and peduncles which contain bitter tannins.

The history of treading, of course, dates back to the beginnings of wine itself. The oldest attested treading facilities in the archaeological record are our *palmenti*. In the Levant they date from c. 1500 BCE, are typically rectangular with walls 30-60 cm (1-2′) high, and are equipped with a channel to another, lower collecting vat in which collecting jars were placed. Where native stone there was permeable, these palmenti were plastered. Some so-called 'cup marks' on the sides of the vats have been interpreted as sockets for poles which treaders at either end of a line held to steady themselves, as intermediate treaders linked arms with them and with each other. Square sockets in the middle of several vats have been interpreted as sockets for a central stanchion for a screw press, but if this is so, this is a much later development.[34] As previously noted, similar installations dating from the Iron Age appear in Malta, Bulgaria, Calabria, Sicily, and parts of Magna Graecia.[35] The standard treading vat in Old and Middle Kingdom Egypt, judging by the iconography, appears to have been a mud-brick vat, either round or rectangular, finished with gesso

32 P. Forbes (1967): 292.

33 Sannino (2009): 210.

34 Walsh (2000): 139-65; Ahlström (1978): 19-49. Note the unfortunate use in the literature of the term 'wine press' for these vats.

35 Anthony Bonanno, "Maltese Wine Pressing in Antiquity," *Melita Historica* 15.1 (2008); Sculli (2002): passim.

plaster. James[36] finds very implausible Lutz's[37] suggestion of the use of wooden treading trough here. Based on the iconography, treaders stood in the vat in line, holding straps suspended from a horizontal crossbeam overhead and stepping in time to a rhythm sounded by two or more men marking the beat with batons or clappers. Otherwise, clay vats are common in Phoenician-Punic Egyptian contexts. In Egypt they are often in an elevate position with receptacles below, often fed by channels. Some of these can be dated to the seventh and sixth centuries BCE. Masonry vats in stone or cement block, coated with plaster, have now been found, but so far the earliest date only to the late Republic.[38]

In the Aegean, the 'bathtub' vats previously referenced appear to be the earliest vessels, and they are now attested archaeologically from Crete as early as the Late Bronze Age.[39] The use of these vessels as treading vats is now confirmed analytically for one such vat from the acropolis of mid-third millennium Titris Höyük, along the upper Euphrates in eastern Turkey. Molecular archaeological sampling revealed that a reddish residue in the vat contained the markers of grape must.[40]

Nearer to home, Brun[41] thinks that the reason none of the apparatus for treading from archaic Magna Graecia have been found, despite the fact that we know well that viniculture was practiced here on a wide scale, is that the apparatus were made of wood. The exception he finds in the village of Punta Chiarito on the southern coast of Ischia, dating from the sixth century BCE. The western part of an oval hut contained pithoi and amphorae, and outside, 3 m from the door, was a stone-cut *palmento* of an irregular oval shape which Brun thinks was likely the treading vat. As will be apparent, my own conjecture is that such *palmenti* were the rule rather than the exception in Magna Graecia. The same can be said, I believe, of Iron Age Etruria, where *palmenti* are found at San Leo (Urbino), San Sepolcro (Arezzo), Abbadia San Salvatore (Sienna), Castello di Vitozza (Grosetto), as well as the islands of Giglio, Elba and Capraia.

An analogue for the Greek 'bathtub' vats has now been found at the Etruscan Fattoria di Poggio Bacherina, now in the Museo Vivico Archeologico of

36 James in McGovern, Fleming and Katz 1996): 206.

37 Lutz (1922): 53.

38 Jean-Pierre Brun, "Le techniche di spremitura dell'uva: origini e sviluppo dell'uso del torchio nel Mediterraneo occidentale," in Ciacci, Rendini and Zifferero (2007): 57-8.

39 Leonoard in McGovern, Fleming and Katz (1996): 244-8.

40 McGovern (2003): 182.

41 Brun in Ciacci, Rendini and Zifferero (2007): 56-7.

Chianciano Terme and dating to the second or first century BCE.[42] Additionally, Fatucchi[43] asserts that Cato's reference to *lintres* in his list of equipment for the winery is to be interpreted as reference to portable wooden treading vats which were taken to the vineyard. More telling than Cato's reference is that of Tibullus (1.5.23-5 (Delia aut mihi servabit plenis in lintribus uvas/ pressagne veloci candida musta pede): Delia will save for me grapes in full troughs or clear must pressed by nimble feet." But Tibullus' use of the adversative (*aut*) suggests here two distinct possibilities, namely, table grapes and wine. Thus it is more likely that these little implements (also called *navia*, 'dugouts' by analogy with the shape of small, dugout boats: Paulus Festus 169M) are no more nor less than wooden troughs, 75-120 cm (2 1/2'-4') long, with a flattened base and upcurved ends, used for carrying bulk items and for kneading bread, and still common in practically every museum of farm life throughout Italy. If such troughs were used as treading vats, they were almost certainly not taken to the vineyard; there is no possible advantage for transporting trodden must from the vineyard. But such implements were indeed used to transport harvested clusters to the winery, installed on carts, as we have noted.

Finally, a case has been made[44] that an 'unusual' number of fragmentary pips found at treading installations in France suggests the possibility that pestles were used to crush grapes, as was often done in later times in France and Italy. On the Greek representations of treading, two of the satyrs who tread in a 'bathtub' basin clutch short wooden stakes, either to steady themselves or to aid in crushing the grapes. But there is a more likely explanation for such fragmentary pips. As we will see, after press wine had been expressed and before the afterwines were made, presscake was removed from frails and thoroughly hacked apart with axes before being mixed with water and put back into the frails to be pressed again.

On the villa farm, treading took place in the *torcularium*, either in a separate vat, the *calcatorium*, as at the urban vineyard at Pompeii (II.ix) or in the *forus* of the press, as at the Villa Regina and other installations. Unfortunately the literary evidence is not as explicit as we could wish. The only direct evidence for treading (*calcatio*) in Cato's manual, for example, occurs in reference to the

---

42 Giulio Paolucci, "L'occupazione dello spazio agrario in Etruria: vici e fattorie," in Consiglio regionale della Toscana, *Il mondo rurale etrusco* (Grosetto, 2009): 44-5 and Figs. 5 and 8.

43 A. Fatucchi, "Palmenti romani da uva dell'Etruria nord-orientale," *Rivista di storia dell'agricoltura* 27.2 (1987): 21.

44 I. Figueiral et al., "Archaeobotany, Vine Growing and Wine Producing in Roman Southern France: The Site of Gasquinoy (Béziers, Hérault)" *Journal of Archaeological Science* 37 (2010): 147.

making of 'Coan' wine (*Agr.*11.2.3). Varro (*RR* 1.54) says simply that 'when the grapes have been trodden (*calcatae*) their stalks and stems should be pressed so that any must they retain may be expressed in the same vat." Does this imply that by Varro's time, treading was done in the press vat? Billiard[45] says that proceeding directly to the straining out of woody elements (*égrappage*) occurs only when the vintager is making an afterwine called *vinum acinaticum* (cf. Palladius *RR* 1.6), i.e., an astringent product indeed. But treading is never omitted, the antiquity of which is attributed by Tibullus (2.1.5.45) to Bacchus himself. Thus Billiard thinks that pressing must have been a relatively late development. The treading vat was located in the winery for convenience (Vitruvius *de Arch* 6.6) and according to Palladius (*RR* 1.6) is composed of a rectangular basin (note that Palladius uses the term *lacus*, however) of masonry with raised sides. Curtel[46] had thought Cato's remark about treading in the pressbed (*torcularium*: *Agr.* 112.3) was a confusion of pressing and treading, but Rossiter[47] has convincingly demonstrated that archaeological evidence bears out Drachmann's[48] conjecture that treading often took place in the press bed. Raised curbs on the pressbed were perfectly adequate for both treading and pressing, and treading here eliminated the necessity of moving the grape solids from a separate vat. This, then, is the feature of the room which Varro (*RR* 1.54.2) and Columella (*RR* 11.2.71) refer to as the *forus*. Rossiter shows that where the press bed is circular rather than rectangular (e.g., Granaraccio) the treading took place in the press room, but in a separate *calcatorium* nearby. The etymology of the term *forus*, deriving from the raised gangplank on the trireme above and between the ranks of thranite rowers, suggests that the *forus* was also typically raised and behind two *lacus*, as is attested in several of the villa sites.

The actual process of treading (*calcatio*) is virtually timeless.[49] Treaders (*calcatores*) of varying numbers between two and seven, strip to their loincloths and carefully wash their feet. They tread the grapes with naked feet (Virgil, *Georg.* 2.5.1ff.). Needless to say they are required to maintain strict bodily cleanliness and are enjoined from letting fall during the process any food or sweat (Apuleius in *Geopon.* 6.11) and are forbidden to tread if they have wounds on their feet, not for hygienic reasons but since wine thus made could not serve for religious libations (Pliny, *NH* 14.119). Treaders worked to a strong

---

45 Billiard (1913): 438-40.

46 G. Curtel, *La vigne et le vin chez les Romans* (Paris, 1903): 107.

47 Rossiter (1981): 349-51.

48 A. G. Drachmann, *Ancient Oil Mills and Presses* (Copenhagen, 1932): 87.

49 Billiard's [(1913): 440-2] is still the best synthesis.

FIGURE 22 *Treading the grapes. Based on a mosaic found at Saint-Romain-en-Gal, now at the Louvre. After Billiard (1913): Fig. 148.*

cadence (Calpurnius, *Egl.* 10.5.44), accompanied by rustic airs played on the double flute (*aulos*) or the pipes (*syrinx*). Frequently this 'march' becomes an actual dance (Theocritus, *Idyl.* 7.5.24 writing in the 3rd c. BCE), called *epilénion*, from the Greek, a sort of pantomime of the tasks of the vintage. The dance might be accompanied by a chant, also called *epilénion* (Oppien, *de Venatione* 1.5.1270) or Latin *celeuma* (Martial, *Epigr.* 4.64.5.21-22). The bruised pulp of the grapes consist of a viscous mass on which it is easy to slip; in order to maintain equilibrium, treaders support themselves on crutches or curved staffs or hold to cables hung from the ceiling or horizontal beams or lock arms with others, as we see in the iconography.

White vs. Red Wine

We now confront one of those areas where our literary sources fail us almost completely, namely, in distinguishing the making of white wines or red, which require different procedures at this point in the process. We know that wines of various hues were made by the Romans, since we hear Pliny (*NH* 14.80) speak of white wine (*album*) tawny wine (*fulvum*), red wine (*sanguineum*), and even 'black' wine (*nigrum*). It is a common misconception that white wines derive exclusively from white grape cultivars and red only from red cultivars, when in fact, with the exception of a few red cultivars which have red juice, grape juice for both is basically clear. It is therefore quite possible and in many cases desirable to make a white wine from a dark grape. The red color, in fact, derives from the hypodermis of the grape skin, and it requires maceration of the skin with the juice for several hours to several days, depending on how thoroughly the skins are bruised, to extract color. Additionally, less astringent tannins and polyphenols that include flavanoids also derive from skins, both desirable qualities in wine. We know from artistic representations and from literary evidence (e.g., Pliny *NH* 14.42) that Roman grapes ranged from greenish to yellow to rosy to red to purple, and the Romans therefore had the same basic options as modern winemakers in this area.[50]

Fortunately, once we understand the possibilities, many of the apparent contradictions and ambiguities in the literature begin to make perfect sense. Thus Cato (*Agr.* 25) says that must from treading is directed immediately to the fermentation vessels, in this case dolia, or to a settling tank (*lacus vinarius*) and then to the dolia (*Agr.*11.3.1). Columella (*RR* 12.27.1, 12.19.3, 12.41.1) mentions the same options. Pliny (*NH* 14.59, 14.83) has must from both treading and pressing flowing into the *lacus*, and thence to dolia (*NH* 14.133) or to *cupae* (14.132). Likewise, in several artistic representations we see must from treading vats flowing through spouts directly into what appear to be *dolia defossa*.

It therefore seems clear that musts which went directly to fermentation vessels were intended for white wines. In this regard the ancients made another distinction as well. Free-run must, *mustum lixivium* (Fr. *moût de mère-goutte*) (Columella *RR* 12.27), also known as *protropum* (Pliny *NH* 14.85), i.e., must which flowed from the bruised grapes under their own weight, was typically

50 As early as eighth-century Assyria we can recognize basic wine styles. David Stronach ["The Imagery of the Wine Bowl: Wine in Assyria in the Early First Millennium B.C.," in McGovern, Fleming and Katz (1996): 180] refers to Assyrian wine lists which categorize wines as 'sweet/good white wine', 'tannic wine' and 'strong wine'. Presumably some of the latter will have been reds.

mixed with must of treading (*mustum calcatum*) and of pressing, but sometimes was decanted immediately into amphorae to be fermented as a luxury product. The wine was fermented out, sealed, and then left under the sun for 40 days to prematurely 'age', after which it was ready for consumption. This will have been an intensely sweet white wine.[51]

More typically the treading must, now undergoing a vigorous fermentation, was conducted from the *calcatorium* or *forus* to one or more (two is somewhat typical) *lacus*, settling tanks. This must, lightly alcoholic and still quite sweet (the Greek term is *gleukos*) was also drunk as a special treat, though medical writers warn that it causes gas, as of course it would.[52] The *lacus* served as a mixing tank as additional must came from the presses, and as a settling tank to allow gross lees to precipitate (Fr. *débourbage*). Several sites in Italy such as the Villa della Pisanella at Boscoreale demonstrate the system.[53] Here must from one *forus* was conducted to a *lacus* by an underground pipe. Other outlet pipes conducted must directly to *dolia defossa* embedded in the floor of the *torcularium*. Rossiter thinks that these may have been for separate storage of trimmings, but it is equally likely, given the smaller capacity of dolia, that musts for afterwines were segregated here. The *lacus* was typically fitted with a spigot or spout, several inches above the bottom, which was closed until the gross lees had settled below its level. If the must was designated for a white wine, we should imagine that it was left here for several hours or perhaps overnight,[54] and then the spout opened and the must conducted or decanted to the dolia in the *cella* for fermentation and aging. At the urban vineyard of Pompeii (II.v), must from the *forus* flowed to a rectangular *lacus* (.9 m × .65 m × .36 m/ 3′ × 2′ × 14′), which was conducted by an overflow channel to a second *lacus* and thence by a covered gutter along the wall to the *cella*.[55] Then the gross lees, which still contain a great deal of must, will have been filtered through the *colum*, the must which drained away transported to the cellar, and the grape solids, the marc or pomace, taken to the press.

51 Estreicher [*Wine: From Neolithic Times to the 21st Century* (New York, 2006): 28, n. 19] compares this to French Vin de goutte and Hungarian Tokaji Esszencia.

52 H. R. Immerwahr, "New Wine in Ancient Wineskins: The Evidence from the Attic Vases" *Hesperia* 61 (1992): 128-30. In Greece wineskins partially filled with *gleukos* and inflated by the CO_2 were favorite objects in various games, especially *askoliasmos*, 'jump on the wineskin'.

53 Rossiter (1981): 351-3.

54 Bird (2010): 60.

55 Parallels at Grotta del Malconsiglia, Rm 41, Settefinestre (where a screen over the plug hole evidently trapped gross lees), San Giovanni di Ruoti, Guidonia (where the *forus* was connected to the lacus by a covered gutter).

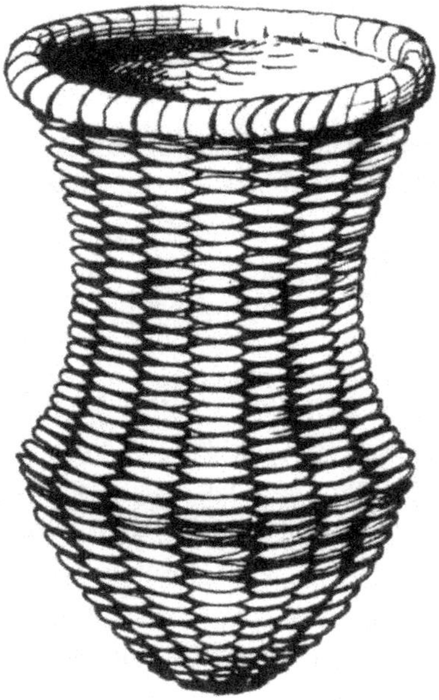

FIGURE 23
The wine strainer, colum, *used for straining gross lees. From White (1975): Fig. 31.* COURTESY OF CAMBRIDGE U. PRESS.

For musts designated for red wines, the *lacus* will also have served for the extraction of color, tannin and flavor compounds, as well as the site of much of primary-phase fermentation. Fermentation at this stage does not expose the must to infection as much as might be expected, since it produces such massive quantities of CO_2, so much heavier than air that it floats over the surface of the must as a protective layer. In modern winemaking the 'cap' (Fr. *chapeau*), i.e., the mass of solids which float to the top of the must, is periodically 'punched down' (Fr. *pigeage*) under the liquid, using various implements, or is stirred back into the mixture to increase extraction of desirable skin elements.[56] A floating cap not only prevents extraction of desired elements but is also dangerous to the must because it harbors swarms of acetic acid (vinegar) bacteria. Traditionally the cap was punched down several times a day. The process is tedious but gentle, causing minimal damage to skins, damage which abrades them so much that they release harsher tannins from their epidermeis.[57] After several days of maceration, the must will have been filtered and decanted in the same way as for white wine, and the pomace taken to the press.

56 Vine et al. (1977): 105-06.

57 Bird (2010): Today it is more typical to use a variety of electrical mechanism to incorporate the cap back into the must.

Pressing

Types of Presses

At this point, as noted, the free-run must (*lixivium*) may be transported directly to fermentation vats or may be reserved to be mixed with the must of treading or that of first and second pressings, though this last does not appear to have been standard procedure. It is important to remember that treading alone will extract some 80% of potential must and that this was considered the highest quality must, as it still is today. Then again, pressing equipment cannot have been cheap.[58] Thus at one end of the economic spectrum our wealthy villa farmer will have been faced here again with the choice between quality and quantity. At the other end, the subsistence farmer may have had no such choice since he had no access to pressing equipment at all, though we needn't overestimate the economic hardship this will have posed. For what it is worth, Cato says that grape pulp and skins, the marc, are to be placed under a mechanical press, in this case the standard beam press, to extract more must, though it is unclear whether he means for this must to be mixed with the *mustum calcatum* or, as Columella and others advise, be stored separately as inferior product (Cato, *Agr.* 23.4). Cato also implies a second pressing. Varro also implies two pressings and is the first to explicitly recommend separate storage of treading must and must of first pressing (*RR* 1.54.2-3). He further recommends mixing the remaining pomace with water to provide a drink for the farm hands in winter. Columella (*RR* 12.36.1; 12.37.1) also advises two pressings and separate storage. Pliny (*NH* 14.59)[59] has musts from treading and pressings flowing into a common reservoir.

Roman presses were of four types: torsion presses, wedge presses, beam or lever presses, and screw presses. Beam and screw presses included a number of variations.[60] Traces of archaic presses are rare, first because the presses themselves were doubtless also rare, second because they were made of perishable materials, especially wood, and they therefore leave no evidence. The bag or torsion press[61] was a standard press only in Egypt. As early as the Old Kingdom, hieroglyphics appear which seem to show a combination of treading vat and torsion press. The marc is wrapped in a flexible bag of some permeable

58 Cf. Rossiter (1981): 346-48.

59 Cf. Rossiter (1981): 4-48.

60 Billiard (1913): 442-55; White (1975): Figs 60-66; Brun in Tchernia and Brun (1999): 48-107; Brun in Ciacci, Rendini and Zifferero (2007): 58-65; Drachmann (1937).

61 Billiard (1913): 442-3; James in McGovern, Fleming and Katz (1996): 200-07; Lesko in McGovern, Fleming and Katz 1996): 217-18; Brun (1999): 49 and Fig. 42.

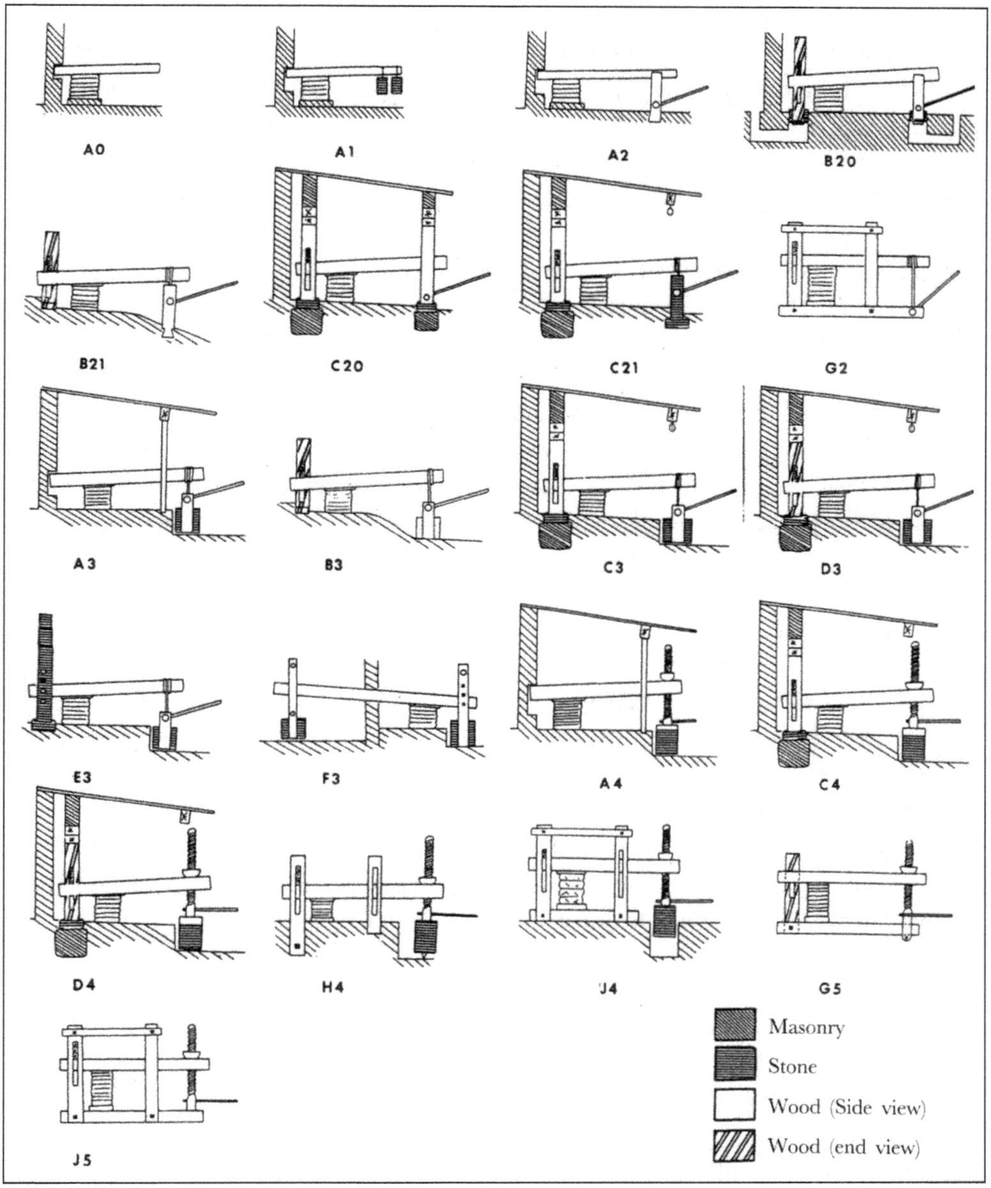

FIGURE 24 *Typology of Roman presses. From Brun (1986): Fig. 28.* COURTESY OF CNRS EDITIONS.

material to which metal loops are attached at each end. One of these loops is attached to a stationary frame and through the other end is passed a stick which is gradually rotated by several men until the torsion wrings the liquid from the bag.[62] The procedure will be familiar to anyone who has wrung out a sopping towel by twisting its ends in opposite directions. The wine extracted falls into a reservoir. Thus in the Middle Kingdom Tomb of Baqet from the 11th dynasty (ca. 2050-2000) a group of men holds a piece of cloth over a jar and pours liquid into the cloth. James conjectures that here we have a representation of the step in which must is filtered from pomace before the latter is placed in the bag press, a logical surmise since the pressing scene is immediately to the right.[63] On a black or red-figure Attic vase of the sixth or fifth century BCE, we see two satyrs treading in a portable vessel which Brun[64] interprets as a wooden vessel. A third satyr presses pomace in a bag of sorts (perhaps an animal skin) on top of a table. He raises his left knee to bring his weight onto the sack as he steadies himself with his right leg on the floor. Liquid flows from the front of the bag into a waiting receptacle at the foot of the table. We have no attestation for such a device in Rome, though its use there is not unlikely.

The wedge press is cited in no texts; in fact, we have only two paintings, one in the House of the Vetii at Pompeii and the other from Herculaneum. The paintings are so similar in style and content it is probable they are by the same artist. They depict two vertical spars solidly fixed in the earth over a stone press table with a depression to hold the olive pulp or herbs to be pressed and a spout to one side to conduct the liquid expressed to a receptacle. Above the mass to be pressed is a press board, a single plank in this case, and above this alternating tiers of conical wedges and additional planks, the uppermost of which is directly under a massive crossbeam between the two spars. Erotes are depicted on either side, driving in the wedges with sledge hammers and thus exerting increasing pressure on the mass. Pompeii was famous for its perfumeries, and perhaps the expression of essential oils from aromatic herbs is what is depicted here; we have no evidence that culinary olive oil or grape must was expressed in this way, but it is not improbable.[65]

Lever presses appear in archaic Greek iconography, for example the skyphos of the Forman collection in the Boston Museum of Fine Arts. The beam press is basically a huge lever, the fulcrum of which may be a socket in the wall of the

---

62 Billiard (1913): 442-44; Brun in Tchernia and Brun (1999): 49 and Fig. 42.

63 James in McGovern, Fleming and Katz (1996): 212.

64 Brun in Ciacci, Rendini and Zifferero (2007): 55-66 and Fig. 1.

65 Forbes III (1965): 144; Billiard (1913): 444-45; D. J. Mattingly, "Paintings, Presses and Perfume Production at Pompeii," *Oxford Journal of Archaeology* 9,1 (1990): 71-90.

torcularium or may be provided by massive piers. Because almost the entire mechanism is wooden, the only archaeological evidence for them is typically the socket in the wall, or the cuttings in the floor where the piers, called *arbores*, were located. Otherwise, occasionally the foundation of a press bed is found, as at Chersonese Taurico and at Punta Chiarito, Ischia, and at the Greek colony of Agathe (Agde, Hérault). The first well documented presses come from the beginning of the second century BCE, corresponding to Cato's famous description, as at Capena and at the Villa Prato at Sperlonga. Here the press is part of a well equipped torcularium.

From Cato's description and the comments of others, as well as from comparative evidence (such beam presses were used in Italy well into the twentieth century and are still used in parts of the world, especially for olive pressing) we have an excellence idea of the parts of the mechanism. The engaged end of the press beam (*prelum*) is called the *lingua* (Vitruvius *de Arch* 10.6.3), the posts on either side of it the *arbores* (Cato *Agr* 21.1; Pliny *NH* 18.317), the anchor sockets in the ground the *pedicini* (Cato *Agr* 18.4). The other end of the press beam is stabilized between two uprights (*stipites*: Cato *Agr* 18.2) and is attached to a pulley (*trochlea*: Cato *Agr* 3.5) with a heavy esparto rope (*funis subductarius* or *capistrum*: Cato *Agr* 12). Underneath it is attached to a capstan (*sucula*: Cato *Agr* 19.2), in the middle of which was a metal loop (*porculus*: Cato *Agr*. 19.2) to which was attached the capstan rope (*funis torculus* or *lorcus*: Cato *Agr* 63, 135.5) made of heavy leather. On the sides of the capstan at either end were sockets into which were placed long iron levers (*vectes*) to crank down the *prelum*.

Cato says that the pressbeam should be made of hornbeam (*Carpinus betulus* L). Wood of this species is still used for tool handles and other applications where extreme durability is required, but is not otherwise used because its extreme hardness makes it difficult to work. The press beams are huge, up to 16 m (50′) long, massively heavy. *Arbores* were typically a pair of slotted piers across which was a heavy lintel. The advantage of the system is that wooden balks can be placed above and below the press beam so that it is more nearly level with the top of the press stack, in which position it develops the greatest force. As the press stack shrinks with the exudation of liquid, the press beam can be lifted and balks from below repositioned above and pressure reapplied. This procedure can be repeated several times as the press stack grows shorter.[66] In either case, the ends of the beams opposite the fulcrum, where

66 Drachmann (1932): Figs 40-41; Raphael Frankel, "Western Galilee: Oil Presses," in *Excavations and Surveys in Israel* 4 (1985): Fig. 2; *idem*, "Some Oil Presses from Western Galilee" *BASOR* 286 (1992): Fig. 7; *idem*, "Oil Presses in Western Galilee and Judaea: A Comparison,"

maximum force is developed, are harnessed by ropes to windlasses by means of which the beams are drawn down with all the massive force which the windlass's mechanical advantage permits.[67]

At some unknown point a refinement was introduced to the conventional beam-and-winch press.[68] The disadvantage of the lever is that it is extremely difficult to exert a constant pressure using it unless a ratchet and pawl mechanism is built into the windlass. The screw is a simple machine which does not have this disadvantage and has considerable mechanical advantage as well. The screw itself dates to the generation of Archimedes (300 BCE) and by the time of Heron of Alexandria (first century CE) there were screw cutters available. The screw was introduced to Italy at the end of the first century BCE. At first the screw simply replaced the windlass to haul down the press beam. The screw was fixed in the floor such that it could freely rotate. The press beam in this case was chosen because the end opposite the roots had two massive branches which forked. Typically we have a tree trunk, often of oak instead of hornbeam, with roots attached, crudely worked, and the truncated ends of whose forked branches provided a yoke for a huge screw.[69] A fixed nut threaded on the screw was placed over this fork. The screw was turned by handspakes and the nut forced the beam down.[70] Alternately, the screw is used to draw up a stone weight or a chest filled with stones. This latter device is described by Heron and is still used in many parts of the Mediterranean. The advantage of this so-called beam and counterweight press is that once the weight has been lifted from the ground it will continue to exert constant pressure on the press stack as it shrinks, whereas the beam and capstan press and the beam and screw press must be periodically ratcheted or screwed down as this happens.

Obviously this is simply a modification of a lever press, but true screw presses (*torcula graecanica*: Pliny *NH* 18.317) were also used by the ancients beginning from c. 50 CE, and are described by Heron and Pliny. In one type two screws at opposite ends of a short horizontal beam go through them and are dovetailed into a massive press bed. As the two screws are turned simultane-

---

in David Eitam and Michael Heltzer, edd. *Olive Oil in Antiquity: Israel and Neighbouring Countries from the Neolithic to the Early Arab Period* (Padua, 1996): 199 and Fig. 1; Brun in Tchernia and Brun (1999): 48-107 (especially good for the archaeology and iconography).

67 Drachmann (1963): 110-15.

68 For a Greek example, S. C. Bakhuizen, "*Torcula Graecanica*: A Note on the Archaeology of Olive and Grape Pressing," in M. Gnada, ed., *Stips Votiva: Papers Presented to C. M. Stibbe* (Amsterdam, 1991): 1-6.

69 E.g., Angelo Pesce, "Un testimone d'eccezione: Axel Boëthius," in Guzzo (2000): 105-11.

70 Drachmann (1963): 115-26.

FIGURE 25 *Lever and screw press.* DRAWING BY G. SILVESTRO IN SANNINO (2009): 209.

ously, the pressure increases on press boards and frails. This appears to be a refinement of the wedge press. Alternately, one screw has a nut which is actually cut into a solid press beam, and both are fixed to the press bed by two stout uprights, after the fashion of the later copying press. The press lid, *tympanum*, is drawn down by turning the screw, *cochlea*, which is forced downward by the fixed nut.[71]

The press bed itself was typically a circular stone or masonry pedestal, often raised several feet from the floor. On this the frails filled with grapes were stacked. Heron describes an alternative to press baskets called the *galeagra*, which Drachmann[72] has reconstructed from the garbled text of Heron as a lattice of half-lapped timbers arranged on the floor of the *forus* to form a square, into the middle of which the marc is placed. On top of it is placed a square board to act as a 'follower'. A chock (or series of chocks) smaller than the hole in the lattice so that the press beam can drive it downward without striking the sides of the *galeagra*, is placed between the follower and the press beam.

71 Drachmann (1963): 126-35; Forbes III (1965): 142-3.

72 Drachmann (1963): 126. For modern pressing using a reconstructed *galeagra*, Brun in Brun and Tchernia (1999): 101-2.

The Process

Cato's description of the pressing technique is typically cryptic: "Press [the marc] every day. Divide into equal parts the must from the second pressing after the cutting up [of the marc] and put one part into each storage jar" (Cato, *Agr.* 23.3, Brehaut's translation). From that unpromising beginning, Billiard has synthesized this procedure. A quantity of marc which can be handled in a single pressing (*factus* or *pressura*) is put by the pressmen (*factores*) inside the laths (*regulae*) of the *galeagra* or into large press baskets or frails (*fisci, fiscellae*)[73] made of withies, bramble, esparto grass, or rush, dense enough to contain the marc but porous enough to allow the must to exude. If frails are used, between them are placed circular wooden *tympana* or *orbes*, four feet (1.2m) in diameter. The press beam is lowered and the lever-men (*vectiarii*) crank the windlass in the case of the standard beam press, the screw in the case of lever and screw and screw presses. The must flows from the interstices of the frails or lattice and collects in the reservoir of the press bed, whence it is conducted or decanted in one of the ways described above to the settling vats (*lacus*) or fermentation vessels in the cellar.[74]

Pressing often demanded considerable time. When the day did not suffice, the process continued into the night and perhaps for several days. This fact explains Cato's recommendation (*Agr.* 13) of beds for workers as well as bolsters and lamps. Excavations of *villae rusticae* such as the Villa Pisanella at Boscoreale confirm Cato's remarks; clustered about both olive and grape press rooms are cubicles which can only be bedrooms.[75] Pliny (*NH* 18.317) says that on a well appointed villa one pressing (*pressura*) ought to produce 20 *cullei* (2400 gal. / 9085 L).[76]

A single pressing was not sufficient to completely extract all the potential must; therefore, when all the must of the first pressing had stopped flowing, the marc was removed from the press bed and extracted from the frails if these

---

73 Frails are evidently depicted on the Getty *psykter* attributed to the vase painter Smikros [= J. Paul Getty Museum 83.AE.285 = J. Frel, "Euphronios and his fellows," in W. G. Moon, ed., *Ancient Greek Art and Iconography* (Madison, WI, 1983): 147, Fig. 10: 1a-c]. Cf. Immerwahr (1992): 124 and n. 14.

74 Billiard (1913): 451. For the *lacus*, White (1975): 157-60.

75 Pasqui (1877).

76 For a carefully considered reassessment of the yield and potential profitability of Columella's prototype vineyard, see A. Carandini, "Columella's Vineyard and the Rationality of the Roman Economy," *Opus* 2 (1983): 177-204. Jancis Robinson [*The Oxford Companion to Wine* (New York, 1999): App. 2] says that modern Italian estates have recently been producing wine yields of c. 690 gallons per acre. Columella's prescription for a cost-effective vineyard of 3 *cullei* per *iugerum* is approximately 660 gallons per acre.

were used. It was then hacked apart (*circumcidere*) with hatchets (*dolabellae, secures*: Columella, *RR* 12.36), put back onto the press in the same way as before and pressed again. But the resulting product was not necessarily an inferior product; indeed, it was valuable in its own right. The must of the second pressing (*tortivum mustum circumcidaneum, vinum circumcisitum*: Cato, *Agr*. 23; Varro, *RR* 1.54; Columella, *RR* 12.36) was rich in tannin from the crushing of stems and seeds, as was obvious to the ancients; this wine was kept separate according to Varro "because it tastes of iron (*RR* 1.54)." That this tannic product could be useful in palliating the acid deficiency we suspect in ancient wines appears to have been known to ancient winemakers as well; Varro recommends that it be equally apportioned to the free-flow wine and wine of first pressing. When we recall that Varro had recommended separate storage of free-flow and first-pressed must to maintain quality, it becomes evident that the product of the second pressing is not an adulterant here but an additive to *improve* quality. It is well established that it is not the absolute amount of sugars, acids, phenols and tannins in a wine that determines flavor but rather their relative amounts. A given level of acidity or tannin may be quite objectionable in an extremely dry wine but quite pleasing to the palate in one with some residual sugar.[77]

Marc (*vinacea, vinacium, pes*: Varro *RR* 2.2.19; Col *RR* 12.43.3, 12.36, *Arb* 4.5; Pliny *NH* 17.197), composed of stalks (*scopio*: Cato *Agr* 112.3; Col *RR* 11.3.46, 12.39.3, 12.34.10) or *scopius*: Varro 1.54.2, 2.4.17), the skins of the grapes (*pellicula*: Varro 2.4.17; *folliculus*: Col *RR* 12.45; *cutis*: Col *RR* 3.2), the pips (*nuclei, acini*: Pliny *NH* 23.13) and residual must, makes an excellent fertilizer and fodder for farm animals (Anatolius in *Geopon* 6.13) and is therefore preserved in Jars, of which Cato (*Agr* 11) prescribes 20 for his prototype 100-iugera farm.

As must filled the treading vats, pressbeds, and settling tanks, workers will have hastened to decant it and transport it to the cellar. These workers may have been among the pressmen (*torcularii, factores*) or may have constituted a separate group of workers, perhaps called *capulatores* as for olive pressing.[78] Workers, perhaps so named or perhaps called *haustores*, if named separately, placed a jug or pail (*hama, urceus*) under the foot of the pedestal of the vat and filled these pails to be transported to fermentation vessels in the cellar. Workers may also have used pitched baskets of the sort used in the vintage; a marble bas-relief from the Villa Albani at Rome dating from the second century CE

---

77 Winkler et al. (1974): 171.

78 The latter formed a separate guild in Rome with their own *schola* situated in Rome's third Augustan region and inhabiting a quarter of the fifth region called the *vicus Capulatorum*. The name derives from the ansate vessel, *capis* or *capula*, with which oil was decanted.

shows men treading grapes in a *calcatorium* which empties into a smaller *lacus* at a lower level. Another worker scoops must from this receptacle with a flagon and empties it into a tall, somewhat conical basket, while another worker is emptying must from the same type basket into a dolium. Unless these grape solids are being reserved as fodder, we are not seeing here the *colum* used for filtration, but rather another vessel for decantation.[79] Overflow from the *lacus* sometimes fell into a tiled trough and was returned to the reservoir.[80] Alternately, the overflow fell into a tiled gutter or pipe and was conducted thence by gravity to a *lacus* in the cellar. In either case workers filtered the raw must through a colander of withies or rattan (*colum* or *saccus vinarius:* Cato, *Agr.* 11; Columella, *RR* 9.15),[81] conical in shape, which retained gross impurities. Thence the wine made its way to the fermentation vessels, the dolia and/or wooden *cupae.*

The Cella Vinaria

The cellar was a masterpiece of rational design; on that the literary and archaeological evidence is in accord. We have previously noted its orientation to take advantage of ambient temperature and to minimize exposure to foul odors. Such a disposition economized time, work, and supervision. The floor of the typical cellar is composed of a deep bed of sand or, in some cases, of some impermeable material. That sometimes wine jars were sealed prematurely is suggested by Varro's citation (1.13.6) of Fundanius' advice that the cellar floor be sloped to a reservoir because secondary fermentation often bursts both the butts (*orcae*) used in Spain and the dolia used in Italy, and some of this wine can be recovered. Alternately, we may have some evidence here of springtime malolactic fermentation. Judging by the archaeological sites explored by Rossiter, the former arrangement was used exclusively, and Varro may be recording a curiosity.

The crowning glory of the cellar was, of course, its huge fermentation and storage vessels. The two standard vessels, as will be clear by now, were the dolium and the barrique, the *cupa* or *orca*. The dolium, otherwise *tinia* (Varro

79 Brun in Tchernia and Brun (1999): 73 and Fig. 87.

80 Billiard (1913): 456-57.

81 Brun [Tchernia and Brun (1999): 73] interprets these baskets as decanters and remarks that they must have been thoroughly pitched to make them waterproof, but I think Billiard's interpretation is more likely, given the ancient testimonia.

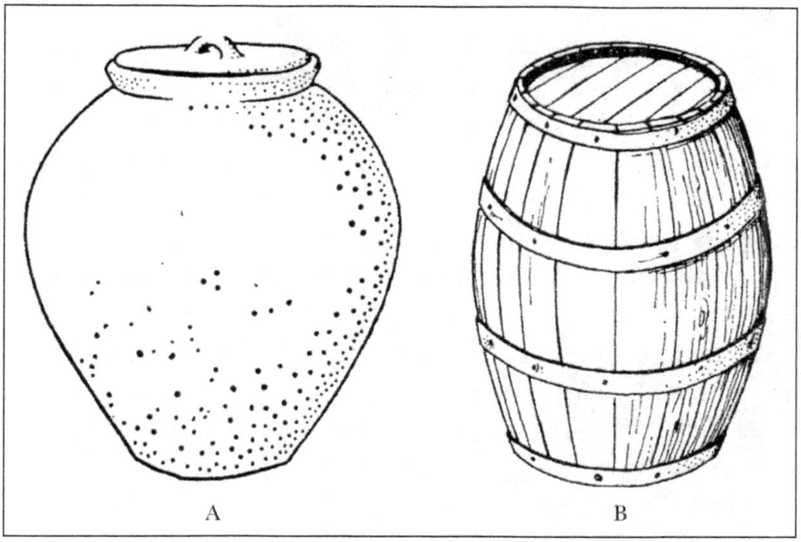

FIGURE 26 Dolium *and* cupa, *the two standard wine vessels. From White (1975): Figs. 144 and 131.* COURTESY OF CAMBRIDGE U. PRESS.

apud Nonnius 544.6; Festus p. 365M)[82] is a large, potbellied, unglazed terra cotta vessel.[83] Dolia have relatively large mouths, a flat base, and walls as thick as 6-7 cm (2 1/2″-3″). There is considerable variation in size and capacity; Billiard cites an example from Marseilles 1.6 m (5′ 2″) tall. 1.22 m (4′) in diameter, and 3.8 m (12′ 6″) in circumference; another from Nîmes 1.9 m (6′ 3″) tall and 4.45m (14′ 6″) in circumference of some 8 hl (211 gal) capacity. At Bougie he found a colossus of some 11 hl (290 gal) capacity. The Villa della Pisanella at Boscoreale had 84 dolia, found *in situ,* holding on average 11.9 hl (315 gal). Seventy-two seem to have held wine.[84] Those figures accord roughly with the testimonia; Columella (*RR* 12.18) speaks of a dolium of 1 1/2 *culleus,* about 7.8 hl (206 gal), and Palladius (10.11) one of 200 *conges*, some 6.25 hl (165 gal). The dolium is the equivalent of the Greek *pithos*, and it is little wonder that Diogenes the Cynic could be said to have lived in one and the Athenians, driven to straits during the Peloponessian War to have resorted to the same stopgap.

82 Also *tinum*: *Notae Tironianae:* a late collection of abbreviations ascribed to Cicero's freedman. Cf. Spanish *tinaja* and Italian *tino*. The *tinaja* is still used extensively in Spain, especially in Andalusia and in the sherry regions, especially Montilla. There have been a number of recent experiments in Italy using dolia as fermentation vessels.

83 Billiard (1913): 465-73.

84 Stefani in Guzzo (2000): 67.

FIGURE 27 *The* cella vinaria *at the large urban vineyard at Pompeii.* PHOTO BY THE AUTHOR.

Form and size of dolia influenced the quality of the wine; preferable were those with narrower mouths and smaller overall size (Pliny, *NH* 14.134. Anatolius (*Geopon*) says that vessels which are too openmouthed promote loss of bouquet and the attack of diseases, both perfectly true.

FIGURE 28 *The* cella vinaria *at the Villa Regina.* PHOTO BY SANDY THURMOND.

Dolia were typically ranged parallel to the longer wall of the *cella vinaria* in ranks. Occasionally they were simply placed on the floor of the cella but far more often were buried from 1/2 to 2/3 their height in the soil of the cella. To reiterate, both Cato (*Agr.* 11) and Palladius (1.18) recommend that a number of dolia beyond the average vintage size be bought so that wines from multiple vintages may be cellared to take advantage of market fluctuations. Cato, in fact, prescribes enough dolia for 5 vintages. Our evidence is exiguous here, but this certainly does not appear from the archaeology to have been the norm; in every case where the size of the vineyard is known or can be estimated, it is more or less consistent with the capacity of the dolia in the cella vinaria.

Other ceramic vessels were also used, most notably the *seria*, a vessel of the same shape as the dolium but of much smaller capacity. Columella informs us that the standard form held 7 amphorae (182 l / 48 gal), roughly ¼ dolium, and also mentions an even smaller vessel, the *seriola*. Some villa farms, in years of superabundance, resorted to a cement *lacus* sunk into the soil of the *cella* (Cato *Agr* 67) and Pliny (*NH* 18.319) mentions a vintner who, in straits, poured his new wine into a *piscina*.

All such vessels were provided with terra cotta covers (*opercula*) which could be sealed with gypsum or plaster. In open-air cellars, very much the norm in Campania, extra protection from the elements was provided by a second cover, the *tectorium*, a sort of terra cotta shield with three ansate feet on the bottom to keep it elevated several centimeters above the *operculum*, and convex on its top to shed rainwater.[85] We should note that a fragment of a sarcophagus conserved at Ince Blundell Hall, Liverpool, dating from the third century CE, shows a scene of a wine sale. Nine embedded dolia are shown, covered with circular lids made from parallel planks with two transverse boards to which they were apparently nailed and which also functioned as handles.[86] But surely this scene shows a wine wholesaler and these dolia have already been broached, since we see two workers filling an amphorae from one of the dolia.

We are now in a position, especially thanks to INAA, to analyze the clays used to make terra cotta wine vessels, and it seems clear that, whenever practicable, such vessels were bought locally. But analysis of the clay fabric of sixth and fifth century BCE vessels in the Albegna valley of central coastal Etruria reveals the presence in many of black augite, and the closest source of this volcanic mineral is some 70 km (50 miles) away near Lake Bolsena. Since augite has no known advantage in the production of ceramics, it seems likely that the mineral itself was mixed with local clays, and it is therefore likely that the vessels themselves were produced near Lake Bolsena. Thus we may infer a systematic commerce in such wine vessels from this early date. Transfer of the technology of ceramics to Etruria in the tenth century allowed for a rapid expansion of viticulture in the area on a commercial scale. But the majority of amphorae analyzed were made in ateliers in Doganella, strongly suggesting that the commercial export of wine was organized at this urban center. Perhaps wine was transported to the city from the valley's hinterland in skin *cullei*, then bottled in Doganella for transshipment.[87]

Terra cotta wine vessels, including dolia and the amphorae in which wine was typically bottled and shipped, were made waterproof by having their inner surfaces coated with pitch, the process known as *picatio*. Strictly speaking, clay bodies with the right amount of 'grog' can be made largely impermeable. For example, Staffordshire pottery designed for potting butter for commercial sale

85 E.g., at the Villa Regina. Cf. de Caro (1994): 224 and Tav. 7 and 8a.

86 Brun in Tchernia and Brun (1999): 76 and Fig. 94.

87 Daniele Vitali, Fanette Laubenheimer and Laurence Benquet, "La produzione e il commercio del vino nell'Etruria romana: Le fornaci di Albinia (Orbetello, Gr)," in Ciacci, Rendini and Zifferero (2007): 191-200.

in London was made of a coarseware, unglazed but impervious clay to prevent dealers from increasing the pot's weight by soaking it in water before sale.[88] Something of the sort seems to have been operant in the ancient Greek world as well. Researchers found a correlation between the permeability of three styles of Corinthian amphorae and whether they were hand-formed (impermeable), wheel-thrown (permeable), or a combination of both (semi-permeable). The grog of the hand-formed vessels was adjusted with the addition of potash flux and a large proportion of coarse 'temper' and by manipulation of the firing cycle by firing rapidly at lower temperatures in a reductive environment. Authors of the study attribute differences to differences in the products the vessels were designed to transport, as well as ease and cost effectiveness of manufacture.[89]

Dolia are far too large to be exclusively or even largely wheel-thrown, so it is theoretically possible to make one impermeable, but that does not seem to have been the case in the Roman world, based on current evidence. Dolia, seria and amphorae for wine were all pitched. Pitch is made by heating various tree resins, especially of evergreens, in a reductive environment, above 350°C (660°F) to produce a substance which is solid at normal temperatures and must therefore be heated to be applied. Equally important, heating resins to such high temperatures removes volatile constituents. It is these volatiles which are bacteriostatic and give to resinated wine (e.g., Greek *retsina*) its characteristic resinous flavor. Thus pitch is an excellent waterproofing substance, but it is not strongly antiseptic nor will it impart a great deal of flavor to a liquid, even one with an alcohol component.[90] Resins, as we have seen, have been used in wine for bacteriostatic purposes and perhaps also for flavor as far back as we can trace wine. But it is critical to make a distinction between pitch and resin in this regard. Pitch and resins are radically different products, both physically and chemically. We will have more to say about the use of resins as bacteriostatics later.

As to the efficacy of the procedure, molecular analysis of pitch from the linings of three vessels (a small pitcher and two Rhodian amphorae) from the

88 Peter Brears, "Pots for Potting: English Pottery and Its Role in Food Preservation in the Post-Medieval Period," in C. Anne Wilson, ed., *Waste Not, Want Not* (Edinburgh, 1991): 51.

89 P. Vandiver and C. G. Koehler, "Structure, Processing, Properties and Styles of Corinthian Amphoras," in W. D. Kingery, ed., *Ceramics and Civilization, vol. 2* (Columbus, OH, 1986): 173-215.

90 Curt W. Beck, Douglas R. Stewart and Edith C. Stout, "Appendix D: Analysis of Naval Stores from the Late-Roman Ship," in Anna Marurite McCann and Joann Freed, edd., *Deep Water Archaeology: A Late-Roman Ship from Carthage and the Ancient Trade Route Near Skerki Bank Off Northwest Sicily* [= JRA *Suppl. Series* 13] (Ann Arbor, MI, 1994): 122-6.

Hellenistic Kyrenia wreck off Cyprus showed that all the linings derived from pine pitch, pitch which had been made made from resinous wood as opposed to raw resin alone, in accordance with Pliny's (*NH* 16.21-2) recommendation. These pitch linings were so 'hydrophobic' that their esters have survived 2,400 years of immersion in the sea. The presence of several alkanes in the pitch cannot be explained by the transformation of the original resin compounds during the heat of manufacture or by degradation, and so must be deliberate additives. Incredibly, these alkanes derive from petroleum, and thus a petroleum product was added to pitches in Hellenistic times, at least in this instance.[91]

There were numerous sources of commercial pitch available to the Roman vintager. The abundance and quality of pitch from the forests of Bruttium (Abruzzo) and Lucania (Basilicata) were famous (Pliny, *NH* 14.127), especially that from the Sila Forest and from the hinterlands between Lokroi (Locri) and Rhegium (Reggio Calabria) (Strabo 6.1.9). The Locrian archives mention the sale of pitch along with that of pithoi and specify the use of this pitch in *kadoi* (amphorae) (*Locrian Archive,* Table 15). Commercial *pix bruttia* in its shipping containers has been found at Pompeii and in Calabria in the plain of Lamezia and at Trebisacce.[92] Strabo (5.2) also speaks of excellent pitchworks in Aemilia.

A contract with an Egyptian potter specifies that he is to provide amphorae which are well fired and carefully pitched, suggesting that, here at least, amphorae could be bought previously pitched.[93] The Roman agronomists, on the other hand, suggest that this is the work of farm hands on slow days at their villa farm. Cato (*Agr.*) says that scrubbing and pitching wine dolia is rainy-day work, and Columella (*RR* 11.2.70) advises that a month before the vintage all the vintage equipment should be cleaned and some dolia should be repitched and others scoured with brine or sea water and allowed to dry for at least 15 days. This seems to suggest that dolia were repitched only when old pitch had become brittle and cracked. Otherwise the dolia were simply scrubbed and disinfected with salt water. In the same passage Columella says that the *forus* is to be pitched if necessary, presumably if it is not made with a waterproof material such as tile or *opus signinum,* Roman waterproof concrete. The procedure for pitching the dolia is not described by the authors but is illustrated on a

91 C. W. Beck and C. Borromeo, "Ancient Pine Pitch: Technological Perspectives from a Hellensitic Wreck," in W. R. Biers and P. E. McGovern, edd., *Organic Contents of Ancient Vessels: Materials Analysis and Archaeological Investigation* [= MASCA *Research Papers in Science and Archaeology* 7] (Philadelphia, 1990): 51-8.

92 S. de Caro, "Anfore per pece del Bruzio," *Klearchos* 105-08 (1985): 21-32; Vandermersch (1994): 46-7.

93 Brun (2004a): 148, citing *Pap. Egypt. Expl. Soc. 364889/J* of 243 CE.

FIGURE 29 *Pitching of wine vessels. From the Seasons Mosaic at Vienne. After Billiard (1913): Fig. 149.*

mosaic found at Saint-Romain-en-Gal and now in the Louvre;[94] one farm laborer tends to a pot of heated pitch over a small fire while another has placed a *seria* on its side and is using a long-handled mop to swab fresh pitch on the interior walls of the vessel. Removing the old pitch would necessitate heating it as well and, unless we are to imagine these huge vessels being heated over large fires, we must imagine workers small enough to crawl inside them with torches or lamps to heat small areas before scraping off the softened pitch. In the case of the *dolia defossa,* I think that we must imagine that the vessels remained *in situ* during cleaning and repitching, and this seems to be the

94 Billiard (1913): Fig. 166.

import of Pliny's warning (*NH* 23.63) to workers descending into dolia for the work to exercise extreme caution to avoid asphyxiation. Pliny suggests that these workers lower a lamp into the vessel first; if the flame is extinguished, the danger is great.[95] In the case of smaller vessels such as amphorae, liquefied pitch will have been poured in and the vessel rotated in order to entirely coat the walls.

There is also a suggestion that the ancients fumigated their fermentation vessels with sulfur matches, as was standard practice afterwards.[96] Unfortunately I have not seen the attestation for the procedure. Certainly the ancients from of old were well aware of the power of burning sulfur to 'purify' a location. There is, for example, the famous scene in the *Odyssey* (22.450-2) in which Odysseus has his housemaids purify his court of the blood-taint of the slaughter of the suitors with strong brine and burning sulfur. But there does not seem to be attestation of sulfur matches used in either Greek or Roman viniculture.

The loss of one of these giant vessels would represent a significant monetary loss, and the thrifty Cato advises that cracked dolia be repaired during slack periods (*Agr.* 39). Cato prescribes a 'mending putty' (*medicamentum*) composed of 1 pound wax, 1 pound resin, 2/3 pound sulfur. These are well mixed and then enough gypsum is added to achieve the consistency of plaster and then smeared over the crack. Afterwards a plaster of 2 parts chalk and 1 part lime is smeared over the repair to make the color uniform. The mend may be reinforced and protected with hoops of lead or oak or with loops of vine shoots.

As viniculture moved out of the Mediterranean and to more northerly climates, wooden barrels (*cupae* and *orcae*) came into use.[97] Ancient barrels were much like our own, with wooden staves (*laminae, tabulae*) retained by metal or wooden hoops. Unfortunately, some of the Roman authors use the term *dolium* when they clearly refer to barrels, as for example, Pliny's (*NH* 18) advice to repair 'dolia' in winter by scraping staves and making new ones. Earliest analytical evidence of contact of wine with a wooden vessel comes from Middle Minoan Monastiraki (ca. 1900-1700 BCE), but we don't know whether the must

95 For a modern recreation of the procedure, Brun in Tchernia and Brun (1999): 106-07.

96 Singleton in Solomon, Fleming and Katz (2000): 76, citing S. M. Hammond and J. G. Carr, "The Antimicrobial Activity of SO_2 with Particular Reference to fermented and Non-Fermented Fruit Juices," *Society for Applied Bacteriology Symposium Series* 5 (1976): 89-110.

97 Billiard (1913): 472; Brun in Tchernia and Brun (1999): 58-9; Étienne and Mayet (2000): 23-4; Élise Marlière, *L'outre et le tonneau dans l'Occident romain* (Montagnac, 2002); *idem*, "Outres et tonneaux," in Brun, Poux and Tchernia (2009): 292-303; Margalit (2004): 140-50; Piccoli (2004): 68-76; Brun (2005): 10-12 and 66-8; McGovern in Tzedakis et al (2008): 201-3. For modern cooperage, Margalit (2004): 127-40.

was trodden in a wooden vessel or the wine was stored in oak, perhaps even for the express purpose for imparting to the wine the lactones which give many modern wines their 'oaky' flavor. The earliest literary evidence for wooden tuns may come from Herodotus (1.194) who describes Armenians bringing wine down the Tigris and Euphrates Rivers to Babylon in the *bikos phoinikeiou*, perhaps 'Phoenician wine jar' but more likely 'palm wood vessels'.[98] In the Roman ambit Strabo (5.2), in speaking of Gallia Cisalpina in what is now Emilia-Romagna, mentions wooden 'jars' (*pithoi*) larger than houses. In Gallia Transalpina the use of wooden vessels began earlier than previously thought, in the Augustan period in the middle valley of the Rhone, if not before. As we have noted before, this may well be the reason for the precipitous decline in the local production of wine amphorae here at this time. Tchernia [99] adds that the impetus may have come from the Roman army itself, since wine was offloaded from river craft at Lyon for transshipment to the Rhine frontier. In any case, by the second century the barrel had become the standard container in Gaul. Thus we have a dedication of the *cupari Vocronenses* at Rochemaure (Ardèche), a combination hatchet and hammer of a cooper at Sainte-Colombe (Rhône), and a bas-relief from Colonzelle (Drôme) representing a scene of a barge laden with barrels. In fact, north of Tricastin there is no attestation of vinification in dolia nor of bottling in amphorae, and the famous *vinum allobrogicum* was vinified and wholesaled in barrels. By the beginning of the third century, wood is making inroads into the Mediterranean parts of Gaul. This is the period when Palladius is recommending the 'mixed' cellaring system previously discussed. At Prés-Bas (Auvergne) there is evidence that dolia were taken up and replaced with barrels, and at Saint-Martin de Taradeau (Var), the truncated walls at one end of the cella vinaria may be to accommodate the new vessels. At Gramière (Gard) we have a beautiful example of a long, rectangular *cella* designed for barriques.[100] The barrels were held securely in place between wooden balks, probably Palladius' *podia* (1.18).[101]

In speaking of Spanish wine Varro mentions *orcae* (Agr. 1.13.6), by which he apparently designates the ancient equivalent of the barrique, since he equates them with dolia and designates them as fermentation vessels. Given the scant evidence it is perhaps reckless to suggest that *orcae* and *cupae* are vessels of

98 McGovern et al. in Tzedakis (2008): 201-3.

99 Tchernia in Brun, Poux and Tchernia (2009): 10.

100 Brun (2005): 66-8.

101 Brun in Tchernia and Brun (1999): 58-9.

different size. On the other hand, iconographic evidence shows transport vessels of markedly different sizes, so there is at least the possibility.[102]

Barrels were made of various evergreen woods such as white fir (*Abies concolor*), larch (*Larix decidua*), and Norway spruce (*Picea abies*), but oak is an ideal wood for the purpose. Wine aged in raw oak will have had two advantages: slow oxidation to soften tannins by slow polymerization and to enhance bouquet; and the addition of oak phenols to add 'oakiness' to the flavor and aroma[103] Strabo's passage cited above (5.1.12) has been interpreted as indicating that wooden barrels were pitched just as dolia and amphorae, but Strabo uses the word *pithoi* when he is obviously referring to these butts, and when he says that the local pitch is excellent for smearing the 'jars' (*pithoi* again), he may mean precisely that. In any case, it seems likely that the barrels will have been pitched only between the staves, where leakage occurs, so perhaps we can still salvage some oakiness in Gallic wines.

Fermentation

Fermentation is the process by which must becomes wine. Fermentation is defined as an enzyme-catalyzed reaction in which a substrate, usually sugars, is broken down into metabolites. The primary, and often the exclusive, form of fermentation in wine is alcoholic, by which sugars in must are catalyzed to alcohol through the agency of yeasts. Today pressed must is left in a bulk container at ca. 5°C (50°F) for at least 24 hours to precipitate solids before fermentation commences. Fermentable sugars are typically measured as degrees Brix, and musts from 14° to 23° Brix represent potential alcohol of some 7-12%. Before their resting period musts are treated with a form of sulfur dioxide, which kills almost all microorganisms and then volatilizes, and then are inoculated with a cultured yeast strain. Fermentation ideally proceeds at 13°-16°C (55°-60°F). The primary and often exclusive yeast in modern winemaking is a cultivar of *Saccahromyces cerevisiae ellipsoedius,* commonly known as beer yeast. This species is a facultative anaerobe, that it, it can metabolize sugars with or without the presence of oxygen. It is also a vigorous reproducer which typically 'overgrows' other species.

A typical wine fermentation will proceed in two phases, a primary, aerobic phase in which sucrose is broken down into glucose and fructose and the latter

102 Ètienne and Mayet (2004): Fig. 2, a bas-relief from Augsburg and a section of Trajan's column.

103 Margalit (2004): 127, 140-50.

into water and carbon dioxide gas. Simultaneously, yeast cells will reproduce exponentially by a process known as budding. During the secondary, anaerobic phase, sugars are metabolized to alcohol and carbon dioxide. A good wine yeast will ferment vigorously and rapidly to overgrow and inhibit spoilage organisms, will have a high tolerance for its own 'waste product', alcohol, until its concentration inhibits or kills it, and then will rapidly flocculate, precipitate, and form a firm yeast sediment, the lees or dregs.[104] As fermentation slows to a crawl,[105] wine is racked (decanted) from its 'gross lees', composed of grape solids and spent yeast cells. Today fermentation is sometimes stopped chemically or by microfiltration of yeast so that the wine will retain residual sweetness. Vessels in which secondary fermentation occurs are typically hermetically sealed to exclude oxygen and equipped with an airlock, a simple device that allows CO_2 to bubble through a liquid medium such as water but does not allow oxygen (air) to invade the fermentation vessel; spoilage organisms are aerobic. Primary fermentation lasts from 3 days to 3 weeks, depending on ambient temperatures and yeast vigor, whereas secondary fermentation takes from 2 days to 2 weeks. After secondary fermentation is complete, the wine is typically racked again (unless the wine is designated for a *sur lies* fermentation and aging) either to stainless steel aging vessels or to oak casks, both completely filled to minimize 'ullage', air at the top of the vessel, and then sealed. It is at this point that wines may be blended, modified, and/or 'fined' to precipitate haze particles and potassium tartrate ions in solution which will eventually crystalize as potassium bitartrate, cream of tartar.

There are some 150 species of yeasts active in winemaking, most subspecies of *S. cerevisiae* and *S. bayanus*, all classified according to alcohol tolerance, temperature and SO_2 tolerance, generation of byproducts, ability to ferment to complete dryness, and flocculation capacity.[106] Ancient fermentation relied on ambient yeasts on the skins of grape, the 'bloom', as well as those in the air and on winery equipment, and comparative evidence suggests that the primary species operant besides *S. cerevisiae* were *S. apiculata, S. beticus,* and *Brettanomyces, Kloeckera* and *Pichia* species. But we now have analytical evidence that *S. cerevisiae* was, in fact, the typical ancient wine yeast. DNA analysis of

104 Vine et al (2002): 211-29; Margalit (2004): 57-61; Bird (2010): 71-80.

105 N.B.: The term 'primary fermentation' refers in this context to the primary phase of alcoholic fermentation, as does 'secondary fermentation', but may also refer to alcoholic fermentation as opposed to 'a secondary fermentation' typically a later, malolactic fermentation.

106 Kenneth C. Fugelsang, *Wine Microbiology* (New York, 1997): 68-115; Margalit (2004): 54-6; Goode (2005): 129-35; Vine et al (2002): 13-19.

residue from a wine jar in Tomb U-j of the pre dynastic Egyptian king Scorpion I at Abydos, a jar which derived from the southern Levant, revealed materials compatible with a precursor of *S. cerevisiae.* Since this material derives from ca. 3100 BCE, the technique promises exciting new prospects for determining the active agent or agents in any number of fermentations, alcoholic and otherwise.[107]

Special mention should also be given to a distinct class of yeasts still prominent in the fermentation of fine sherries, namely, so-called 'flor' yeasts. The Romans called all yeasts colonies blooming on the surface of a fermenting wine *flores*, but they clearly recognized there was one particular 'flor' which conferred wonderful taste, bouquet and longevity on some lucky wines. Flor is a film of yeast cells which covers the surface of a fermenting must, in most cases some two months after the vintage, but not uncommonly in April or May. There are four flor yeasts, the traditional names of which are *Saccharomyces beticus, S. montuliensis, S. cheresiensis,* and *S. rouxii. S. cheresiensis* and *S. beticus* are now considered anamorphs (an anamorph is an asexual reproductive stage) of *S. cerevisiae, S. montuliensis* is now classified as *Torulaspora delbrueckii,* and *S. rouxii* relates to *Zygosaccharomyces rouxii.*[108] All four yeasts have common characteristics including resistance to high alcohol levels and acetaldehyde, but their most important common feature is their lipid content which allows them to float on the surface of wine to create the characteristic *velum* (L. 'veil' = Sp. *velo*). Flor develops when four conditions are present: 1) all sugars have been used up, 2) oxygen is present, 3) temperature is between 15-21° C (59-70° F), and 4) sulfur concentration is less than .018% and tannin less than .01%. Sulfur concentrations will not have been an issue for the ancients. Spring flowering is triggered by proper temperatures during that period. The veil remains constant in thickness at a given time but dead cells constantly fall to the bottom of the fermentation vessel and are replaced by new ones. The veil grows noticeably thinner during hottest and coldest months and the color changes from white to grey. Vessels that are not blended generally breed flor for 6 to 8 years but have been known to do so for as many as 15. When nutrients in the wine are exhausted and alcohol levels grow too high, the flor darkens, dies and precipitates. Thus the practice in Jerez of drawing off portions of old wine to be replaced by fresh nutrient in the form of new wine, the *solera* system. Most importantly, flor is the deadly enemy of vinegar bacteria since it overgrows the weaker vinegar cells which also must grow on the

---

107 McGovern (2003): 103-06; Duccio Cavalieri, Patrick E. McGovern, Daniel C. Hartl, Robert Mortimer and Mario Polsinelli, "Evidence for *S. cerevisiae* Fermentation in Ancient Wine," *Journal of Molecular Evolution* 57 (2003): S226-S232.

108 Jeffs (1992): 182-7; Hornsey (2007): 332-46.

surface of the wine, and metabolizes the oxygen upon which vinegar metabolysis depends. A properly nourished flor will easily maintain the health of a wine for decades.

Flor yeasts produce a flavor compound called sotolon, a powerful eugenol, 4,5-dimethyl-3-hydroxy-2 (5H)-furanone, described as "a complex fragrance with touches of walnuts, curry, toasted fenugreek seeds, caramel, and maple syrup."[109] This complex combines with Maillard reactions in which sugars and amino acids under rising heat create colored, aromatic substances called melanoids which are described as caramel, vanillin and cocoa. When we consider that Roman authors recommend conferring to cheaper wines the artifical age of a vintage wine by adding fenugreek, itself high in sotolon, and by heating the wines in a *cella superior* (Maillard reactions), it is almost certain that Roman vintage wines had something of the flavor of a modern, flor-seasoned sherry. And when we consider that the Jerez region also still practices some of the traditional Roman viticultural practices, who can doubt that Columella, who of course derived from this area, would be proud of his country's traditions? And those of us who cherish the unique flavor and bouquet of this lovingly crafted wine style can perhaps derive more enjoyment in knowing that connoisseurs in various parts of the world have also cherished them for well over 2,000 years.

Vigorous fermentation of *S. cerevisiae* requires basic yeast nutrients in their substrate and optimal temperatures in the environment. Primary yeast nutrients are nitrogen, phosphorous and sulfur. Wine yeast has an optimum temperature range of 8°-14°C (46°-57°F) for white wines, 22°-30°C (72°-86°F) for reds. Above 35°C (95°F) the yeast dies, and below 15°C (59°F) it works very slowly or becomes dormant. But avoiding radical shifts in temperature appears to be more important than exact temperatures, at least within these parameters. Temperatures above 30°C (85°F) cause wine to lose bouquet and allow vinegar bacteria to thrive in the presence of oxygen. Excessive heat is still the bane of the vintner, especially since fermentation itself produces large amounts of heat. We are reminded that the term fermentation derives from the Latin for 'boiling'.[110] Today proper temperature control is considered so important that more and more wines are fermented in temperature-controlled vessels. Our ancient Roman vintner had fewer options, of course, but he did have two advantages. First, dolia and the *cupae* that replaced them were relatively small

---

109 François Chatier, *Taste Buds and Molecules: The Art and Science of Food, Wine and Flavor* (Hoboken, NJ, 2009): 82.

110 For the connection of 'ferment' with cooking in the classical world, Obbone Lonzo, "Techiche di vinificazione nel mondo greco-romano," *Quaderni di Storia* 29 (2003): 197-200.

vessels, and smaller vessels equate to greater heat transfer relative to volume. Second, dolia were typically embedded in soil, promoting even greater heat transfer as well as far greater thermal stability. Indeed Rossiter[111] thinks the primary function of embedded jars was to stabilize temperature, though he asserts that they insulated against heat loss during fermentation. Far more likely, such embedded dolia dispersed heat during fermentation and provided ideal winter temperatures for aging.

Since Roman fermentation was almost exclusively spontaneous, fermentation was not really a distinct step in the winemaking process, as is modern cuvage.[112] In effect, fermentation proceeded from the moment the harvest began and grapeskins were broken and ambient yeasts gained access to sugars. Relatively warm temperatures during transport to the winery in open containers will have incited the rapid multiplication of yeast cells, and by the time treading began, will have provided a vigorous inoculum for the must. Fermentation continued during treading, multiple pressings, settling and maceration of red wines, filtration, and transport to cellars, as well as in the fermentation vessels themselves. For a typical dolium, we should imagine that the vessel was filled only to some 3/4 capacity, and the foam of vigorous primary fermentation as well as gross lees brought to the surface were skimmed twice a day (Cato, *Agr*. 26). The duration of Roman fermentation varied widely, as today, but Pliny (*NH* 14.124) estimates an average of 9 days and Cato (*Agr*. 26) suggests a maximum of 30 days, both estimates perfectly congruent with modern experience. Elsewhere (*Agr*. 152) Cato has prescribed the making of elm-twig whisks to brush the inner sides of dolia so that lees do not adhere to them, excellent advice since they will have provided an ideal environment for spoilage organisms. Varro (1.65) cautions that once fermentation has proceeded to the dolia, the must should not be racked (*promere*) until fermentation is complete.

Enrichment

Two potential disasters can occur at this point. If grape sugars or ambient yeasts are deficient, a weak fermentation can allow vinegar bacteria to flourish

111 Rossiter (1981): 359, n. 39.

112 Billiard (1913): 459-62; Goode (2005): 130-3. Goode remarks that it has been estimated that in the late 1990s some 80% of fermentation worldwide was still spontaneous, though the percentage was changing rapidly. It has been interesting to note a small but significant trend to return to spontaneous ferments on grounds that some of the 'wild' yeasts, including the apiculate yeasts, give wine a more complex flavor.

in this oxygen-rich environment. Achieving sufficient alcohol levels to ensure biological stability in a wine is a function of sufficient fermentable sugar in the must as well as of yeast vigor. Obtaining musts with sufficient fermentable is not likely to have been a problem often in the warm Mediterranean milieu, particularly when grapes were apparently harvested late by modern standards. But it is clear from literary evidence that this was a problem on occasion even here, and perhaps more so in the northern reaches of the Empire. Today's wine maker knows precisely how much fermentable sugar his grapes will yield. He may use a refractometer, or he may simply check the specific gravity of a sample of grape juice; because sugar is much denser than water or ethanol, the specific gravity of a must is an excellent indicator of total fermentable available. Though Archimedes had long since discovered the principles of specific gravity, it would be several centuries before the invention of the precursor to the modern hydrometer. It was Synesios, sixth-century CE commentator on Democritus (*Epist.*15), who first mentioned the *hydroscopium*, a true hydrometer composed of a cylindrical vessel, graduated and weighted at the bottom with a small, conical weight. Apparently the instrument was used exactly as the modern hydrometer and was capable of determining minute differences in density. The fact that none of the agronomists mentioned this instrument suggests that it was used only rarely, if at all, by ancient winemakers. In the absence of such an instrument, they doubtless resorted to empirical evidence from tasting the ripe fruit and/or the must to estimate fermentability of must. Even today winemakers rely heavily on taste to complement technical evidence in this regard.

Today deficiencies in total sugars in musts are corrected by the addition of sucrose, a process, still highly controversial in some quarters, formerly known as chaptalization, now called enrichment.[113] Since yeast easily converts sucrose to glucose and fructose, this does not appreciably affect the taste or quality of the final outcome. Cane sugar was only just known to the ancient Romans as an exotic from Persia, and was far too precious for this use, as was honey, an

113 Boothroyd (1986): 50-52; Tchernia in Tchernia and Brun (1999): 114; Halliday and Johnson (2007): 199; Bird (2010): 65-6. The name derives from Napolean's minister of agriculture Jean-Antoine Chaptal, who formally approved the procedure in 1801. Increased alcohol obviously has a bacteriostatic effect, but it also contributes 'apparent' sweetness in the form of a more glycerine 'mouth feel' and therefore, used judicously, improves wine quality. Chaptal recommended sucrose from beets, which acids in must quickly invert to glucose and fructose, but EU regulations now dictate the use of concentrated grape musts and thus the oldest method of adjusting wine sugars is once again *de rigueur*. There is some evidence that wines enriched in this way were considered inferior in antiquity: *CIL* IV.5588, a Surrentinum with the titulus: Ab Clodio Clemente de superiore sine defrito.

expensive commodity in ancient Rome. Thus, grape concentrates, *defrutum* or *sapa*, were universally used for the purpose and were ideal from a technical standpoint.[114] Hence Cato's (*Agr*. 23.2) recommendation: "If necessary put *defrutum* boiled down from free-run must (*mustum lixivium*) into the must; add a fortieth part of the concentrated juice per *culleus*." Elsewhere (*Agr* 24) he recommends 1/30 part of well aged *defrutum* and coins a term for the process, *defrutare*. He cautions the vilicus that only the best *defrutum* be used lest it ruin the very wine it is meant to ameliorate. Elsewhere we find the prescription for wine too low in fermentables because of the dilution of too much rain; it should be corrected by the addition of 'cooked wine', *vinum decoctum* (Palladius 11.14; cf. Columella, *RR* 12.19). Since wine per se will have had little if any residual sugar in it, our sources apparently refer to cooked musts, that is, grape concentrates, by this same term. According to Columella, when must to be enriched has been removed from the press vat it is cooled and clarified for two days and *defrutum* is added on the third day (*RR* 12.21.2) at the ratio of one sextarius per amphora of wine (*RR* 12.20.1). Boiled musts were an important food product in their own right. Unfermented or lightly fermented musts were taken to the kitchen or to a separate facility called the *defrutarium* and boiled in leaden vessels to concentrate their sugars, the use of lead being prescribed because they did not leave a 'taint' in the must as did the more typical bronze. But when acidic products are heated in lead, significant amounts of lead are leached out, and the dangers of the procedure were recognized; Dioscorides (40-80 CE) warned of the dangers of sweetening wines with grape concentrates, describing symptoms that we clearly recognize today as lead poisoning (*de Mat Med* 5.6-7).[115] There may have been another consideration as well; Hugh Johnson[116] asserts that the Romans had discovered that lead gave wine a sweet taste and succulent mouth feel. Lead also will have had a preservative effect because lead ions have a "drastically inhibitory effect on enzyme growth, and hence on all organisms." His analogy is the use of lead oxide, litharge, to sweeten wines in seventeenth-century France and Germany.

There are two main considerations for the wine maker here, neither, perhaps not surprisingly, mentioned by ancient sources: first, sugar in high concentrations will itself inhibit yeast reproduction and metabolism and so in making wines of higher alcohol concentration must today be added in stages; second, sugar's high density causes it when poured into a fermenting must to

114 Thurmond (2006): 132-4.

115 J. R. Riddle, *Dioscorides on Pharmacy and Medicine* (Austin, TX, 1985): 144, citing Jerome O. Nriagu, *Lead and Lead Poisoning in Antiquity* (New York, 1983): 350.

116 Johnson (2004): 156.

sink quickly to the bottom of the solution; thus the mixture must be vigorously stirred.[117]

Did the ancient vintner have recourse when yeast vigor was deficient? There is suggestive, if inconclusive, evidence that he did. Columella (*RR* 12.26) says that on estates where wine habitually sours, raw must should be taken from the treading before pressing, 1/10 part water added to it, and the mixture boiled until reduced by the same 1/10 part. But, of course, this will have created a sterile environment, and unless an inoculum is added, the same vinegar bacteria will recolonize. But one of the additives prescribed for defective ferments is wine lees from a vigorous fermentation, and such lees contain large numbers of living but dormant yeast cells. Boiling musts in this way will have stripped it of much of its bouquet and varietal character, but at least the procedure has a chance of producing a stable, potable product.

The second potential disaster occurs when a fermentation simply stops in the middle of the process, a so-called 'stuck' fermentation.[118] Reasons are lack of oxygen during yeasts' growth phase, lack or depletion of yeast nutrients, especially nitrogen, temperatures either too high or too low, excessively high sugar concentrations, accumulation of fatty acids, as well as the unviable yeasts we have already addressed. Short of racking the new wine to aerate it and inoculating with wine lees, our ancient vintner was probably at a loss.

Today airlocks are typically used to exclude oxygen from the developing wine. The Egyptians had invented a simple airlock by sealing the necks of their fermentation vessels with clay and using a straw to punch a tiny hole in it, big enough to allow CO_2 to escape but small enough to exclude air currents which would disturb the CO_2 'blanket' floating above the wine.[119] Something of the sort has also been found in the Agora excavations in Athens, in this case on amphorae.[120] We have no direct evidence of the use of such locks by the Romans, but they may have found another solution. After all visible evidence of fermentation had ceased and no more gas bubbles rose to the surface, bulk wine designated for quick consumption remained on its lees, wine to be aged was racked off its lees to clean dolia, and in both cases Roman dolia received their *opercula*, circular terra cotta covers. Judging by Varro's citation of

117 Boothroyd (1986): 51-52.

118 Fugelsang (1997): 137-41; Margalit (2004): 61-5.

119 James in McGovern, Fleming and Katz (1996): 207. In later times Champenois winemakers often used a cornstalk (maize) as a crude lock. The porous pith of the stalk allowed carbon dioxide to escape without allowing oxygen or impurities to enter. Cf. P Forbes (1967): 301.

120 Koehler (1996): 329.

Fundanius, it is obvious that the vessels were often sealed prematurely, and one wonders how many burst vessels were required until the practical Roman vintner learned not to seal the *opercula* with gypsum or plaster, as was typically done, for several weeks. More typically during this phase the dolia are covered with *opercula*, pitched in the same fashion as the dolia, but not sealed, so as to allow excess carbon dioxide to be emitted (Macrobius, *Sat.* 7.12.15).[121] Even the smallest fissures between the neck rims of the dolia and the *opercula* would provide ample space for gases under pressure to escape, although there may have been some residual 'sparkle' in the wine, what Italian winemakers call a *vino frizzante*.

At the end of primary fermentation bulk wine, *vin ordinnaire,* has been allowed to remain on its lees, designed as it is for daily consumption and for sale to taverns. Wines to be aged have been racked, that is, physically transferred, to clean vessels after the primary ferment and perhaps again before being sealed or modified in one of the ways to be discussed later. The ancient winemaker knew perfectly well that wines allowed to remain on their lees (French *sur lie*) are more prone to infection. Today, though some varietals are sometimes aged *sur lie* to increase flavor extraction, it is understood that lees provide an ideal growth medium, primarily in the form of amino acids from autolyzing dead yeast cells, for any number of spoilage organisms.[122] In either case, after all apparent evidence of fermentation has subsided and no more bubbles rise to the surface, the wine may be racked and sealed.

The huge vats are covered with the same terracotta covers, but this time they are sealed (*oblinere*) with the same clay or plaster attested for sealing wine stored in amphorae (Persius 4.5.29; Columella, *RR* 12.39.2). After the wine is sealed there will be far less work for the cellarman. Columella explains that after the sealing of the vessels until the spring equinox it will suffice for him to examine the wine every thirty-six days, though wines aged at higher temperatures will require more frequent monitoring. Each time the jars' seals are broken, he advises, the mouths of these vessels are to be scoured with pine cones (*RR* 12.30)

In these bulk vessels in the cella vinaria, the wine is allowed to mature and age. Maturation is defined as a bulk storage period, aging as storage in smaller containers, today, of course, the glass bottle. Maturation is all about oxidative reactions, including those with wood in many cases, whereas aging does not involve oxygen or wood extractives. Generally only those wines with a high phenolic and/or acid content will improve dramatically with age. Maturation

121 Cf. White (1975): 179-80.

122 Vine et al. (2002): 213.

and aging both involve subtraction of undesirables such as CO_2 and excessive or excessively harsh tannins, addition of wood extractives or oxidative flavors of flor wines, retention of desirable features such as grape character, and multiplication of the complexity of wine.[123]

Modification

It is at this point, when wine is being racked off its lees to be aged or to be sold in bulk, that the majority of modifications will have been made to the wine. Modification is simply the addition of various substances to wine to enhance its flavor, aroma and bouquet, color, or biological stability.[124] Billiard distinguishes between 'amelioration' which improves organoleptic qualities or facilitates the preservation of wine and 'adulteration', modifications designed to mask the deficiencies of wine.[125] A nice distinction, but one man's amelioration is another's adulteration. The ancients thought that a sound, unmodified wine was most healthful (Pliny *NH* 23.39), but many ancient modifications were harmless if ineffective, and most quite beneficial to the wine. And we should bear in mind that modern wines are subject to a great number of modifications to exactly the same ends and sometimes even using the same substances.

One such modification has already been discussed, namely enrichment (chaptalization) to increase the concentration of fermentable in a must and thereby ensure a reliable fermentation. But, of course, that modification by its nature must occur before or during fermentation. Perhaps the most common modification otherwise was the addition of salt to must or wine.[126] The practice was apparently borrowed by the Romans from the Greeks, witness the many references to salted wine as 'Coan' or 'Greek' wine (Columella *RR* 12.25; Pliny *NH* 14.120). The ancients recognized that salts promoted biological stability, suppleness, and clarity (Cato, *Agr.* 23; Col. *RR* 12.21; Pliny *NH* 14.120). It was thought particularly helpful in weak musts such as those from rain-diluted grapes, in conjunction with, or as a substitute for, enrichment. Columella prescribes it as a preventative of the wine disease the Romans called *mucor*,

123 Hornsey (2007): 293-308.

124 Margalit [(2004): 38-46] has a lucid discussion. For modern experiments with modification using Columella's formulae, Tchernia in Tchernia and Brun (1999): 113-17.

125 Billiard (1913): 489-91.

126 Billiard (1913): 500-02; Tchernia in Tchernia and Brun (1999): 114-15.

'ropiness', to be discussed later. Salt, incidentally, is still widely used to stabilize and clarify wines.[127]

Salt was added either as rock salt or as brine. For his 'Coan' wine, Cato (*Agr* 112) prescribes treated sea water, taken from the open sea 70 days before the vintage, left in a dolium with ample ullage for 30 days without being agitated, racked into another vessel, and racked yet again before use. Presumably the proportions used were the same as for his 'Greek wine' (*vinum Graecum*: *Agr* 24), for which dead ripe Apician grapes are used and 2 quadrantals (45 l. / 12 gal.) of the aged sea water are added per *culleus* of must or one modius (8.7 l / 8 qts) of rock salt per *culleus*. In the latter case the rock salt is to be placed in a frail which is suspended in the must until the salt dissolves. For standard wines Cato recommends (*Agr* 23.3-4) 1 1/2 pounds of rock salt per *culleus* of must, along with 1 pound of marble dust, and 3 pounds of resin. The vintner may also mix the salt with water to create a brine (Cato *Agr*. 105; Col. *RR* 12.25). Elaborate methods were prescribed to ensure that the water was pure, particularly if rainwater or spring water was used.

Columella's standard prescription calls for a heaping spoonful or a 1/2 oz (15 gr) measure of roasted, pounded white salt per urna (13 l / 3.4 gal) of must, as well as roasted, pounded fenugreek, 1/8 cyathus (5 ml / 2 oz) per urna. Then, after fermentation has subsided, gypsum is added in the same proportion as salt (*RR* 12.21.2-3). Columella adds that his uncle, the talented viniculturist, used brine instead, but boiled it down to half its original volume from sea water brought from far out at sea (*RR* 12.25). Nriagu[128] notes that Columella's advice to boil down brine in "in the same way as for boiled musts" implies that this is done in leaden cauldrons and cautions that "lead forms highly soluble complexes with the chloride ions. Thus, boiling down the solution in a leaden vessel would certainly result in a high-level contamination of the liquor with lead. The preservative action of the 'treated' brine solutions most likely can be attributed to the lead acquired from the boiling vessels."[129] Doubtless Nriagu is correct, if the Romans used leaden *cortinae* for boiling brines. But we should remember that the standard cauldron of the Romans was bronze, and boiling a liquid to concentrations in leaden vessels poses significant challenges.[130]

127 Longo (2002): 207.

128 Nriagu (1983): 343.

129 For the use of lead to sweeten and preserve wine, Estreicher (2006): 32. Use of lead continued in Europe well into the eighteenth century, despite increasing evidence of its toxicity.

130 Thurmond (2006): 251.

There is a good chance that Columella's prescription does not necessarily imply a leaden cauldron.

The Romans had recourse to standard treatments to correct acidity in wine. Acidity contributes significantly to both the taste and the stability of wine[131] Acid has an obvious preservative effect, but low acid wines also taste flat and vapid. Acids also contribute to bouquet, since they react with alcohol during fermentation to create volatile esters. Additionally, yeasts flourish best in an acidic environment. Grape acids are primarily tartaric and malic, with much less citric. Wines which undergo a malolactic fermentation also contain lactic acid in significant concentrations. Major adjustment to acidity of musts is best made during fermentation, today usually in the form of added tartaric acid, while fine adjustments are made immediately before bottling as tartaric, malic, and/or citric acids. For ancient deficiency in acidity, the standard, almost universal, procedure was the addition of calcium sulfate in the form of gypsum or plaster (*gypsatio*).[132] Plastering of wine is attested in the Mediterranean from the fourth century BCE, and was common in Rome. Columella (*RR* 12.20) recommends it as a substitute for enrichment if wood for boiling musts is scarce. Palladius (11.14) and Democritus (in *Geopon.* 7.4) recommend it especially for diluted musts, naturally weak from rainy-day harvests. Frontinus (in *Geopon.* 7.12) is struck by its ability to increase acidity. Quantities were carefully modulated, since overly plastered wines had a well deserved reputation as insalubrious (Pliny, *NH* 23.45). Columella recommends 1 ounce per amphora (28 gr / 26 l), but says that the proportions should be adjusted based on the quality of the must, and gives prescriptions for various scenarios. Palladius' prescriptions range from proportions of 1:50 to 1:100, otherwise 3 sextarii per 10 amphorae (1.4 l / 3 pt per 260 l / 70 gal).

The chemical action of plastering is simple, if unpredictable in its results: calcium sulfate is added to must, calcium tartrate (cream of tartar) precipitates, and sulfuric acid, which has a lower pH, replaces tartaric acid.[133] But acid and pH changes are not exact, and sulfuric acid also adds bitterness. About the use of the precursor of sulfuric acid, sulfur dioxide, among the Romans, we know painfully little. Sulfur dioxide in very small quantities is a natural byproduct of fermentation, but the health implications of higher concentrations are still highly controversial.[134] Apparently sulfur dioxide was used by Galen's father in the mid-second century CE as an antiseptic, but its use does not seem

131 Boothroyd (1986): 37-8; Halliday and Johnson (2007): 200-01.

132 Billiard (1913): 495-500; Tchernia in Tchernia and Brun (1999): 116-7.

133 Philip Jackisch, *Modern Winemaking* (Ithica, NY, 1985): 104.

134 Goode (2005): 120-1.

to have caught on; the ancients were well aware that sulfites had inimical effects of their own.[135] Finally, Pliny remarks (*NH* 14.129) that Cato had recommended the use of sulfur in wine, but the reference is not in the extant manuscripts, nor does Pliny explain Cato's procedure. Was this elemental sulfur, a sulfide or a sulfate? We simply cannot say.

We might mention in passing that plaster is still added to traditional sherry grapes before they are crushed, in proportions of 1 kg (2.2 lb) per 770 kg (1,540 lbs.) of grapes. The practice is thought to encourage the growth of the flor yeasts which give sherry its distinctive flor character.[136] This is yet another reason that the practice was efficacious for Roman vintage wines, since these were almost surely fermented by the same yeasts in some cases, as discussed earlier.

Added to wine to decrease acidity were marble dust or chalk, i.e., calcium carbonate (Cato *Agr* 23; Colum. *RR* 12.20; Pliny *NH* 14.120). Calcium carbonate is a strong base which neutralizes acids. Specifically, the compound precipitates acid ions and leaves a much weaker acid. Today excessive acid can be physically removed from musts and wines, but the chemical process is still used as well.[137] Cato recommends a pound of marble dust per *culleus*, to be added to raw must and stirred periodically.

Also used frequently were various aromatic substances to create *vina ficticia* (Pliny *NH* 14.98). By far the most commonly mentioned are tree resins and pitches made from them. There is little doubt that resins in particular were sometimes added to wine to modify flavor (e.g., Pliny *NH* 14.124). But, again, it must be emphasized that these were often used as preservatives, not to modify flavor, since our authors are at pains to recommend concentrations beneath the taste threshold. Aromatics do, in fact, have a marked antiseptic power.[138] Pitch (*pix*) and resin (*resina*) were obtained from a number of evergreens, including terebinth, lentisk, cypress, several species of pine, especially umbrella pine (*Pinus pinaster*) and Scots pine (*Pinus sylvestris*), cedar, spruce, larch, and mugho pine (*Pinus mugo*). Pitches and resins were imported from the Orient, from Asia Minor, the Levant, Greece, and Spain, and were also obtained from many parts of Italy, in particular from Bruttium and Liguria.

Resin is a common find on ancient shipwrecks, both in vessels for transport of wine and as a cargo in its own right, apparently for modification of wines. As cargo it was transported as oval blocks (e.g., Chretienne I shipwreck) and in

135 V. Nutton, "Medicine and the Roman Army," *Medical History* 13.3 (1969): 260-70.

136 Halliday and Johnson (2007): 188.

137 Jackisch (1985): 107-8.

138 Majno (1975): 207-27; Thurmond (2006): 264-7.

amphorae. Pitch was obtained almost exclusively from resins of conifers.[139] Koehler[140] is typical of those who think that the coatings of vessels will have given a resinous taste to wine, though she acknowledges that there is no consensus that pine pitch, as opposed to pine resin, is even soluble in wine. Wilkins and Hill[141] in the same vein quote Strabo's (4.6.2) notice of mountain dwellers of the Ligurian Alps who bring sheep, fleeces and honey to Genoa to trade for olive oil and 'Italian' wine, since they produce only resinated wine. "Strabo probably means that Ligurian wine was so flavored to a pronounced degree." More likely Strabo means exactly what he says. In the Ligurian Alps, the custom was to use high concentrations of resin to preserve weak wine, whereas the wines of central Italy did not have an artificially resinous taste. Thus we have Columella's (*RR* 12.22-4) rather elaborate method for producing a pitch to preserve wine (his sense is unambiguous) without imparting a resinous taste, in this case by leaching lye through it. Cato's procedure (*Agr* 113) is even more elaborate: a piece of earthenware or roof tile is covered with pitch, then covered with melilotus, rush, or palm branches, suspended in an empty dolium which is then sealed. The next day the fumigant is removed and the wine added. To reiterate, pitch used to coat interiors of wine vessels was highly inert.

Resins were typically introduced in the must during fermentation either raw, cooked, liquefied, powdered, or dissolved in *defrutum* (Cato *Agr.* 23; Colum *RR* 12.20; Pliny *NH* 14.125). Concentrations prescribed are 1 pound per *culleus* (1 kg per 524 l / 2.2 lb per 138 gal) or 10 sextarii per 90 amphorae of *defrutum* (in conjunction with enrichment) (5.5 l per 23.4 hl / 1.4 gal per 618 gal); or 6 scruples per 48 sextarii of must (6.75 gr per 26.2 l / 2.4 oz per 6.9 gal). Medical writers think resinated wine unhealthful, causing headaches and vertigo as well as the malady known as *crapula*, the word thus designating both the resin itself and the 'crapulous' hangover it was reputed to cause (Pliny, *NH* 14.46). Pitch seems to have been brayed and added in powdered form, and one form, used by the Allobroges, is called 'bark pitch' (*pix corticata*) and is especially good, according to Columella (*RR* 12.24) since it is easily brayed to a powder and added to wine after fermentation has subsided. After another 14 days the wine can be sealed. Doubtless the most exotic resin attested is myrrh, which has known bacteriostatic as well as analgesic effects.[142]

---

139 A. J. Parker, *Ancient Shipwrecks of the Mediterranean and Roman Provinces* (Oxford, 1992): 374; Jurisic (2000): 131.

140 Koehler in McGovern, Fleming and Katz (1996): 328.

141 John M. Wilkins and Shaun Hill, *Food in the Ancient World* (Oxford, 2006): 131.

142 McGovern (2003): 71.

Other aromatics prescribed by Columella (*RR* 12.20.3-6) are iris, the fenugreek previously discussed, leaf of spikenard, Illyrian sword lily, Gallic spikenard, costus, date, angular rush, sweetrush, sweet reed, cinnamon, balsam, saffron, and the unidentified *cripa*, to which may be added aloes, mastic, and pepper (Pliny, *NH* 14.25; Palladius 11.14). Used to smooth and mellow harsh wines were cinders of vineshoots or oak; pulverized shells; and a mixture of bitter vetch flour, *sapa* and wine, formed into lozenges and stored for use, then soaked in wine for 1 1/2 days and mixed with the wine. Used to remove off odors (acetaldehyde?) from wine was a variation of the pitched roof tile (Cato *Agr* 109-10). We have seen that wine lees were used to aid a weak fermentation, but they are also recommended after fermentation has ceased to mellow a harsh wine (Col. *RR* 12.30; Pliny *NH* 14.121; Palladius 11.14). Columella (*RR* 12.28) and Frontinus (in *Geopon.* 7.12) also recommend wine lees for 'defective' wines, though they do not specify the defect. Also not specified is the agent for artificially coloring wine, "so many poisons are employed to force wine to suit our tastes—and we are surprised that it is not wholesome! (Pliny, *NH* 14.130).," in the case of red wines, the leading suspect is elderberry (*Sambucus nigra*), used almost continuously in many parts of Europe to add depth of color to red wines and reputedly still used surreptitiously in some. Billiard is rightly astonished by Pliny's (*NH* 14.120) mention of the use of vinegar to 'preserve' (*condire*) wine, and wonders if resort to the commercial sale of *posca*, sour wine, had inured Roman wine drinkers to sour wines.[143]

An additive which was common in Mediterranean winemaking well into the twentieth century,[144] but for which I can find no attestation, is willow bark. Clearly the ancients knew of the powerful medicinal powers of the active ingredient in willow bark, salicylic acid (aspirin). In wine, salicylic acid inhibits bacterial growth but not yeast growth and thus can be added to a fermenting wine as well as to finished wine. In fact, Koch used salicylic acid to prove that bacteria and not yeasts were responsible for malolactic fermentation.[145]

A modification which clearly qualifies as adulteration is the dilution of wine with water, as practiced in Greece (Pliny *NH* 14.119). There were several prescriptions for detecting such wines, presumably at the point of sale. For example, throw into it a cricket or cicada; if it swims on the surface, the wine is pure, but if it sinks, the wine is diluted. Better, decant the wine into a new amphora, and if it acetifies, it is diluted (*Geopon.* 7.8). Here there may be some

143 Billiard (1913): 506-07.

144 E.g., in the southern Campanian region called the Cilento. (Professor Fernando La Greca, personal communication).

145 Hornsey (2007): 224.

truth; dilute wines are deficient in alcohol and are therefore conducive to acetification. But the process will take several days to weeks, and is therefore useless to the buyer. Another recommendation: make a cup of ivy wood, pour some of the suspect wine into it; if diluted, the wine will soak through the wood and the water will remain, since a cup of ivy wood will not hold wine (Cato, *Agr.* 111). Incredibly, this technique has now been experimentally proved effective by the Laboratorio di Ricerche at Pompeii, though the exact mechanisms are still not understood.[146]

Aging

Choice of Wines for Cellaring

Given the numerous infections to which ancient wine, in the absence of sulfur dioxide and microfiltration, must have been susceptible, most notably various acetic fermentations, the vast majority of Roman wines were probably consumed at the point of origin or sold in bulk to the *négociant* after minimal aging, to be consumed quickly by the urban masses. But the Romans were aware that certain grape cultivars were capable of producing powerful wines that were capable of sustaining age and might not, in fact, be potable (because of their powerful tannins) or at their peak for ten years or more. They also knew that such vintage wines were more valuable the longer they were kept before sale. Thus the vintner was faced with the decision to age or not to age wines. Varro (1.65) suggests a compromise, advising that all healthy wines should be cellared for at least a year before consumption, but that wines from grape cultivars prone to acetify must be consumed within the year. Cato (*Agr* 108) suggests a test for judging the aging potential of a wine: a small quantity of barley groats is put in a clean pan, a sextarius of wine poured over it, the mixture heated to boiling two or three times, the groats strained out and the wine set out until the next morning. If the wine tastes like the original, it will keep; but if it has a slight sourness, it will acetify. His advice, unfortunately, is completely useless. Boiling wine even once will have killed any spoilage organisms, nor is it possible that a sterile wine will acetify perceptibly overnight, under the influence of ambient organisms.

146 Maurizio Landi, "Così i Romani scoprivano il vino annacquato," *Storica* 20 (Oct. 2010): 14. Cf. Annamaria Ciarallo, "Le vite e il vino a Pompei," in Di Pasquale (2010): 190 and Figs. 7a-d.

Cellaring

The purpose of aging or cellaring (Fr. *cuvage*)[147] is to allow wine to mature by completely merging its elements. Cellaring must be done with minimal but not negligible exposure to oxygen, to prevent excessive conversion of alcohol to acetaldehyde (small concentrations are considered desirable in many red wines), with minimal exposure to light to maintain color, and at optimal temperatures between 10° and 15°C (50° and 60°F). Changes that occur are extremely complex and interrelated and depend in part on temperature, humidity, and other environmental factors. But, in general, oxidation, either from the oxygen of ullage or from that invading semipermeable closures such as corks for wines aged in bottle, softens and mellows flavors; the interactions (polymerization) of tannins and anthocyanins intensify colors of both red and white wines and soften the astringency of tannins, and esterification of many wine components develops the aroma of a mature wine. Cellaring of prestige wines goes back to the beginnings of historical records.

Filtration

Today wines are aged in bulk, then in bottle, and the same was true of vintage wines in ancient Rome. Modern wines are typically coarse filtered before being pumped into stainless steel or oak aging vessels.[148] They may be microfiltered as well before bottling. Filtration in this case refers to the transmission of wine through one or more more or less semipermeable materials to remove solids. Roman filtration (*liquatio*) was designed primarily to remove lees (*defaecare*: Pliny *NH* 14.138, 19.53, 23.45),[149] though they recognized that filtration stripped wine of some of its flavor and bouquet (Pliny *NH* 19.53; Horace *Sat* 2.4.53-4). But the Romans had no effective way to fine filter large quantities of wine. Today, for comparison, wine is forced under pressure and at great speed through filters, often arranged in series. Romans coarse filtered wines at the winery using the *colum*, as we have seen, but fine filtered them at the point of service. Cato seems to suggest that coarse filtering could also occur in the dolia when he specifies (*Agr.* 26) that lids not be placed on dolia for at least 30 days, when the wines have rejected the skins ('bene deacinata erunt'). This implies that the

147 Vine et al (2002): 126-37; Glen L. Creasy and Leroy L. Creasy, "Grape-derived Flavanoids and Stilbenes," in Sandler and Pinder (2003): 199-227; Halliday and Johnson (2007): 202-09; Boothroyd (1986): 81-3.

148 Vine et al (2002): 137-9.

149 Billiard (1913): 508.

lacus was not standard in Cato's time. Alternately, perhaps some skins have passed through the *colum*.[150]

Filtration of fine lees was effected with a *saccus vinarius* or *saccus linteus,* made of linen (Pliny *NH* 14.138, 19.53). Sometimes this was infused with aromatics such as myrtle oil to refresh the taste (Pliny *NH* 15.124), since myrtle oil was thought to have the taste of a sound wine. Others soaked the bag in brine or sea water or coated its interior with plaster, sand, *defrutum*, honey, or wine lees of a good vintage (Martial *Ep.* 14.104; Florentinus in *Geopon.* 7.37). Alternately, a small metal strainer, also called *colum*, was used. These are depicted in a number of artistic representations, suggesting that they were part of the standard wine service along with drinking vessels, mixing bowls, ladles, and serving pitchers. For example, a luxurious silver service is depicted on a wall painting from the tomb of Vestorius Priscus at Pompeii. A serving pitcher (*lagoena*) and a strainer are shown at the foot of a table on which is displayed the elaborate silverware. The confusing double sense of the word is perfectly illustrated in Cato's list of vineyard equipment (*Agr.* 11) in which he recommends 3 basketwork strainers (*cola vitilia*) as well as 3 'strainers to dip up the flor' (*cola quae florem demant*). Presumably the former reference is to our conical baskets for straining gross lees and the latter to a metal strainer of the sort we see in the artwork, but in this case used for removing a vinegar flor when it had begun to form, or a flor veil when a wine of this fortunate type was ready to be bottled.

Clarification

Clarification refers to the removal of fine particles in wine that make it cloudy.[151] The most basic form of clarification is simply transporting wine off its lees to a clean vessel. As noted earlier, if wine is left in contact with gross lees it often develops a bitter taste and so is physically removed to a clean vessel, a process called racking.[152] We have seen that this, in combination with the wine strainer, was standard procedure before the wine went to the fermentation vessel. In many cases gravity flow moved the wine, but often it was laboriously decanted by workers, using simple implements. As noted earlier, implements were the *hama*, 'bucket', although here again the attestation is for oil, or the *urceus mustarius,* the 'must pitcher' (Cato, *Agr.* 11).

As to the frequency of racking once secondary fermentation was completed and bulk vessels had been sealed for aging, we have only Pliny's (*NH* 18.232) citation of Hyginus that wines may be racked (*defaecari*) at the winter solstice

150 Billiard (1913): 461.

151 Margalit (2004): 89-91.

152 Bird (2010): 143-56. The word derives from O.E. *rakken*, 'marc'.

or may be bottled (*diffundi*) to be further aged. In the latter case the implement was the ladle (*simpulum* or *trua* or *trulla*). Notice Hyginus' 'may', but not 'should'; surely this lack of enthusiasm for winter racking represents a well founded fear of the potential effects. Today wines are pumped in closed systems from one container to another to avoid oxygenation, but once the Roman wine was in the dolium or the butt, racking will have oxygenated the wine and exposed it to any number of disease organisms. But our ancient vigneron was between Scylla and Charybdis; as previously discussed, aging *sur lies* also promotes infections, although it is important to make a distinction here. Storage of wine on gross lees is disastrous, but storage on fine lees has advantages. It contributes to structural balance, complexity, and stability, especially protein stability. Fine lees are strongly antioxidant and therefore keep wine fresh and crisp. Contact with fine lees also integrates yeast character and fruit with wood aromas, imparts a creamy mouth feel, and introduces smoky, toasty, and cheesy flavors. Nitrogenous compounds in fine lees act as yeast nourishment and also promote the LAB responsible for malolactic fermentation.[153] Since ancient racking, unlike modern microfiltering, will have left fine lees, we may assume that ancient wines derived these advantages from fine lees.

It is just possible that some ancient vintners had discovered the advantage of the siphon in this procedure; the Romans certainly knew the siphon (*sipho*), and Piccoli[154] has interpreted an implement depicted on the stele of a *negotiator vinarius*, Publius Tenacius Esimnus, a stele found in the bed of the River Inn in Southern Bavaria, as a siphon. Esimnus resided at Iulia Tridentum (Trento). On the left, the *négociant* is represented with two vessels in his hands and two small barrels at his feet. On the right another man holds the suspected siphon in a vessel and has three small barrels near his right leg. But this is not a siphon in the modern sense; the implement bears an uncanny resemblance to a wine thief, strictly speaking a type of siphon, one which would be ideal for decanting small quantities of wine without roiling the lees. But such an implement would be completely impractical for racking large quantities of wine.

Precipitation of lees and the development of a firm lees sediment is also facilitated by cold temperatures,[155] and here the Romans' understanding of passive solar heating came into play. The ancient agronomists, as we have seen, recommended that the cella vinaria be located to the north and east of the winery. Once wine in a cool environment has 'fallen bright' it can be 'topped up' if necessary to eliminate ullage. We are reminded that the globular shape of

153 Hornsey (2007): 296-8.

154 Piccoli (2004): 128 and Fig. 25.

155 Boothroyd (1986): 63-66 and Margalit (2004): 95-101 for the science.

the dolium and its relatively narrow mouth was an attempt to minimize ullage, though a limit was imposed here by the necessity of accessing the interior for cleaning and pitching.

In the vast majority of cases, cold stabilization and racking of wine are sufficient to ensure that it clarifies. But sometimes fine solids remain in suspension, and various methods may be used to clarify the wine, that is, rid it of these fine solids, a process called 'fining' (Fr. *collage*).[156] Turbidity in wine is caused by an optical effect called the Tyndall effect. A new wine will contain two fractions which will cause this turbidity, namely gross particles such as grape debris and yeast cells, and macromolecules such as proteins and tannins, generally referred to as colloids, subdivided into stable and unstable colloids. Stable colloids remain invisible but unstable colloids often cause turbidity, especially when wine is chilled ('chill haze'). The main agents of unstable colloids are protein pectins, gums, metallo-colloids, and polyphenol degradation products. Most fining agents work by electrostatic attraction, the fining agent carrying a positive or negative charge, the particulate the opposite charge. Types of fining agents available to the ancients were proteins such as gelatin, isinglass, casein, albumen (especially from eggs, though why Horace specifies pigeon eggs escapes us), and yeast protein; earths (bentonite and kaolin); and polysaccharides such as alginates and gum arabic (Horace *Sat.* 2.2.58; Frontinus in *Geopon.* 7.27).[157] Typically the fining agent is mixed with a small quantity of wine, poured into the bulk vessel, and allowed to combine with the colloids and precipitate, carrying the offending substances to the bottom with them. Among the astonishing variety of additives prescribed for ancient wines was a type of clay, an agent the Romans had learned from the Greeks (Pliny, *NH* 14.24; *Geopon.* 7.12). This is nothing more than our bentonite. Bentonite is a natural argillaceous material of volcanic origin, geologically a form of the smectic clay, montmorillonite. Soaked in water, plates of montmorillonite in bentonite separate to form a colloidal suspension with an enormous surface area for absorption, and the fact that these conglomerates tend to clump together and then precipitate makes bentonite an excellent fining agent. Bentonite is not as effective as proteinaceous agents but cannot over-fine a wine.[158] The chemical action of fining agents was probably recognized by the ancients, though scarcely understood; the physical action was well known.[159]

---

156 Billiard (1913): 508-09; Boothroyd (1986): 68-9; Margalit (2004): 101-12; Byrd (2010): 146-56.

157 Hornsey (2007): 242-92.

158 Austin (1968); Jackisch (1985): 83-86; Hornsey (2007): 255-8.

159 Cf. Billiard (1913): 508-09.

Oddly, there is no ancient attestation for the practice of sealing out oxygen by adding a small amount of oil to the vessel, oil which floats on the top of the wine because of its lower density. The practice is age-old, witness the Talmudic injunction: "Let no man send his neighbor wine with oil upon its surface (*Chullin*, fol. 94, col. 1)." It was common in Italy well into the twentieth century.

Artificial 'Aging'

Odd as it may seem to modern sensibilities, the Romans frequently subjected their vintage wines to procedures designed to impart on them artificial 'age'.[160] The most common method was heating the wine, in or out of the presence of smoke. Columella (*RR* 1.6.20) advises storing vintage wine in lofts so that smoke rising from kitchen hearths will age the wine, though he cautions that too much smoke will also taint it. Obviously, this will have necessitated bottling the wine and placing the amphorae in a special loft. A loft in the Greek household, the *apotheca*, became common in Rome under its Greek name as well as under the Latin term *cella superior*. Originally for all valuables in its Greek guise, the Roman *apotheca* becomes the storage cellar for vintage wines specifically. Thus the poets speak of bringing *down* a fine wine from the cellar for celebrations. The shrewd businessman Cato does not mention the loft, confirming Pliny's assertion (*NH* 14.94) that Rome adopted the form in 120 BCE. In some cases this 'cellar' represented a considerable fortune in vintage wines (Pliny *NH* 17.96). If specifically sited to receive smoke, it was sometimes called the *fumarium* (Martial 10.36.1; Colum. *RR* 1.6). Wine 'aged' in this way was considered insalubrious (Pliny *NH* 23.39), but was nevertheless valued by gourmands (Juvenal *Sat.* 5.33-5; Tibullus 2.1.27-8). There is also the possibility that the process effected a sort of crude pasteurization.[161] If liquids are brought to and maintained at temperatures as low as 60°C (140°F), almost all pathogenic microorganisms will be killed and, provided the vessel is hermetically sealed, the liquid will remain sterile indefinitely. Billiard thinks that during Mediterranean summers the temperature in a loft might easily reach 60° or more. Otherwise, wine amphorae could simply be placed in bright summer sun for several days (Pliny *NH* 14.85, 14.136; Cato *Agr* 105).

The *apotheca* survived in Italy well into the twentieth century.[162] In Barolo, vessels were glass demijohns in a loft above the wine cellar, exposed to the chill of winter and the heat of summer. The modern analogue of smoking and heat-

160 Billiard (1913): 521-4; Tchernia in Tchernia and Brun (1999): 135-8.

161 Billiard (1913): 526-8.

162 Piccoli (2004): 71.

ing is maderization, specifically in Madeira itself, where the wine 'oven' is the *estufa*.[163] Wines here are exposed to temperatures of 50°C (122°F) for three months or 40°C (104°F) for six months, the latter the preferred treatment for finer wines. The effect is twofold; sugars in the wine are caramelized and the wine is simultaneously oxidized to produce the *rancio* effect, literally 'rancid' but in fact the characteristic nutty, caramel taste of Madeira and maderized sherries. Agitation of wine was also known to prematurely 'age' it. Pliny (*NH* 14.118) says that navigation does this. Again, we are reminded that the combination of heat and agitation of Madeira wine in the holds of East India merchant vessels was thought to confer a desired effect on this wine. The effect was discovered when a shipment of wine from the island to the East Indies returned unsold and was discovered to have matured to a remarkable degree.

As white wine, especially maderized wine, ages and slowly oxidizes it gradually darkens, and Cato (*Agr*. 24) has a recipe for conferring a tawny color of an aged wine (*vinum helvolum*) by adding 1/30 part of *defrutum* to an immature blended wine. Otherwise, some additives were used to confer the taste of age on wine, and these include bitter almonds, absinthe, gum of sloe, fenugreek, licorice, Celtic nard, and aloes (Palladius 11.14; Damogeron in *Geopon*. 7.24).

Specialty Wines

The preceding accounts for most facets of the production of standard white and red wines, but the Romans produced in addition a number of specialty wines which ranged from the sublime raisin wines to the execrable after wines. To begin with the latter, afterwines were known as *lora* or *vinum operarium* (Cato *Agr* 57, 147; Varro 1.54; Columella *RR* 12.40; Pliny *NH* 14.86). Cato specifies in his model contract that the contract vintagers are to leave the marc and any lees. The marc is then hacked apart with axes, mixed with various proportions of water, then trodden and pressed and the thin must thus obtained allowed to ferment to provide a 'thin wine' for the farm workers, as one of its names implies. This will have been a thin wine indeed, very deficient in alcohol, though quite tannic, and subject to rapid acetification. Thus it had to be consumed within a year, though Columella claims that the shelf life could be extended by the addition of the foam skimmed from the manufacture of concentrated musts and of good wine lees. The standard formula seems to have been 1 part marc to 10 parts water. Such afterwines have a long history; they are attested in ancient Greece as the *deutereios oinos* of Dioscorides 5.6.15[164] and in

163 Halliday and Johnson (2007): 184-9.

164 Cf. Marie-Claire Amouretti, "Les sous-produits de la fabrication de l'huile et du vin," in Amouretti and Brun (1993): 85.

latter-day Italy and France until quite recently. The traditional Tuscan analogue is *acquarello*,[165] the French is *piquette*.[166]

Cato mentions another afterwine, *vinum faecatum* (*Agr* 113; cf. Pliny *NH* 14.86), 'lees wine', which is exactly what the name suggests, a 'wine' made by soaking wine lees of fermentation in a proportion of water and allowing it to ferment.

At the other end of the spectrum were the exquisite raisin wines of antiquity, made by partially raisining select grape clusters for a more or less extended time, then treading and pressing the grapes in the usual fashion, or after first macerating them in high quality must and then treading and pressing them, to obtain a must of extremely high sugar concentration. Again, such wines have a long history. Gorny[167] conjectures that the Hittites made raisin wine, though he acknowledges that the only literary evidence for this is a curious phrase that refers to the raisin "which holds the wine in its heart." If he is correct, raisin wine was made as early as 1600 BCE. Likewise, the Babylonian Talmud mentions *hilistom* and *simmiqim*, both raisin wines.[168] Some of the most prestigious imported Greek wines during the Roman period were also raisin wines, including the famous Pramnian. And the tradition continued in the West after the Roman period, right down to the present; the modern Italian product is *vino passito* or *vin santo*, the French analogue *vin de paille*.[169]

The Roman product, *passum*,[170] was made by selecting the choicest clusters and rasining them for several days to several weeks. As previously noted, ancient viniculturists seem to have harvested late by modern standards, especially in the earlier times, doubtless to concentrate the sugars, foremost to ensure a vigorous fermentation and thereby a stable product, but secondarily to produce a wine with residual sweetness. Today an off-dry or sweet wine is produced by allowing fermentation to proceed to the point at which the desired residual sweetness remains and then permanently halting it by chemical means or microfiltration of yeasts. But, in the absence of either of these, the Roman either sweetened his wine at the point of consumption, or started with a must of such high Brix that, once alcohol level had reached some 14%, yeasts would die or become dormant while residual sweetness remained. The best

165 Calabresi (1987): 131.

166 Billiard (1913): 557.

167 Gorny in McGovern, Fleming and Katz (1996): 158-9.

168 Lutz (1922): 26.

169 Calabresi (1987): 122-34. Today in Tuscany the grapes are allowed to raisin for over a month, until early December, at which time they are pressed and the must put in *barilli*, small oval barrels.

170 Billiard (1913): 434-5, 491-2.

way to obtain such a must was by raisining grapes to concentrate their sugars. Certain grape cultivars were known to conform better to the practice: Psithian, Melampsithian, Helvolan, and Scripula, for example (Vergil *Geo* 2.93; Pliny *NH* 14.80). Whole clusters of choicest quality were selected before the general vintage, suspended on hurdles (*uvae pensiles*) or from roof beams to raisin in a dry, airy room, or under the autumn sun, or were left on the vine to raisin naturally (*uvae passae*). Columella (*RR* 12.39) advises that raisins left outdoors be covered at night to keep off dews.

Columella, citing Mago as his authority (*RR* 12.39-40), describes the method for making raisin wine. Grapes are picked dead ripe and dried outside on hurdles, then are plucked from their pedicels, placed in a vessel and covered with best-quality must and allowed to absorb the liquid for six days, at which point they will have absorbed all of it. They are then put in frails and pressed. Afterwards the marc is put in the treading vat, covered with fresh must, trodden, pressed, and fermented for 20 to 30 days, after which it is strained and racked, covered and sealed. Obviously, this wine of second pressing will be of lower quality. That account accords in most details with Pliny's description (*NH* 14.81-2). Early-ripening grapes are dried in the sun until reduced in weight by about half, then are trodden (the only way I can make sense of Pliny's *tunsa*) and gently pressed. Afterwards to the marc is added the same quantity of well water as has been expressed as must, and this mixture is trodden and pressed for a raisin wine of second quality. But, Pliny adds, more fastidious vintners pick the grapes from the pedicels, soak them in excellent wine until they swell, and only then press them to make *passum* of the highest quality. This marc is also soaked in water and processed for a second-quality wine.

Finally, it should be noted that Manzi[171] long ago cited Vergil's reference at *Aeneid* 1.738-9 (Ille impiger hausit spumentem pateram) as evidence that the Romans knew a sweet sparkling wine, a claim which still enjoys considerable currency in Italian popular culture. But surely this is far too literal a reading of the poet's words. It is certainly true that in Manzi's time Champagne wines were often sweet and were universally so up until the 1850s. It is also true that *blanquette de Limoux*, the *vin de paille* to which he compares this Roman wine, was *cremant*, i.e., slightly effervescent, but hardly a sparkler in the style of Champagne. That style has one prerequisite which the Romans did not have: bottles and closures stout enough to withstand several atmospheres of pressure.

171 Manzi (1883): 167-8.

Wine Diseases

Whatever the style of wine produced, after secondary fermentation has subsided, modifications have been made, and the wine has been sealed in its aging vessels, our ancient vintner will be content to monitor his aging wines much less often, every 36 days if he follows Columella's prescription. But periodically he will unseal his vessels and remove the *operculum* from the dolium or the bung (presumably) from the *cupa* to look for signs of dreaded wine disease. Wines are living entities and are therefore subject to infections no less than their human counterparts.[172] Obviously it is impossible without samples to identify with any certainty the diseases discussed by the ancient agronomists, but in general it is clear that they struggled with the same acetic and lactic infections as their modern brothers and sisters. Most feared were (and are) acetic species such as *Acetobacter aceti, A. pasteurinus, A. peroxydans* and the *Gluconobacter* species, and lactic acid bacteria such as *Pediococcus cerevisiae* and *Leconostoc oenos*. Even today the most important factors in maintaining the health of bulk wines in storage are maintenance of containers with minimum ullage to reduce oxygen exposure, and maintenance of low temperatures in the cellar, as long as they are above freezing. Did the ancients know about microbes? Apparently the most intelligent and educated did. Varro, for example, advises (1.12.3) that, in establishing the plantation, the farmstead itself should be placed on elevated ground in a well ventilated locale which receives ample sunlight "so that microbes (*animiculae*) which are bred in the vicinity are blown away or die from lack of humidity." Nor can we doubt that the ancients recognized the role of contagion. Thus Pliny's injunction (*NH* 14.134; cf. Florentinus in *Geopon.* 6.2) that dolia be arranged with sufficient space between them, not only to facilitate the movement of the cellarmen, but to prevent contagion which is especially rapid in wine.

It also seems clear that the ancients recognized both acetic and lactic diseases. In speaking of the contract for buying a vintage, Cato (*Agr.* 148) stipulates that the seller will deliver wine that is "neither sour nor ropy (*neque aceat neque muceat*)." The former obviously refers to acetification and, as I will argue below, 'moldiness' per se is unlikely to have been a problem in wine or in vinegar where the malady is also mentioned, so it is most probable that this is a reference to lactic bacterial infection which causes off odors and ropiness. Likewise Pliny (*NH* 14.131) characterizes acetic infections as the most typical wine infections, a fact which explains the "volumes of treatments" for the problem. Sadly, we are not privy to those volumes.

---

172 Vine et al. (2002): 80-88 for an excellent overview.

Today, in speaking of acetic infections[173] it is more typical to talk in terms of volatile acidity, the two components of which are acetic acid (vinegar) and ethyl acetate. Both are present in minute quantities in all wines, even the finest and most expertly made, but the extent to which volatile acidity should be detectable has recently become a subject of considerable debate among even expert tasters. Many contend that a slightly elevated level of volatile acidity accentuates the bouquet of some wines.[174] Normal volatile acidity develops from primary alcoholic and malolactic fermentations, but abnormal levels from bacterial activity from acetic and lactic bacteria, especially from species of *Acetobacter*. At some point the smell and piquancy of acetic acid and the chemical smell of ethyl acetate (nail polish smell) become so strong that the wine becomes not just unpleasant but undrinkable.

The best 'cure' for acetification was (and still is) prevention. Acetic fermentations are aerobic and highly contagious, and thus the obsessive care of the Roman vintner to maintain cleanliness and ample spacing of his wine vessels and minimum ullage to exclude oxygen. Acetification which has progressed to the level at which it is detectable is impossible to cure; nevertheless the geoponics provide some bizarre and totally ineffective prescriptions for the malady (Pliny *NH* 14.136; Tarentinus in *Geopon.* 7.16). And thus we hear of the rather alarming variety of aromatic bactericides that the Romans used as (largely effective) prophylactics.

Acetic infections typically occur soon after alcoholic fermentation has ceased, particularly in wine deficient in alcohol, but other secondary fermentations are more likely to occur in spring. Vitruvius, for example (*de Arch.* 6.6), says that wine stored in cellars exposed to excessive heat from southern exposures becomes "roiled and flaccid" (*confusum et imbecillum*), and Pliny (*NH* 14.125) adds that secondary ferments strip wine of taste, after which it is called *vappa*. Today, as previously noted, one such secondary lactic fermentation leads to the desired malolactic fermentation in red wines, in which lactic acid replaces malic acid and the harsh acidity and fiery tannins of some 'big' red wines are mellowed. Several organisms are implicated, but *Leconostoc oenos* (now reclassified as a *Dekkera*) is the desired organism, since it does not produce the disagreeable odors that other species do. But in wines of lower acidity, all such species strip wines of flavor and bouquet, just as Vitruvius and Pliny say.[175] Almost all other Spring infections are caused by so-called LAB (lactic

173 Billiard (1913): 531-4; Fugelsang (1997): 48-67.

174 Halliday and Johnson [(2007): 211] have a dispassionate discussion.

175 For the pros and cons of malolactic fermentation, A. Versari, G. P. Parpinello and M. Cattaneo, "*Leuconostic oenos* and Malolactic Fermentation: A Review," *Journal of Industrial Microbiology and Biotechnology* 23.6 (1999): 447-55.

acid bacteria),[176] which convert alcohol to lactic acid but also produce unpleasant odors. Pliny (*NH* 19.135) mentions a wine malady which he calls 'cabbage' (*brassica*). This may, in fact, be a defective alcoholic fermentation: when yeast has limited access to its preferred nitrogen sources (ammonium and amino acids) it will resort to sulfur-containing amino acids and release sulfur dioxide in the process. Sulfur dioxide, as we have seen, has powerful bacteriostatic and bactericidal effects, but, unfortunately, it may combine into more complex compounds called mercaptans which have odors described as 'cabbage' 'rubber' or 'burnt rubber'. But malolactic bacteria such as *Lactobacilli* and *Pediococci* also produce unpleasant odors which are sometimes described as 'krauty'.[177] *Brettanomyces* species produce odors described as 'ammonia', 'mouse droppings', 'manure', 'burnt leaves', 'barnyard', 'stable', 'horsey', 'wet dog', 'sweaty', and 'cheesy', and may be implicated as well, though 'Bretty' flavor is now regarded as adding complexity to some of the world's most famous red wines.[178] Working symbiotically with *Brettanomyces* species are certain LAB to produce a 'mousy' smell. Today the standard preventative for taint is SO_2, but once a smell is present it cannot be removed. Sadly, Pliny's assurance that 'cabbage' taint in wine can be removed by soaking in it beet leaves is desperate indeed.

Another frequently mentioned malady is *flos*, 'flower'. Columella (*RR* 12.30) says that during the aging, if the cellarmen in the periodic checks of the vessels notice a flor developing, they must check the vessels more frequently than the typical 36 days before the Spring equinox and 18 days thereafter and must skim off the flor to prevent it from sinking to the bottom where it will ruin the flavor of the wine. But Pliny (*NH* 14. (27.3) considers a white flor a good sign, a red flor a sign that the wine will soon acetify. Tchernia[179] had conjectured that the white flor was nothing more than the veil of *Saccharomyces cerevisiae,* the sign of a healthy fermentation, whereas the red flor is a vinegar flor. A gray vinegar flor becomes rosy as it thickens and begins to turn the wine to vinegar. In the latter case he is surely correct, but in the former we must make a distinction. Both Columella and Pliny mention flor in the context of aging, not alcoholic fermentation, and thus we are probably dealing here with our sherry flor, created by the other *Saccharomyces* species. And Pliny's reference to a 'good' flor is well taken; as one modern authority has aptly noted, a wine aging under a vigorously growing protective layer of these unique yeasts is completely pro-

176 Fugelsang (1997): 3-47.

177 Goode (2005): 121-5; Margalit (2004): 80-6.

178 Goode (2005): 136-43.

179 Tchernia in Tchernia and Brun (1999): 128-32.

tected from the invasion of oxygen and the attacks of acetic acid bacteria.[180] Unfortunately, the flor yeasts need regular replenishment with fermentable to continue to grow, and thus the *solera* system of the sherry bodegas in which older wines are refreshed with proportionately younger ones. A Roman wine of high sugar concentration, particularly a *passum*, will have had enough fermentable to sustain flor yeasts for considerable time, and as dolia were 'topped up' they may have received more sustenance still, but it is impossible to say how long this fortunate condition might last. Archestratus (H 59-60) describes a Lesbian wine with "liquid locks thickly overgrown with white flower," an obvious description of sherry flor. Furthermore, a vase painting shows a long dipper which strongly suggests the *venencia*, the sherry-shipper's instrument for plunging through the flor on the surface to extract the wine underneath without seriously disrupting the veil. On the other hand, sherry makers long ago discovered that if they replenished the flor less often, then the veil would not completely cover the surface of the wine and slow oxidation would maderize the wine and create the characteristic nutty *rancio* taste.[181]

As previously mentioned, I believe that the ancient *mucor* is a lactic infection. *Lactobacillus trichodes* is a spoilage bacterium that forms mannitols from fructose as well as CO_2, ethanol, acetic and lactic acids. It is especially prevalent in wines deficient in acid. Because the bacterium causes increased viscosity of the wine due to polysaccharides formed as a protective covering, wines thus spoiled become so thick that one can dip a finger into the liquid and lift a mucilaginous 'rope' from it, hence the name 'ropiness'. The condition is mentioned by several of the ancients, though only Columella offers a preventative in the form of salt (Cato, *Agr.* 148; Pliny, *NH* 14.26; Columella *RR* 12.33).

Other wine 'diseases' are not so much infections as deficiencies in organoleptic qualities. Pliny (*NH* 23.40) and Columella (*RR* 3.2.17) mention *caries*, 'flaccidity', typical of wines aged beyond their potential and perhaps victims of LAB infection as well. Otherwise we hear of *amaritudo* (Pliny *NH* 14.6.3; 23.22.3), literally 'bitterness', and said to be a characteristic of generous wines as they age. 'Bitterness' in big wines immediately suggests excessive tannic astringency, except that aging tends to soften, not accentuate such tannic notes. Perhaps these wines are victims of bacterial infection as well. By far the most alarming 'infection' mention by the ancients, at least to modern sensibilities, is that from small animals such as rodents and snakes which have fallen into the wine. But far less alarming than Columella's (*RR* 12.31) cure: the animal's body should be removed, burned, and the ashes stirred back into the wine.

---

180 Halliday and Johnson (2007): 189.

181 Johnson (2004): 25.

Blending

Blending (Fr. *assemblage*) is the proportional mixing of wines from different cultivars, vineyards, crús, lots, or even fermentation vessels to add complexity or correct deficiencies in acidity, color, tannic structure, or organoleptic qualities in the base wine.[182] The ancients were not as sophisticated at the practice as their modern counterparts, but they certainly practiced the rudiments.[183] Cato, for example (*Agr.* 24), recommends a half-and-half blend of musts from the Helvola and Apicia cultivars, to which is added 1/30 part *defrutum*, presumably during the secondary fermentation. Pliny (*NH* 14.68) mentions a wine from Marseilles so 'fat' that it was used to blend (*condire*) with other wines. Elsewhere (*NH* 14.74) he advises that a sweet wine from Mount Tmolus may be blended with dry wines and a well-aged wine with pronounced bouquet be blended with young wine to add 'age' to it. We also hear of a big, hearty Falernian being blended with a sweeter Chian (Horace *Sat.* 1.10.24); wines from Erythaea being blended with a Hecutaean, and a small fraction of Falernian or Formian being mixed with a cheap red to confer on it the semblance of nobility (Horace *C.* 1.20.11). A flaccid new wine was refreshed with an infusion of another, more vigorous wine (Pliny *NH* 23.40). Obviously, the key to blending astutely is recognition of the strengths and weaknesses of individual crús, and the development of a vocabulary to describe them is a rough indicator of the degree of sophistication achieved in blending itself. Billiard has created a glossary of Latin oenological terms which indicates that the Romans could speak with some precision about qualities of sweetness, astringency, smoothness, 'fatness', 'flabbiness', bouquet and aroma, though they never resort to the descriptive, metaphorical language used by modern oenophiles to describe the subtleties of the latter two. And as to color, we have seen that the Romans recognized white (*album*), tawny (*fulvum*), 'black' (*nigrum*), blood-red (*sanguineum*), and rosé (*medium*) wine (Aulus Gellius, *NA* 13.13; Pliny *NH* 14.80).

Bottling

Roman bottling (*diffundere, in amphoras diffundere*: Juvenal *Sat.* 5.30; *Digest* 33.6.15)[184] corresponds to modern practice except in choice of materials.

182 Vine et al. (2002): 117-26; Margalit (2004): 118-21. Michel Bouvier ["Recherches sur les goûts des vins antiques," *Pallas* 53 (2000): 115-33] has made a valiant attempt, including several experiments, to recover to some extent the tastes of ancient wine.

183 Billiard (1913): 509-12.

184 Billiard (1913): 514-7.

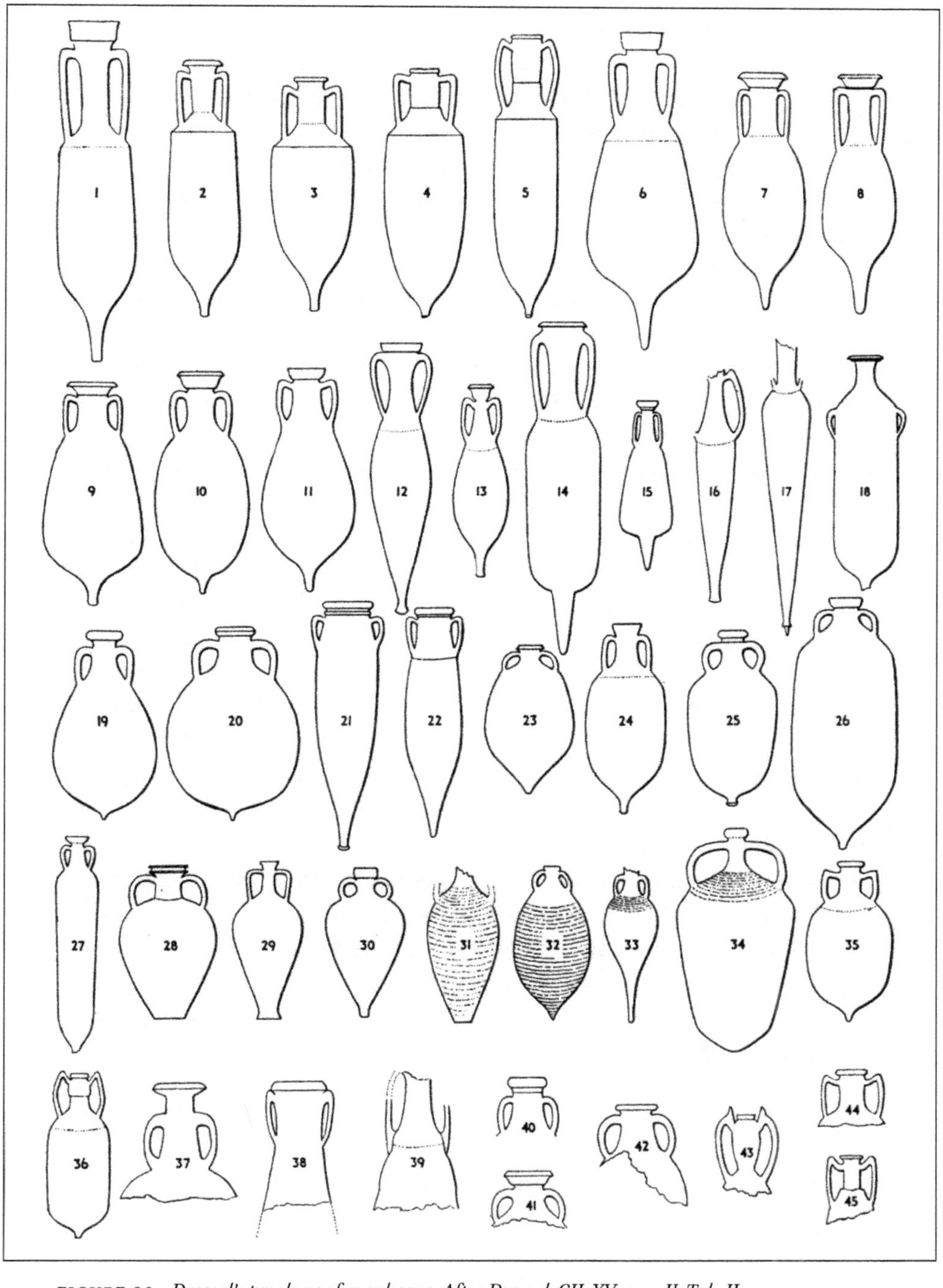

FIGURE 30 *Dressel's typology of amphorae. After Dressel, CIL XV, pars II, Tab. II.*

Whereas today wine goes from wood or stainless steel to glass, the ancients bottled in the same terra cotta as their standard fermentation vat, the dolium, but in smaller increments. And whereas today most wine is sold, if not aged, in bottle, we have every reason to believe that the vast majority of Roman vintages were sold and transported in bulk by the *négociant* (*negotiator vinarius*). For vintage wines, the Roman bottle was the *amphora* or *cadus*. We are reminded that the Greek term *amphora* refers to the two 'ears', i.e., loop handles attached to the neck or shoulder of the vessel; hence Manzi[185] calls attention to Horace's (*C.* 1.9.8) use of *diotas*, 'jar with two ears', obviously a reference to the amphora and used for metrical purposes. Gras[186] thinks the Greek term *kados*, whence Latin *cadus*, originally designated an amphora from the Orient and perhaps more specifically from Phoenicia. Ultimately the two terms are used interchangeably by the Romans.

Amphorae might be bought from a local or regional atelier or produced on site, as were the dolia. In some villas, such as the Stabian Villa of the Fondo Bracaccio, the presence of numerous amphorae shows that bottling was done on the estate. In others, such as the Fondo d'Acunzo and Villa XVIII at Stabia, access to a public road has led to the establishment on the estate of a caupona, from which the wine of the estate could be sold at retail directly from bulk vessels. But at the Villa Regina at Boscoreale, as in the majority of cases, absence of amphorae suggests that the wine was sold to the *négociant* and transported to the city in skins, the smaller *utres* or the giant *cullei.*[187] Villa B at Oplontis (Torre Annunziata) may have belonged to just such a *negotiator vinarius,* as evidenced by the partial inscription *negot* ... and by the presence of hundreds of wine amphorae for transport but the unusual lack of dolia.[188] In Roman Gaul, local potteries begin producing amphorae, first imitations of Dressel 1 and 2/4, then the local styles Gauloise 1-5. Here the height of Gallic amphora production is the second half of the first century CE to the end of the third. Wine was bottled on estate (e.g., Prés-Bas, Pertissol, Pardigon), at sea ports (Fréjus, Saint-Lambert), and at river ports (many places along the Rhone and its affluents). To limit the chance of infection, evidently here the wine traveled

185 Manzi (1883): 153.

186 Gras (1985): 263-5.

187 Cf. Immerwahr (1992): 121-32, though here the wineskins are filled with freshly pressed must, either for immediate consumption or as a fermentation vessel. If the skin was not completely filled and was freshly tanned, the ullage and the elasticity of the skin would permit considerable gas pressure to build up before it was purged.

188 De Caro (1994): 226.

to its places of embarkation in skins, either loaded on wagons or on the backs of donkeys.[189]

On the Italian mainland, we are told that the best amphorae were to be had at Capua (Cato, *Agr.* 135). Often, when destined for long-distance travel, they were wrapped in wickerwork of esparto grass, broom, withies, or straw, to absorb shocks, and these prototype *fiasci* were then called *amphorae spartae* (Cato, *Agr.* 11). The practice may have been especially prominent in Gaul. Obviously, for long-distance shipment, the greatest possible ratio of contents to container weight which would still protect the contents reliably from breakage would be a desideratum, and this may explain the popularity of the so-called Gauloise 4 type of amphora which is much more bulbous and thin-walled than the classic southern Italian types such as Dressel 1. For example, a typical ratio of Dressel 1 for content to container weights is 1 to 1, whereas that of Gauloise 4 is more like 3 to 1. But the thinner walls may have led to considerable breakage, which in turn may have led to the enclosure of these vessels in our prototype *fiasci* as illustrated in the top range of a relief from Neumagen, Germany.[190]

The shape of the amphora made it ideal for transport in holds of ships but rather awkward to stand upright and pour from, so a stand called the *incitega*[191] may have been used to keep it upright during service. Alternately, the *incitega* may have been a wooden block, a sort of fulcrum on which the belly of the jar was rested and the neck lowered to pour wine into smaller vessels, as we see in a fresco in the House of the Vetii where two putti are serving wine.[192]

Amphorae were filled with wine and sealed with stoppers which were then pitched and sealed with clay or plaster (Horace *C.* 3.8.10; Persius 4.29). The name of the sealing operation is *oblinere* (Col. *RR* 12.32) or *gypsare* (Col. *RR* 12.39, 12.41-2). The closure of choice in antiquity, as today, was cork.[193] Cork is

---

189 Brun (2005): 65-6.

190 F. Widemann et al., "Analytical and Typological Studies of Gallo-Roman Workshops Producing Amphorae in the Area of Narbonne," *Archaeophysika* 10 (1978): 317-41; F. Widemann, F. Laubenheimer and J. Leblanc, "Amphorae Workshops in Western Narbonensis: The Non-resolution Space Problems," in *XIXth Symposium on Archaeometry and Archaeological Prospection* (London, 1979): 57-71.

191 Manzi (1883): 147, n. 2 cites Paulus Festus 107M: incitega machinula in qua constituebatur in convivio vini amphora.

192 Cf. Athenaeus E.5. Casaubon in Du Cange: a device by which an amphora is held.

193 Avashalom Zemer, *Storage Jars in the Ancient Sea Trade* (Haifa, 1978); Lloris Beltrán, *Las ánforas romanas en España* (Zaragoza, 1970); Peacock and Williams (1986): 50-1. Tchernia in Tchernia and Brun (1999): 142-5 for a hypothesis that a change in closures from pitched stoppers to more permeable materials may have led to slow but perceptible differential

the outer cambium (bark) layer of the cork oak, *Quercus suber,* which is common in the Mediterranean climates of Spain, Portugal, France and Italy as well as some parts of Eastern Europe. When the tree reaches maturity at about 20 years, the outer cambium layer can be cut off and not only will the tree not be harmed, but a new layer of cambium will start to grow. Afterwards, about every nine to ten years, the cork can be safely harvested again. Because of the cell structure of cork and the fact that these cells are filled with gas, cork can be compressed to about one half its normal size without losing elasticity. Furthermore, when compressed in one direction it does not distend in the other. It will resist moisture for decades and will remain safely compressed for almost as long, about thirty years. But cork also harbors TCA, i.e., 2,4,6-trichloroanisole, which gives about one in twenty modern wine bottles 'cork taint', a musty odor. But the odor is strictly an organoleptic defect, since TCA poses no health risk.[194] Other stoppers used were fennel stalk and wood.

Once Roman bottles were sealed they were typically labeled. As we have seen, this was a common practice in Egypt from the eighteenth dynasty (ca. 1540-1300 BCE).[195] Here hieroglyphic labels were impressed in the still-wet mud of the stoppers with which Egyptian vessels were sealed during fermentation and storage. It may be assumed that such labels were present during the Old and Middle Kingdoms as well. The labels included the date (vintage), maker (vigneron) and source, just as today. Roman wine labels identify the vigneron and vintage and often descriptive information about the quality, sweetness, even adherence to kosher law. Labels were impressed on seals and also painted.

Koehler[196] thinks that amphorae of wine were stored upright but also on their sides in racks, and for the same reason as corked bottles today: the wine keeps the stopper of the bottle moist so that it doesn't dry out, shrink and crack, thus admitting spoilage organisms. As we have seen, such a practice may have precedent as early as 3500 BCE at Godin Tepe in central western Iran, where a red residue was found on one interior side of presumed wine jars. Additionally, an appliquéd rope motif running across the shoulder of the vessel and along two sides indicate where real ropes are to be applied has been interpreted as chocks to keep the vessel stable on the rack.[197] In the

evaporation of water, a phenomenon which will have left the wine more alcoholic and pungent.

194 Goode (2005): 144-58.

195 Lesko in McGovern, Fleming and Katz (1996): 221-3; James in McGovern, Fleming and Katz (1996): 198-9.

196 Koehler (1995): 330.

197 Badler in McGovern, Fleming and Katz (1996): 45-56, esp. p. 50 and Fig. 4.2.

FIGURE 31 *Wine shop at the House of Triton and Amphitrite, Herculaneum, showing a wine rack and various amphorae.* PHOTO BY SANDY THURMOND.

Roman world, we have the shop of the House of Neptune and Amphitrite at Herculaneum, where a wooden wine rack still contained amphorae on their sides at the time of excavation.

CHAPTER 5

Distribution and Consumption

Distribution

The final stage of the winemaking process was the spring tapping of bulk wine not previously bottled as described above, and preparation for shipment.[1] At the beginning of spring when aging had progressed and the winter chill of the cellar had 'cold-stabilized' the wine, the ancient vintager sampled his product to evaluate its merits and decide on further aging or sale to the *négociant*. In fact, a religious festival, the *Vinalia* of April 15, marks this important event. Tapping (*diffusio*) also separates the wine from any residual lees.[2] The ancients, thinks Billiard, were especially anxious to rack the spring wine from lees before it was exposed to summer winds, obviously because the heat associated with these southerlies encouraged secondary spoilage fermentations. Wine was thus racked from dolia to amphorae and sealed in the way described above. At this point wine might also be transferred in bulk to the *négociant*, and Cato describes an ingenious method for transferring fixed measures to the *négociant*'s huge wineskins. A vat (*lacus*) is made with four handles at its corners so that it can be lifted. There is also a hole at the bottom of one side into which a stoppered pipe is fitted, and the *culleus* level (520 l.) marked on an inside wall. The vat is placed on a platform, filled to the *culleus* level with wine, then unplugged and piped into the *négociant*'s bulk vessel. In such ways does bulk wine make its way to and from all parts of Italy and the Roman world, and especially to the city of Rome itself, where it is drunk in huge quantities.

The vigneron has a number of options for sale of his product. Cato (*Agr* 147-8) gives us terms for sale of the grapes on the vine, presumably as insurance against late-summer weather disasters, as well as for wine in the cask. The vigneron could also contrive to distribute the wine himself. For example, the villa itself could be the site of markets (*nundinae*), though the owner needed Senatorial authorization for this, and we have a number of examples where this permission was granted (e.g., Pliny *Ep*. 5.4, 5.13).[3]

The *négociant*, *negotiator vinariarius* (CIL 6.9676.7, 6.9992.7, 10.1931.5) or *negotiator vinarius* (CIL 6.9181.6.14, 9.4680.3, 10.6493.1; Sal. *Hist*. 1.63; Suet. *Cl*.

1 Cf. Billiard (1913): 487-89.

2 Foroaster in *Geopon*. 7.6.

3 Marzano (2007): 182-84.

 | DOI 10.1163/9789004334595_006

FIGURE 32
The culleus, *used for bulk transport of wine. From White (1975): Fig. 38.*
COURTESY OF CAMBRIDGE U. PRESS.

FIGURE 33
The uter *as a serving vessel. After Rich,* A Dictionary of Roman and Greek Antiquities, *s.v.*

40.1), might distribute wine on a purely local level or risk disaster in exchange for the chance of fabulous profits and engage in the transmarine trade. Our best evidence for local trade comes from Pompeii, naturally enough. There wine was transported from the hinterlands of Campania in the *culleus* wine-skin. In an unprovenienced relief now in the British Museum we see the wagon with a full *culleus* being towed by a yoke of oxen. One hard-bitten ox strains at the load while a dog nips at his hocks and a drover beats him with a bastinado.[4] In Pompeii the wine was dispensed directly into amphorae, as we see in another famous Pompeiian artistic representation, or directly into the dolia of wholesalers, wine shops or taverns, through a spout made from one of the ani-

4 Potter (1987): 159.

mal's legs.[5] Wine was delivered in amphorae to buyers by porters (*phalangarii*: CIL 6.7803) who inserted a long pole through one 'ear' of one or more amphorae and then placed the ends of the pole on their shoulders to deliver the amphorae to the consumer. A *vinarius*' home and shop at Pompeii (IX.ix.6-7) on the Via di Nola, about six blocks from the Porta di Nola, gave easy access for deliveries from the countryside. Some 114 amphorae were found in the shop, of which 63 have stamps. Another 42 were found in the garden shared by the house and shop, and 29 empties were stored upside down in a small room off this garden. The labels of several of the amphorae indicated that they had been reused. Five amphorae were discovered lined up in the shop itself. The shop had no street counter of the sort we see so often in Roman towns, suggesting that wine was not served here but sold in bulk, unless we should imagine that wine was served in the atrium of the *vinarius*' house. In the shop there is an amphora of the famed Falernian as well as two Surrentines and a Lampha (or Lympha), i.e., a clear white wine grown on the slopes of Vesuvius. A dolium and a number of vine roots in the garden suggest that some wine was also produced on site.

There is intriguing evidence that the famous Vettius brothers, among their many other business interests, may have been wine merchants as well. Three amphorae (CIL 4.5572, 5573, and 5578), plus a picture of a wine shop found in their home, seem to indicate as much. The amphorae meticulously note the estate and lessee and/or vigneron from whom the wine was purchased and even the individual dolia from which the wine was decanted. One allotment apparently comes from the estate of an Arrius, another from that of an Asinius. A certain C. Arrius Secundus is known to have produced wine in the area, so perhaps this is he. The wine was apparently left in the cellars of the estates until called for, as indicated by the records of the dolia from which the wine was drawn. Alternately, these wines may have been of especially high caliber and the Vettii retained the names of the original estates for prestige value, after the manner of a famous *château*.[6]

What percentage of wine consumed in Pompeii will have been local? A study of inscribed amphorae found at Pompeii[7] indicates that 62% can be provenienced to Pompeii itself by inscriptions on, and style and fabric of, the amphorae. Another 11% is attributable to Herculaneum and some 7% is of

5 Jashemski (1979): 219-26 and Figs 328 and 329.

6 J. Day (1932): 190-2.

7 André Tchernia, "Il vino: produzione e commercio," in Fausto Zevi and Elena Fiorentini, edd., *Pompei 79: Raccolta di studi per il decimonono centenario dell'eruzione vesuviana* (Naples, 1979): 87-96.

Greek extraction. When we consider that much if not most wine arrived in Pompeii in the *culleus* and was transferred to recycled amphorae or directly to dolia in wine shops, we can certainly imagine that some 70% or more of wine consumed in Pompeii was local and probably over 90% was either local or regional.

But how typical are these figures for the rest of Italy, much less the Empire, when we consider that this part of Campania was particularly noted for the quality of its wines? Sadly, we do not know. But we can say with assurance that there was a thriving business in imported wines in other parts of Italy and the Empire. For example, wine produced in Apulia was widely exported across the Adriatic into the Aegean as well as northward along the coast and up the Po valley from that river's ports. Names on the amphorae indicate that some aristocratic individuals, including Appius Claudius Pulcher, consul of 38 BCE, and Gaius Laecanius Bassus, from the prominent family of the first century CE, were involved in this trade. The Laecanii Bassi also manufactured dolia at Pola (Istria). Thus members of eminent senatorial families apparently did not consider it *déclassé* to be involved in a petty industry of a sort typically eschewed by such aristocrats. Presumably the wine trade had something of the same cachet it enjoys today.[8] Already at the end of the Republic, Varro speaks of Apulian wine merchants transporting wine (presumably in skins) on donkeys to port facilities (*RR* 2.6.5). Of the Cisalpine traces of this trade we have some fragmentary evidence. The ancient *pagus Arusnatim,* in what is now Valpollicella, was involved in the local wine trade as well as in imports. Verona had a wine port on the Adige River, as attested by a fragmentary bronze lamina (*CIL* 5.3945) of a *negotiator vinorum* of the gens Gavia, of whose gentilician there are also numerous examples at Aquileia.[9]

It will be apparent here as in the study of the history of wine that amphorae studies are among our most productive archaeological tools for unraveling the commerce in wine. Especially valuable, of course, are 'stamped' (more commonly inscribed) amphorae. Such stamps have a long and distinguished history. Already from the eighteenth dynasty (ca. 3000 BCE) it was common practice in Egypt to label bottles by impressing hieroglyphs on the still-wet

8 Potter (1987): 163. Robert I. Curtis ["Storage and Transport," in Wilkins and Nadeau, *A Companion to Food in the Ancient World* (Oxford, 2015): Ch. 16] has an excellent overview of the logistics of transport. Cf. Fabrizio Paolucci, "Le vie di Dioniso. Commercio e trasporto del vino nel mondo antico," in Di Pasquale (2010): 160-69.

9 Piccoli (2004): 83.

mud of the stoppers with which the vessels were sealed.[10] Such labels included the vintage, maker, and estate, just as today. Greek and Roman amphorae were both stamped at the pottery, usually on the handle, and inscribed (painted, in fact) on their shoulders at bottling or before transport, in much the same way. Unfortunately the only practical way to study these amphorae in international trade is by way of catalogues of the vessels, and all too often the catalogues were made before the systematic study of clay bodies, either visually or by analytical techniques such as INAA, were practical. Thus there is much uncertainty attached to unstamped amphorae catalogued purely on typological grounds, since we know that local potteries across the Mediterranean were wont to imitate international styles. A 'Rhodian' amphora manufactured in Lucania is a real possibility. There are also problems with interpretation of stamps as well, quite apart from the telegraphic format of these inscriptions. We know of instances where elements of provenience were counterfeited, nor is it easy to tell sometimes whether a descriptor refers to place of origin or style of wine. We have 'Chian' wines from Italy, as we have seen, and a 'Naxian' wine which derives from over 100 years after that city was destroyed, unless it refers to the Sicilian city of the same name, and "Naxian' may refer to a style of wine as well, in the same way that bulk American wine producers have the deplorable habit of labeling their plonk as 'Burgundy' or 'Rhine' when it is nothing of the sort. Still, taken as a whole, amphorae studies are a powerful lens on this subject.

The port of Rome, of course, was Ostia, both the old roadstead at the mouth of the Tiber and the new Claudian port facility at the aptly named Portus. From inscriptions and amphora stamps we know of a Forum Vinarium at Ostia (CIL 14.409, 430) as well as a guild of *négociants* there, the *Vinarii importatores negotiantes*.[11] But reference to a patron of the guild *ad quadrigam fori vinarii* (CIL 14.409) might have eluded us were it not for a clever bit of scholarly detective work that has also helped to clarify the topography of the new Trajanic facility at Portus. Comparison of the famous Torlonia Relief (Museo Torlonia No. 430) and the Belvedere Relief (Museo Vaticano)[12] strongly suggests that the Porticus vini at Portus was located at the juncture of the eastern mole of the Trajanic basin and the Tiber canal, in the vicinity of the Temple of the Bona

10 T. G. H. James, "The Earliest History of Wine and Its Importance in Ancient Egypt," in McGovern, Fleming and Katz (1996): 198-99; Leonard H. Lesko in McGovern, Fleming and Katz (1996): 221-3.

11 Micheline Fasciato, "'Ad quadrigam fori vinarii': Au tour du port au vin d'Ostie," *Mélanges d'archéologie et d'histoire* 59.1 (1947): 65-81, citing E. Bloch, "Inedita Ostiensa I" *Epigraphica* 1.1 (1939): 37.

12 Cf. Fasciato (1947): Pl. I and II.

Dea. To the west of the wine port was the site of a temple of Liber Pater/ Bacchus and a triumphal arch above which can clearly be seen a *quadriga* pulled by four elephants and driven by a divinity. The iconography of this figure suggests that it is not, in fact, Bacchus, but the Genius of the Vinarii. We know there was a temple to Liber Pater Commodianus near the Forum Vinarium in old Ostia as well, and doubtless this and the *quadriga* there provided the prototype for those at the new facility. Honorary officials of the guild are among the most prestigious in the town, and the guild itself is described as a *splendidissimum corpus.*

At Rome a Portus Vinarius appears in the record from 68 CE (*CIL* 6.9189-90), doubtless among those in the city's port facilities in the southwest sector near Monte Testaccio, and another further up the river, a Portus Vinarius Superior (*CIL* 6.37807). Wine warehouses (*cellae vinariae*) are attested from the time of Trajan, as well as a center for the wine trade called the Septem Caesares, probably to be located in Transtiberim.[13] Purcell's perception that vineyards in the Tiber Valley were partly supplying a huge growth in urban demand for inexpensive bulk wine reminds us that by way of the the Portus Vinarius Superior and perhaps the Ciconiae Nixae, there was at least as much wine coming down the river as up it. It is in this light that we should read the famous story of the freedman Palaemon's fabulous success with his farm at Nomentum. Palaemon may have been far more the shrewd businessman who anticipated demand than a careful vigneron.

In the late Empire the production and sale of wine was not only regulated by the state but was also a part of the *fiscus*. From the *Codex Theodosianus* (12.6) we hear of functionaries such as the *exasciator*, whom Cogrossi[14] interprets as one who seals barrels, but, given the etymology of the name, may as easily have been an imperial cooper. We also have *haustores*, 'decanters', and a *custos cuparum,* 'guardian of barrels', our *phalangarii* again, and an *ampullarius*, in charge of the 'little amphorae', the flasks from which wine was drunk.

Mention of barrels reminds us of the practical issue of how exactly wine imports worked. The amphora to the modern eye may look like an ungainly transport vessel until we consider its role in sea transport. Early port facilities in the Roman world had no davits and so all loading had to be done piecemeal by stevedores (*baiuli*). The weight of a filled amphora is very nearly at the limit of a muscular man's ability to carry over an extended period of time, and we

13 Purcell (1985): 12.

14 Cornelia Cogrossi, "Il vino nel 'Corpus iuris' e nei glossatori," in Gabriele Archetti, ed., *La civiltà del vino. Fonti, temi e produzioni vitivinicole de Medioevo al Novecento* [= *Atti del convegno Monticelli Brusati, 5-6 Ottobre 2001*] (Brescia, 2003): 499-531.

can imagine our stevedore hoisting the amphora to his shoulder by grabbing the pointed 'foot' with one hand and one 'ear' with the other. For the actual procedure we have an excellent relief from Portus formerly in the Torlonia Museum which shows a naked stevedore carrying what has been interpreted as wine amphorae down a gangplank. On shore is a clerk who, along with his assistant, tallies the merchandise as it is unloaded and gives the stevedore what appears to be a chit or token. Apparently the stevedore is paid piecemeal as well.[15]

Once in the hold of the ship, the amphora design demonstrates its true genius. The pointed feet of the vessels are wedged into the lathing covering the ribs of the hold, and create a solid tier of vessels. Then the feet of upper tiers are wedged into the interstices created by the shoulders of four vessels on the lower. So tightly are amphorae packed into the holds of ships that breakage from shifting must have been quite minimal, and a huge number of vessels can be laden. Already by the fourth century BCE, cargoes of 3,000 amphorae (ca. 100,000-140,000 kg/ 120-150 tons) are not at all unusual. The famous Madrague de Giens shipwreck is the largest yet explored with the single exception of the Albenga wreck, and was capable of carrying some 8,000 amphorae of Dressel 1b type, but was only carrying some 6,000-6,500 at the time of the wreck (much of the cargo was salvaged shortly after the wreck) because it also contained a huge cargo of Campanian ware. The amphorae were loaded at Terracina.[16] We are reminded that Marcus Aemilius Lepidus, Pontifex Maximus and Censor, constructed at public expense a breakwater at Terracina near his own estate there at the same time he was building the wharf on the Tiber in Rome. One is tempted to speculate that Lepidus was himself involved in the wine trade; certainly his construction of the breakwater caused public outcry because of a perceived conflict of interest. The Albenga ship[17] was a large merchant vessel of some 450 tons which may have been carrying upwards of 10,000 amphorae of wine when it foundered, the equivalent of 250,000 liters (66,000 gallons) of wine.

Oddly, however, the wine vessels were never standardized in shape or capacity, despite the fact that the 'amphora' was a standard Roman measure, as we have seen. Amphorae are of widely varying heights and shapes and their capacities are equally variable. For example, amphorae from Hellenistic Magna Graecia and Sicily vary in capacity from 20 to over 30 liters and every *entrepôt*

---

15 Cf. Lionel Casson, *Ships and Seafaring in Ancient Times* (Austin, TX, 1994): Fig. 76.

16 Patrice Pomeroy, "Le navire romaine de la Madrague de Giens," *Comptes-rendus* 126 (1982): 133-54.

17 Nino Lamboglia, "La nave romana di Albenga," *Rivista di Studi Liguri* 18 (1952): 131-236.

seems to have had its own standard. At Selinunte, for example, there was a unit measure called the *ktyle* (.27 l.) based on a *choûs* of c. 3.24 l. Here it seems that the *cadus* (amphora) contained a minimum of 6 chóes.[18] It must have made the haggling between the *vinarius* and wine wholesaler quite colorful.

But there was an evolution in the lading of wine. By the time of the later Empire we have attested in Rome's wine port an unloading point called the Ciconiae Nixae, "The Leaning Storks," because of the cranes used to unload cargo.[19] As we have seen, by the fourth century barrels were commonly used for wine transport and storage, not only in Gallic areas where they originated but in other parts of the Empire as well.[20] We should imagine barrels in slings being raised and lowered through the hatches of transport ships by derricks, just as is done today. We have also seen that dolia were used in Gaul for transport of wine beginning from the first century BCE, and we must assume that such cranes were used there for the same purpose if not in Ostia and Rome at this early date as well.

A fascinating refinement of the loading procedure is reported by Athenaeus (12.519d). It seems that the Sybarites of Magna Graecia had a hilltop wine installation at Thuri during that city's Sybaritic period, from which a closed pipe conducted wine to the port facility below, and archaeological remains of this 'vinoduct' may actually have been found.[21] How wine was decanted or piped into transport vessels at the port we are unfortunately not told.

As we have seen, amphora studies from Ostia have been used to suggest a collapse of Italian viticulture in the early Empire. But, as Purcell points out,[22] there are problems with making Ostia the measure of Italy's wine production. Ostia is strategically located for the importation of products from the western Mediterranean and we should expect a preponderance of such merchandise

18 Vandermersch (1994): 112.

19 Purcell (1985): 12, citing J. Rougé, *REA* 59 (1957): 320. There is a beautiful representation of one such crane on the tomb relief of the famous contractor Quintus Haterius.

20 Sereni (1964): 175. A bas relief conserved in the museum of Langres (in the territory of the Galli Lingones) (cf. Rostovzev Tab 29.3, p. 204); the stela of Quintus Veiquasus Optatus found at Cherasco and conserved in the Museo Archeologico di Torino (cf. Papasogli, *L'agricoltura degli Etruschi e dei Romani* [Rome, 1942] Tab. 14, p. 68) and the bas relielf found at Cabrières d'Aigues and conserved at the Museé Calvet at Avignon (cf. Papasogli Tab 27, p. 128). The Gallic god of cooperage is Sucellus, represented holding a hammer and a butt.

21 E. Galli, "Alla ricerca di Sibari," *Atti e Memorie della Società Magna Grecia* (Rome, 1929): 46-8; *idem*, "Due ville romane in agro Sibaritano," in *Atti del II Congresso Nazionale di Studi Romani*, vol. 1 (Rome, 1931): 267; Vandermersch (1994): 94.

22 Purcell (1985): 10.

FIGURE 34 *A wholesale wine shop. After Bisconti (2003): Fig. 1.*

there. But there will have been huge quantities of wine being brought overland to Rome from the numerous villas in the Suburbium, if not in Latium and northern Campania, and most of this will have traveled in bulk (ie., in skins). This in addition to the huge quantities remarked earlier that were making their way down the Tiber or overland from northern Latium and Etruria.

From such facilities wine made its way either to wholesalers or directly to retail outlets or consumers. We have previously noted the wine wholesaler near the Porta di Nola in Pompeii. We now have one priceless iconographic representation of the wholesale trade as well.[23] A marble bas relief found in a gallery of the catacombs of the ex- Vigna Chiaraviglio, part of the Catacombs of Callistus three kilometers from the city along the Appian Way, depicts such a shop. Here we see many aspects of wine sales at once. On the far left of the relief, amphorae of wine are stacked in ranges along the back wall. Next,we see a young man pouring wine from one of the amphorae into a krater to mix it with water. He uses a tall, upright podium with a type of cradle on top, perhaps our *incitega*, as a fulcrum so that he can control the flow of wine from this heavy vessel. The krater rests on a low table. To the right of the table a mature man raises his hand as a signal for the flow of wine to stop. In his left hand he holds a sack of coins. Obviously wholesale customers were expected to sample the wares before they purchased. To the right of this customer a short wicker stand holds an *urceus* and above this on the back wall is a shelf on which are represented a beaker, a ladle and a bowl. Next to the stand a young man offers his beaker to the proprietress, who sits behind a counter raising her right hand to receive it. Above the beaker is a *tintinnabulum* attached to the head of

23 Fabrizio Bisconti, *Mestieri nelle catacombe romane: appunti sul declino dell'iconografia del reale nei cimiteri cristiani di Roma* (Rome, 2000): Figs. 71-3; *idem*, "Scena di commercio del vino in un relievo inedito della regione dell'ex Vigna Chiaraviglio in S. Sebastiano," *Rivista di Archeologia cristiana-Roma* 79 (2003): 14-44; *idem*, "Una taverna sulla via Appia," *Archeo* 22.9 (2006): 60-5.

FIGURE 35 *Retail wine shop illustrated on a funerary monument.* DRAWING BY C. TOUZEL FROM S. DEYTS, *SCULPTURES GALLO-ROMAINES MYTHOLOGIQUES ET RELIGEUSES*, DIJON—MUSEE ARCHEOLOGIQUE (PARIS, 1976).

Mercury, god of merchants. In her left hand the proprietress holds an object which is difficult to identify. On the bench in front of her, fat coins are represented. To her left a young woman has returned from supervising the delivery of a consignment and proffers a bulging sack of coins while a puppy, obviously the mascot of the shop, jumps up on her dress. To the right of this young woman, two men clad in short tunics, with leather bands encircling their torsos to help them sustain the weight, heft wine amphorae on their right shoulders, holding them by one ear with their right hands. Altogether we see a picture of thriving success.

A retail shop is depicted on a funerary relief now in the archaeological museum in Dijon.[24] The relief shows a man and woman, presumably his wife, who are dispensing wine from a high counter to four customers standing beneath and in front of the counter. On the counter two basins appear, whose drains are attached to funnels, the ends of which appear beneath the counter.

24 Christian Vernon, "À la recherche des outils antiques de vignerons," in Brun, Poux and Tchernia (2009) 316-7.

Under one funnel a child holds a wine skin and under the other a young man holds a *lagoena*, flagon. It would appear that into the stoppered basin a measured quantity of wine was poured from a serving vessel of standard capacity. Then the stopper was unplugged and the wine flowed into the customer's container. On a shelf behind the counter appear these standard vessels, to judge from the size, a *sextarius*. On the right we see an *uter* being transferred from the woman to another youth or vice-versa. This scene comports perfectly with another funerary monument of the same sort, but in this case we see an assortment of small measuring jugs. Here too the vendor pours the wine into a basin and a man holds the neck of his wine pitcher beneath the neck of its funnel.

Consumption

The Romans, as did other ancient cultures, consumed wine in the public sphere, on semiprivate occasions, and in private. Purcell[25] argues that it may have been the state itself which whetted the public's thirst for wine. For example, wine or afterwine was a regular component of the legionary's ration, and we have seen how wine, wine culture, and viticulture followed the legions into western and eastern Europe, stimulating trade and vitiviniculture among the local populations as it progressed. It is worth pausing to remember, for example, that there were typically no fewer than four legions and on occasion as many as eleven in Gaul; at even a meager half-liter ration per day the implied totals are staggering. As veterans returned home or settled in colonial cities they will hardly have abandoned their 'wine habit'.

Wine as a prestige product has always been a part of public feasts and holidays as well.[26] Wine was a regular part of daily household ritual in the form of libations and was also a regular offering to the gods in the public sphere.[27]

25 Purcell (1985): 14.

26 For a stimulating analysis of the role of wine and wine paraphernalia as a diacritical marker of status in feasting, see M. Dietler, "Driven by Drink: The Role of Drinking in the Political Economy and the Case of Early Iron-Age France," *Journal of Anthropological Archaeology* 9 (1990): 352-406; *idem*, "Feasts and Commensal Politics in the Political Economy: Food, Power and Status in Prehistoric Europe," in P. Wiessner and W. Schiefenhovel, edd., *Food and the Status Quest: An Interdisciplinary Perspective* (New York, 1996): 87-125; B. Hayden, "Feasting in Prehistoric and Traditional Societies," *op. cit* (1996): 127-48; C. Bakels and S. Jacomet, "Access to Luxury Foods in Central Europe during the Roman Period: The Archaeological Evidence," *World Archaeology* 34 (2003): 542-7.

27 K. Kirker, *Die sakrale Bedeutung des Weines in Altertum* (Giessen, 1910); F. Bömer, "Juppiter und die römischen Weinfeste" *Rheinisches Museum* 90.1 (1941): 30-58; R. Schilling, "La

Even a cursory examination of all the Roman festivals in which wine played an official role would lead us far afield, but we should mention at least the two wine festivals.[28] The Vinalia Rustica was held on August 19 to celebrate the grape harvest. Gardens and vineyards were dedicated to Venus Obsequens, that is, Venus in her archaic guise as goddess of cultivated crops, and to Jupiter, controller of weather at this critical juncture when torrential rain or especially hail can be disastrous. Jupiter's priest, the Flamen Dialis, picked the first cluster of grapes, blessed the harvest and vinification, and offered propitiatory sacrifice. The Vinalia Urbana (Varro *LL* 6.16.20; Ovid *Fast.* 4.863ff; Pliny 18.287; *Fast. Praen.* Apr. 23 (= *CIL* 1, p. 236); Paulus Festus 374M),[29] celebrated April 23, marked the other critical phase of the vine cycle, florescence and fruit set, and again was dedicated to the god of weather whose good will at this time was critical. In the city it marked the tapping of the autumn's vintage (*vinum spurcum*).

Public banquets financed by officers of the state were a product of the late Republic and early Empire (Varro 3.2.16). We are reminded of Caesar's lavish banquet in celebrating his Spanish triumph (60 BCE) in which he apportioned an amphora of Falernian and one of Chian to each table, a feat he repeated in 46 BCE. It was also in 46 during his third consulship that he provided no fewer than four *grand cru* wines at a public banquet, the first time four different wines had ever been served on such an occasion (Pliny *NH* 14.97). Wine also figures prominently in the communal dinners of the *collegia* of the urban plebs.[30]

Obviously a prestigious wine was *sine qua non* at exclusive banquets of aristocrats for their social peers, and it must surely be in this regard that we see the first lists of *grand cru* vintages. Varro records (by way of Pliny, *NH* 14.96) that in his own day imported Greek wines were still the *premiers crus.* When Lucius Lucullus was a boy, Greek wines were given at formal banquets, but only once, but by the time he himself returned from campaigning in Asia (80 BCE) he

signification du culte des Vinalia," in *La religion romaine de Vénus: depuis les origines jusqu'au temps d'Auguste* (Paris, 1982): 91-155; G. Piccaluga, "Numa e il vino," *SMSR* 33 (1962): 99-103; H. H. Scullard, *Festivals and Ceremonies of the Roman Republic* (London, 1981); 98; Michel Gras, "Vin et société à Rome et dans le Latium à l'epoque archaîque," in *Modes de contact et processus de transformation dans les sociétés anciennes: Actes du Colloque de Cortona (24-30 Mai 1981)* (Pisa-Rome, 1983): 1067-75.

28 Bömer (1941): 30-58; Scullard (1981): 177; Olivier de Cazanove, "Rituels romains dans les vignobles," in Murray and Tecuşan (1995): 214-256.

29 Scullard (1981): 106-08.

30 Jean Pierre Waltzing, *Etude historique sur les corporations professionnelles chez les Romains,* vol 1 (Louvain, 1895).

distributed more than 100,000 amphorae as largesse. Varro also cites the praetor Gaius Sentius who says that the first Greek wine ever to touch his lips was a Chian prescribed by his doctor, but that Hortensius left over a ten thousand amphorae to his heir.

By Caesar's time, Greek prestige wines still have their old cachet, but Italian wines have come of age; it is telling that among the four famous vintages that Caesar served at his banquet are two Greek (a Chian and a Lesbian), one from Magna Graecia (a Mamertine), but the famous Campanian Falernian as well. Pliny himself supplies the most famous list of Roman *grands crus* and even divides them into *premier, deuxième* and *troisième crus* (Pliny *NH* 14.64-73).[31] First, however, practical politician that he is, he nods to authority. It seems that Livia Augusta used nothing but Pucinum into her eighty-second year, a grape from the vineyard of *Praetutianum* (*NH* 14.60, 67; Athen. 1.33a; Diocletian Price Edict 2.2)[32] But right away we are in a quandary. Pliny specifically state that this wine comes from the Adriatic sinus near the river Timavus, the modern Reka in Istria. But elsewhere Pliny has garbled his geography of Umbrian Hadriana/Adria, modern Atria, south of Ancona in Abbruzzo, and Praetutia Interamna is a prominent city in the region, capitol city of the Petrutii, modern Teramo. Praetutia, incidentally, has given us the name Abruzzo. It is at least possible that the Pucinum grape is in fact the Picenum grape and that Livia's wine is a product of Umbrian Picenum, not of Istria. This wine is named in papyri in Egypt where it was shipped as early as the fourth century BCE. It is probably also the Picenum listed among seven *premiers crus* wines in Diocletian's Price Ediict which are entitled to higher prices. Meanwhile, Augustus himself preferred Setinum, wine from Setia, modern Sezze, in the environs of Forum Appi in Latium, overlooking the Pomptine Marshes (Pliny *NH* 14.61, 14.52, 14.64; Martial *Ep.* 13.112, 6.86, 9.3, 10.74; Juvenal 5.34; Silius 8.378; Strabo 5.3.6).[33] Martial counts it as the equal of Massic, but satirizes a Papillus who serves it in a mixing bowl from which he has poisoned four wives.

Pliny lists Caecuban as his *premier cru* wine but quickly acknowledges that it is no longer on a par. Of the true *premiers crus* listed by Pliny, the famous Falernian takes pride of place.[34] In myth, Falernian was the gift of Bacchus to

31 On the possible sources of Pliny's famous classification see André Tchernia, "Le cercle de L. Licinius Crassus et la naissance de la hiérarchie des vins à Rome," *Comptes-rendus* 141 (1997): 1247-59.

32 Tchernia (1986): 167-8, 348-9; D. Rathbone, "Italian Wines in Roman Egypt," *Opus* 2.1 (1983): 81-98; Andrew Dalby, *Food in the Ancient World from A to Z* (London, 2003): 171.

33 Tchernia (1986): 204, 345-7; Dalby (2003): 298.

34 Best general discussion in Andrew Dalby, *Empire of Pleasurers: Luxury and Indulgence in the Roman World* (London, 2000): 48-50.

a certain Falernus, who taught men the art of viticulture (Sil. 7.162-21). First written evidence for the wine is an amphora stamp (*CIL* 15.4554), unless we credit Polybius (39.11.1; cf. Athen. 1.31d), who lived from 205-120 BCE. and says that 'the area of Capua produces a wine called Anadendrite, particularly famous, with which no other can compete." 'Anadendrite' is Greek for a vine trained to the *arbustum*, and we know that Falernian typically was.[35] Production for export goes back at least to the first half of the second century BCE when it achieved international fame. Sinuessa provided the entrepôt for shipment to Rome and overseas, and a large pottery industry developed there to provide amphorae for the purpose.[36] The district, which stretched from the Massic Hills to the River Volturnus (Voltorno) and may have included Massic, Calene and Statan vintages as well,[37] was subdivided. The Ager Faustinianus, in the lowland close to Sinuessa, was drinkable after ten years but best after another ten, then crapulous. Pliny distinguishes three styles, a dry (*austerum*), a sweet (*dulce*) and a thin (*tenue*) (*NH* 23.34). Galen (in Athen. 1.26c) cites only two, a dry and a sweet, the sweet variety made when the south wind blew during the vintage, apparently partially raisining the grapes, the dry variety made otherwise. Another classification is given, such that Caucinian Falernian is grown at the tops of the ridges, Faustinian on the middle slopes, where we would expect the best wine, and a simple *vin ordinaire* on the plain.

The wine could also apparently be made as either a red or white wine since it is described as *niger*, 'black', as red wine does appear when served in a goblet not made of clear glass. But otherwise it is amber (*xanthos*), best quality amber being described as 'Falernian amber'. This style is moderately sweet (Galen, *On Antidotes* 14.77).[38] Pliny tells us that Falernian is the only wine to take fire when a flame is applied to it, and some have asserted that Falernian may have achieved as much as 30 degrees alcohol, but Warner Allen[39] sensibly points out that any reasonably alcoholic wine when heated will volatilize enough alcohol to *flambé*. On the other hand, if we assume that some Falernian was made from partially raisined grapes and from free-run must (the white version), and

---

35 Guadagno (2009): 230.

36 Cotton (1985): xxvii; Clementine Panella, "Retroterra, porti e mercati: l'essempio dell'ager Falernus," in J. H. D'Arms and E. C. Kopff, edd., *The Seaborne Commerce of Ancient Rome: Studies in Archaeology and History* [= *MAAR* 36 (1980)]: 251-60; David Peacock, "Recent Discoveries of Roman Amphora Kilns in Italy," *AntJ* 57 (1977): 262-9.

37 Dalby (2003): 138-9.

38 Cf. Dalby (2000): 48-50.

39 H. Warner Allen, *A History of Wine: Great Vintage Wines from the Homeric Age to the Present Day* (London, 1961): 103-07.

especially if it has undergone a *flor* fermentation (the amber color), we would expect it to be a powerful wine capable of long aging.

During the later years of the Empire, Falernian's reputation declined, perhaps as a result of overproduction and/or the expansion of the vineyard, but it never completely lost its cachet. From the late Empire we have six amphora stamps related to an imperial Falernian distributed in England, southern Spain, Gaul, Ostia, Scafati (CA), and Jerusalem.[40] The tomb of Domitius Primus (*CIL* 14.914) cites the deceased's crowning glories: "I frequented Baiae and I often drank Falernian." When Diocletian in his Price Edict (1.7) declares that the price of 'Falernum Italicum' will be 30 denarii per sextarius, he implies that 'Falernian' is being made elsewhere and being sold under the name, much like American 'burgundy'. But Falernum was still the Château Haut Brion of the Roman world in the late fourth century, witness the notorious case of Theodorus, a senior bureaucrat in the imperial service executed in 371 for plotting against the emperor Valentinian's brother Valens (Ammianus 29.1). During a seance, the plotters asked the name of the next emperor; when the answer came back piecemeal as T-H-E-O-D, the plotters popped a cork on a fine Falernian. A bit precipitately, in the event, since the plot was revealed and the plotters duly executed, and Valens' successor in due time turned out to be Theodosius.

Columella (*DA* 3.8.5) lists Massic as one of the four *premier cru* wines of Rome and yet we have no later mention of it. Doubtless Dalby is correct[41] that the appellation 'Falernum' subsumed all the vintages of the northern border of Campania, specifically Massic, Calene and Statan. Massic (Pliny *NH* 14.62-4; Horace *Sat.* 2.4.51; *C.* 1.1.19; 1.7.21; Mart. *Ep.* 13.111; Sil. 7.207)[42] derived from the Massic Hills on the border between Latium and Campania and is described as 'somewhat harsh'. Calenum is specifically named by Strabo (5.4.3) with Falernum. Cales is modern Calvi Risorta.[43] Athenaeus (1.27a) calls it easy to digest. No one has been able yet to identify the exact provenance of Statanum, though it doubtless is in the vicinity of the others. It is ranked as *premier cru* by Strabo (5.3.6, 5.4.3) but cited as third-rank by Pliny (*NH* 14.65).[44] Athenaeus (1.26c) calls it one of the best, but lighter and less alcoholic than Falernian, which it resembled.[45]

---

40 Daniele Manacorda, "Nel calice dell'Imperatore," *Archeo* 259 (2006): 104-07.

41 Dalby (2003): 209.

42 Cf. Tchernia (1986): 283, 342.

43 Dalby (2003): 71; Tchernia (1986): 159-60; 345-7.

44 Allen (1961): 116-19.

45 Dalby (2003): 311; Tchernia (1986): 160.

Another Republican *premier cru,* Caecuban, had already lost its reputation by the time of Pliny (*NH* 14.61)[46] In the first century BCE the wine had been said to surpass all others, being smoother than Falernian and more full-bodied than Alban. The vineyard was near the poplar swamps bordering the Gulf of Amyclae close to Fundi (Martial *Ep.* 13.115). Warner Allen takes Martial's description "vitis et in media nata palude viret" to mean a wine grown on marshy, alluvial soil and therefore a heavy, full-bodied but not very refined wine as is all wine grown in such campestrine terrain. But it is just as likely that the green reflection of the vines from the neighboring slopes is what Martial references; again, vines love to smile on water. In any case, the wine's reputation was destroyed by Pliny's time due to careless cultivation, overproduction, and the disruption from Nero's infamous canal from Baiae to Ostia. The name thereafter was bastardized for any wine which was generous, full-bodied, strong and age-worthy ('Galen' *apud* Athen. 1.27; Hor. *C.* 1.20.9, 3.23.2). It is described as a white wine which grew tawny with age, made in a sweet style (Diosc. *M.M.* 5.6.7-11). In Martial's time genuine vintage Caecubans were still aging in cellars at Amyclae and Galen (*MB* 6.805, 6.809; *On Ther. Meth.* 10.834) is the last recorded to have tasted it, almost a century after production had lapsed.[47]

It would seem, then, that the wines of northern Campania were the prestige wines, along with the traditional Greek vintages, of the late Republic and Augustan period, *pace* the idiosyncratic tastes of the Emperor and his wife, but that the reputation of these wines was beginning to decline. Pliny's 'third-rank' wines suggest that Latian wines were beginning to establish a reputation of their own. Particularly noted were the wines from the Alban Hills southeast of Rome herself.[48] These wines were made from the famous Sicilian varietal Eugenia, seemingly transplanted to its ideal microclimate since "elsewhere it practically loses its nature" (Col. *RR* 3.2.16). Wines had been grown in some quantity here from at least the fourth century BCE, when the invading Gauls were said to have besotted themselves with Alban wine. It was made in various styles from very sweet (*praedulce*) to sweet (*dulcis/gleukazon*) to dry (*austerum*) to sharp (*omphakías*). It is described as invigorating (*nervis utile*) and best after fifteen years' aging (Mart. *Ep.* 13.109; Pliny *NH* 23.36; Hor. *Sat.* 2.8.14; Juv. 5.33; Athen. 1.26d). Warner Allen[49] posits two harvests, one for the light style and a late, possibly raisined harvest for the intensely sweet, comparable

46 Dalby (2003): 67-8.

47 Dalby (2003): 67-8.

48 Dalby (2000): 115; *idem* (2003): 3-4; Tchernia (1986): 108, 206.

49 Allen (1961): 123-4.

to the lighter styles of Moselle and the various *spätlese* styles. One Alban vintage, Labicum, is mentioned specifically, from ancient Labicum/Labici, lying along the Via Labicana some twenty km (12 m) southeast of Rome, or, more likely, its replacement, the Roman *colonia respublica* Lavicanorum Quintanensium, a bit southeast of the modern village of Colonna. Since there are few amphorae to attest Alban wine, most was probably transported to Rome in *cullei* and bottled there.

The wine of the Sorrentine Peninsula, Surrentinum, falls into the same category (Pliny *NH* 14.64; Athen. 126d; *Diocl. Price Ed.* 2.6).[50] Apparently this was the ideal microclimate for the Gemina minor cultivar because of the persistent westerly breeze. Again we find the wine made in totally different styles, a thin, dry (*alipês psarapharós*) style and a strong, winy style. In both cases the wine was dry and tannic (*austerum*). It was still regarded as a *grand cru* wine at the time of Diocletian's Price Edict. The wine had its detractors. According to Pliny, Caesar called it 'noble plonk' (*nobilis vappa*) and Tiberius 'generous vinegar (*acetum generosum*). Elsewhere we are told that Surrentinum was not fit to drink for twenty-five years, so obviously it was sometimes a big, powerful wine that needed long aging. Warner Allen[51] suggests that perhaps what Tiberius was experiencing as vinegar was not so much acid as the astringent tannins which had not yet mellowed.

From the other side of the Bay of Naples came Gauranum, specifically from the volcanic ridge (Monte Gauro) above Baiae and Puteoli (Pliny *NH* 14.64; Galen *ST* 6.335, *BM* 6.806; Athen 1.26f, 1.27a; Hor. *C.* 1.31.9; Tibullus 2.1.27; Juv. *Sat* 1.69; Mart *Ep* 10.35.13-113).[52] It was probably produced from the local Calventina cultivar. It was produced in small quantities of excellent quality. It is described variously as full-bodied (*eútonos*) and nourishing, a bit crapulent, oily and light. This will have been Athenaeus' 'Cumaean'. A bit farther to the north was the vineyard of Fundanum, close to that where Caecuban had been produced (Pliny *NH* 14.64-5; Strabo 5.3.6).[53] As were the famous Falernians, the vines were trained to the *arbustum*. The wine is described as strong, full-bodied, but a bit crapulent. It was losing its repute by Pliny's time. Martial writes a gift tag for an Opimian vintage Fundanum which purports to be 220 years old, but we needn't take this at face value.

Finally there were the wines of the Volscian Hills of southern Latium, which in Pliny's time had begun to establish a certain repute (*NH* 14.65. Cf. Strabo

50 Dalby (2000): 56; *idem* (2003): 316; Vandermersch (1994): 33.

51 Allen (1961): 111-14.

52 Dalby (2000) 52; *idem* (2003): 158; Tchernia (1986): 202-03.

53 Dalby (2003): 151.

5.3.10; Athen. 1.27b; Mart *Ep* 13.116).[54] Vintages came from Veliternum (Velletri), Privernum (Priverno) and Signinum (Segni). These wines are described as having a strong aftertaste. Privernum is said to have been thin but pleasant, but Signinum strictly medicinal because of its powerful astringency. The emperor Marcus Aurelius (*Letters* [Front. vol. 1 p. 176 Haines]) insists in a jocular letter that he would rather drink the wine than the raw must, so great was its astringency.

Pliny now moves further afield, and we have the distinct impression that his program is more geographic than qualitative. For example, the wines of Sicily have a certain repute, thanks largely to their champion, Julius Caesar, who, as previously noted, had introduced Mamertine at a public banquet (Pliny *NH* 14.66; Strabo 6.2.3; the Sicilian of Dioscorides 5.6.11; Mart. *Ep.* 13.117; Athen. 1.27c).[55] This wine from the environs of Messina, most notably a vintage called Potitianum (the Potalinus of Athenaeus) derived from the vineyards nearest the mainland. It is described as sound, light, medium bodied and healthful. We are also told that Tauromenitanum (Taurmenian) was sometimes fraudulently substituted.[56] After Martial we no longer hear specifically of Mamertinum, and Dalby conjectures that the appellation may have been subsumed by the *Sikelios aminaios* recommended by Galen (*On Ther. Meth.* 10.834; *On Compounding* 13.659). Another Sicilian wine of some repute was Syracusan,[57] made in the style of the famous Bibline of the eastern Mediterranean and sometimes labeled as such.

Pliny now adopts a strictly geographical itinerary and, sadly, often we know little more than the bald name of the vintage and not so much as the specific location of the vineyard. For northeastern Italy, for example, he mentions Praetutian, Anconan, Latiniensian, Caesenatian, Maecenatianum, Reatic, Graviscan, and Statoniensian (*NH* 14.66). Praetutian he has already detailed in regard to the preferences of Livia Augusta. Anconan (Cf. Polyb. 3.88; Strabo 5.4.2; Athen. 1.26f)[58] obviously derives from the environs of the modern city of

54 Dalby (2003): 303.

55 Dalby (2003): 206-7; Allen (1961): 114-15; Vandermersch (1994): 31. Vandermersch has the best discussion of the vintages known from Sicily and Magna Graecia (25-57). From the sixth century the wines of Sybaris and Naxos bore comparison with those of the Greek motherland, and acclimitazation of Greek cultivars to the western Mediterranean seems to have been rapid. Vandermersch also has an excellent list of literary sources as well as a brief introduction and bibliography for the archaeology, paleobotany and iconography of these wines.

56 Dalby (2003): 133-8.

57 Vandermersch (1994): 28.

58 Dalby (2003): 11.

the name, where excellent wine is still made. Apparently that is a tradition which dates back at least 2200 years; by 218 BCE Hannibal discovered so much aged vintage wine in the district, says Polybius, that his cavalry used it to rub down their horses. Athenaeus describes the wine in his own day as having an excellent taste and mouth-feel. Caesenatan presumably derives from the the hills around modern Cesena, northwest of San Marino, though we have no evidence for it except the name. Statoniensian may derive from the Etruscan city of Statonia, but if so Pliny has made a mishmash of his geography again, since this city is in Etruria proper and not in the northeast.[59] The only other wine in this list about which we can say anything substantive is Raetican.[60] This wine, from the hills of the Athesis (Adige) River valley near Verona was loved by Augustus and was recommended by Strabo. The vineyard was later the home of a wine called Acinaticum, and this is perhaps the same wine. Catullus says that Verona also produced a distinctive type of earthenware called *panaca* and boasts of serving local wine (Catullus was a native of Verona) in local cups. Dalby wonders if Augustus' preference anticipated a preference for wine aged in wood, since we know that the district adopted the butt and cask early on. But he notes that such barrels were still sealed with pitch and conjectures that none of the vanillin and eugenol so associated with wood aging will have infused the wine nor will the wood have allowed the slow, subtle oxidation necessary to develop phenols and soften tannins. But, as previously noted, perhaps this Cisalpine pitch used for the purpose was used only between staves, in some of the 'oakiness' may have found its way into the wine.

Of the wines of the northwest, namely northern Lazio, Tuscany, Umbria and Liguria, Pliny mentions only those of Luna and Genoa (*NH* 14.68) but elsewhere we hear of several good vintages. Marsican, for example, was grown by the ethnic Paeligni and Marsi north of Rome. It was said to be austere but digestive, though Martial recommends giving it only to one's freedmen (*Ep.* 1.26, 13.121, 14.116).[61] By the time of Athenaeus (1.27f) it had perhaps a more favorable repute and we hear of two styles, a dry and a sweet. From the hills around Tibur (Tivoli) came Tiburtum, described as light, drinkable after ten years but even better later. In the second century CE it was made in a light and noble style, the noble being described as an astringent, dry white wine. It was still listed as a prestige wine in Diocletian's *Price Edict* (Galen *ST* 6.334-7; *DA* 14.15; *On Ther.*

59 Cf. Enrico Angelo Stanco, "La localizzazione di Statonia: nuove considerazioni in base alle antiche fonti," *MEFR* 106 (1994): 247-58.

60 Dalby (2000): 90; *idem* (2003): 278.

61 Dalby (2003): 208.

Meth. 10.831; Athen. 1.26e; Dio Price Ed 2.2).[62] From the territories of the Sabine Hills came two vintages, one from the local Vinaciola grape, another from the more popular transplant, the Visulla (Strabo 5.3.1; Hor. *Car.* 1.9, 1.20; Athen. 1.27b).[63] It is described as light and drinkable from seven to fifteen years. Horace mentions a 'cheap' (*vile*) style as well that is nevertheless ageworthy. The light style was doubtless made to counter demand from Rome's urban masses, but some growers obviously persisted in growing for quality instead of bulk, witness Diocletian's *Price Edict* (2.3) where it is also one of the seven prestige wines.

From the traditional areas of Latium and Campania we also hear of other good wines. From the vineyards of the Latian coast on the Gulf of Caieta (Gaeta), neighbor to the famous Caecuban and Massic but evidently not as highly regarded, came Formian. Athenaeus describes it as quick-maturing and 'oily', by which he presumably means that it has a good mouth feel (Athen. 1.26e; Hor. *C.* 1.20).[64] From Campania we hear of Trifolinum (Mart *Ep.* 13.14), whose exact site is unknown, though it was perhaps somewhere on the Vesuvian slopes. Marangou-Lerat[65] in her study of Cretan wine amphorae at Pompeii found several Cretan 2 amphorae reused and inscribed *TRI*, which she took to be Trifolinum. Alternately, it may have been from the environs of Capua.[66] The wine was newly celebrated in Pliny's day (*NH* 14.69), in seventh place among Italian wines for Martial (*Ep.* 13.114). Athenaeus (1.26e) describes it as earthier than Surrentine and equally slow to mature. Pompeiian wine was made with another Sicilian transplant, the Murgentina.[67] Trebellian wine from somewhere in the same area was briefly esteemed in the mid-first century CE, but seems to have failed quickly (Galen *SDT* 6.335, *BM* 6.806, *DA* 14.16, *On Ther. Meth.* 10.833; Pliny *NH* 14.69; Athen.1.27c).[68]

Some wines from areas of the deep south also enjoyed a favorable reputation[69] Pliny (*NH* 14.69) lists Tarentine, Servitian, Cosentian, Temsan, Calabrian,

62 Dalby (20003): 328; Tchernia (1986): 206-07, 344.

63 Dalby (2000): 72-3; *idem* (2003): 287.

64 Dalby (2003): 150.

65 A. Merangou-Lerat, "Le vins et les amphores de Crète," *Etudes Crètoises* 30 (1995): 142; *idem*, "Vin et amphores de Crète en Campanie," in *Eukrata: Mélanges offertes à Claude Vatin* (Aix-en-Provence, 1994): 137-43; Tchernia (1986): 203, 276.

66 Gianfranco De Rossi, "Indicatori archeologici della produzione e diffusione del vino della Baia di Napoli in età altomedievale," *insulae Diomedeae* 4 (2003): 541-9.

67 Dalby (2000): 230.

68 Dalby (2003): 230.

69 T. J. Dunbabin, *The Western Greeks: The History of Sicily and South Italy from the Foundation of the Greek Colonies to 480 B. C.* (Oxford, 1948): 93-4, 220-1; Cf. Vandermersch (1994): 29-46.

Lucanian, Thurine, and Lagarian. In his seminal work on the area, T.J. Dunbabin cites Galen's list (in Athen. 1.26e) of the famous vintages of the area as Rhegine, Buxentine, Tarentine, and Mamertine, but adds that this was at a time when Magna Graecia was in ruins. At an earlier time wine had also been exported from Sybaris (Athen. 519d), and from the unknown town mentioned in connection with the Serdaoi whose name begins with Mef- or Ser- (BM Cat. Italy 395). The earliest coins of Galaria, one of the first Sikel towns to mint, had a coin depicting Dionysus holding a *kantharos*, and presumably wine was made here on a commercial scale. Inykon, another Sicilian town, on the southeast coast, was famous for its wine (Steph. Byz., s.v.). The wines of Messina we have already met as Mamertine, and we have seen that the Murgentina (Morgentina) grape was famous both as a native and as a transplant. Export of wine from Akragas (Agrigento) to Carthage was one source of the wealth of this Sicilian city (Diod. 18.81.4-5, 11.25.5) and wine was also grown at Syracuse. The wine of Rhegium (Reggio Calabria) is described as being like Surrentine only smoother. Buxentine[70] in the zone of Greek Laos-Pyxous, is probably the original home of the Italiote Serdaioi, and this would explain our mysterious city whose coin in the British Museum collection depicts the head of Dionysus on the obverse and a fat cluster of grapes on the reverse. Thus it seems that the area had a long tradition of wine growing. The vineyard, located along the Bussento and perhaps Mingardo rivers, is probably related to the Oenotria of the archaic period, and the connection with the Lucanians of nearby Roccagloriosa also raises interesting questions. A coin with the symbol of Sybaris, a locust over a bull, and the legend Sirinos-Pyxoes, suggests that the founding of the Greek city of Pyxous in 471 BCE may have incorporated a native settlement called Sirinos on the analogy of nearby Palinurus-Molpa, also attested on a coin. The Greek town of Pyxous was a subcolony of Rhegium; the Roman colonia was founded as Buxentum in 194 BCE. About the wine itself we hear little, though the town continued to export wine beyond the third century CE.[71] Oddly, there is no modern DOC for the sleepy little town of Policastro Bussentino nor is there much wine grown in the immediate environs, perhaps another victim of bradyseism. Alternately, Buxentum was simply the port of departure for wines

70 Clara Bencivenga Trillmich, "Pyous-Buxentum" *MEFRA* 106 (1988): 701-29; Vandermersch (1994): 46. On the Ager Buxentinus, Helena Fracchia, "The Romanization of the Ager Buxentinus," in E. Lo Cascio and A. Storchi Marino edd., *Modalità insediative e strutture agrarie nell'Italia meridionale in età romana* (Bari, 1994): 55-73; W. Johannowsky, "Appunti su Pyxous-Buxentum," *Atti e Memorie della Societá Magna Grecia* 4 (1992): 173-83.

71 Fracchia (1994): 72; Vandermersch (1994): 154.

made father up the Bussento and Mingardo valleys; today excellent wines derive from these areas.

Pliny's Tarentine wine (Taranto) Galen describes as light and pleasant. It is typical of the wines of extreme southern Italy for both Pliny and Athenaeus: simple, light, pleasant, and easily digestible (Pliny *NH* 14.69; Athen. 1.27e).[72] Servitiana is perhaps to be associated with San Severino Centola (SA), not far from Buxentum, though this is rank speculation. Tempsan is to be associated from modern Temesa (CA), but we hear nothing more of it. In the same passage, of course, Pliny gives the nod to these whole regions of Lucania and Calabria. The wine of Thurii (Strabo 6.1.14; Pliny *NH* 14.69)[73] enjoyed a good repute from the first century BCE. Certain varietals here were not harvested until the first frost, we are told, a sort of prototype *eiswein*, very sweet and strong since the frost desiccates the grapes in the same way as raisining does. The wine of the adjacent Lagaria was reputed for its medicinal value, though what its connection to the health of the Messalae Potiti or the famous statesman of the name, we cannot say. The site of the city has been variously placed at Trebisacce, on the ancient road between Siris and Sybaris, or at Monte Cappolo in the commune of Valsinni (MA) Basilicata.[74] The wines of the area are still said to be especially good. Vinum Balbinum (alternately Barbinum and Babbianum) (Pliny, *NH* 14.69; Athen. 1.48)[75] no less an authority than Dalby can identify only as 'from somewhere in deep southern Italy." It is described as austere and ageworthy, but is not heard of again after Athenaeus. The wine of Cosentia, modern Cosenza (CA), is also cited by Strabo (6.15). Vandermersch[76] notes that the high plateaux situated at the southern extremity of the Crati valley create a distinctive microclimate. Pliny is the first to mention the vintage, but there is reason to believe it derives from at least the third century BCE. Eventually viticulture became more practicable here in the uplands than in the lowlands of Sybaris/Thurii and Temesa. Obviously the Greek colonists were discovering the advantage of hilly terrain for the purpose.

The wines of Roman Gaul, specifically of the provinces of Provincia (Provençe) and Narbonensis (Narbonne) had already by Pliny's day achieved a certain repute (*NH* 14.68) and he mentions specifically those of Massilia

72 Dalby (2003): 322.

73 Dalby (2003): 327.

74 F. Lenormant, *La Grande-Grèce* (Paris, 1881-4): vol. 1, 219; Dunbabin (1948): 158; Richard Talbert, ed., *Barrington Atlas of the Greek and Roman World* (Princeton, NJ, 2000): Map 46.

75 Dalby (2003): 41.

76 Vandermersch (1994): 45-6.

(Marseilles) and Baeterrae (Béziers). Spanish wines were also starting to make their mark (Pliny *NH* 14.71), and we hear of wines of Laeetania (Barcelona), Terraconensia (Tarragona) and Lauronensia (perhaps the western coast of Catalonia) as well as of the Balearic Islands.

Pliny nods to the Homeric Greek wines (*NH* 14.73-4), though we cannot be sure that all of these storied wines were still being exported in his own time: wine of Thasos, Chios, Lesbos, Clazomenae (modern Urla in the Ismir province of Turkey), Mt. Tmolus (modern Bozdag in Lydia), Cicyon (Isthmus of Corinth), Cypros, Telmessus (either the Lycian or the Carian city of the name), Tripolis (Greek Tripoli), Berytus (Beirut), Tyrus (Tyre) and Sebennytus (an unknown Greek city, unless we have a scribal error or Pliny has confused his geography and is referring to the Hellenistic Egyptian city)[77] Of Egyptian wines per se Pliny (*NH* 14.75) mentions Hippomantian, Mystican, Cnidian, Catacaumenitan, Petritan, Myconian, Mesogitan, Ephesian, and Apamenan, all mere toponyms, sadly, unless Pliny has jumbled his geography yet again and the Apameia referred to is in fact that on the southern coast of the Sea of Marmara. The wines of Pontus (*NH* 14.76) are listed as Nasperceniten, Orticum, Oeneaten, Leucadium, Abrioten, and Peparethium. Again, Peparethus is a Greek island of the northern Sporades, but this has nothing to do with Pontus.

Lesbos provides a transition to Pliny's subsequent discussion of specialty wines of the east, since it was the home of the famous 'Pramnian Essence'. From Archestratus, as reported by Athenaeus (1.29) we learn that it was a wine seasoned by *flor*, "a wine which had its liquid locks thickly overgrown with white flower." We know as well that Pramnian was made from *lixivium*, that is, free-run must, the densest and richest of musts. Such a high-density wine fermented under flor will have been incredibly complex and long-lived.

The Greek world was the home of other specialty wines still popular in Pliny's day (*NH* 14.80-1). *Psithium* and *Melampsithium* were raisin wines with distinctive flavors, generic names for the style, made either as a tawny (*Psithium*) or a dark (*Melamsithium*) wine. Scybelite from Galatia tasted like *mulsum*, wine sweetened with honey, as did Sicilian Haluntium. Greek *passum* was still regarded as best, followed by Cilician and African. The Psithia grape, perhaps synonymous with the Latin Apiana as well as the Scripula, were used in Italy for the purpose. Elsewhere we hear that raisined wines, made from

77 Allen (1961): 61-71; cf. Dincer Savas Lenger, "Il mercato del vino nell'antichità in Asia Minore," in *Archeologia e Cibo* [= *Nutrirsi: Rivisti online di scienza della nutrione umana*] (2006): 1-14.

grapes raisined on mats as well as on the vine, were called *retan*, *protropos* or *pramnios*, and were made in black, medium and tawny styles (Dioscorides 5.6.4)[78] Finally we hear of a specialty wine called *conditum*, an aperitif which supplanted *mulsum* at some time in the second century CE. This was a highly spiced wine, a sort of prototype vermouth, to judge by Apicius' recipe (1.1) which includes honey, wine cooked down, pepper, mastic, tejpat leaf, saffron, date stones and date flesh soaked in wine. The emperor Nero refined his *conditum* by boiling the water used to dilute it, then bottling it and chilling the bottle with snow in a wine cooler (Pliny *NH* 31,40; Suetonius *Nero* 48.3; Martial *Ep.* 14.116-18)[79]

Obviously such prestige wines were reserved for the elite, at least in their heavier, more austere and ageworthy forms. The poor drank simple wines, probably young and light in style, and often well on the way to vinegar. The vast majority of Romans drank such wines at home, having brought it there in their own containers from a retail wine shop of the sort we have seen depicted in Gaul, or at one of the thousands of local pubs (*popinae*), food shops (*thermopolia*) or inns (*cauponae*). We suspect that such outlets for food and wine were incredibly common, not least because the majority of urban dwellers had no, or at best very crude, cooking facilities, and therefore often ate cooked food at retail shops. For example, Kleberg in his famous study of such outlets[80] counted some 118 *thermopolia* in Pompeii, not evenly distributed, but rather clustered near city gates, the barracks of gladiators, and the public baths. Wine was carried home or on the road in a wide variety of flasks, one of the most charming of which is donut shaped with a flanged neck at the top which could be stoppered, and typically an incised message around the body of the clever sort so often found on such inexpensive wine vessels and conveniently collected at CIL XIII.10018: 'Hilaris sis!' (May you be giddy: 57); 'Gaudias!' (May you rejoice: 97); 'Copo imple!' (Innkeeper, fill me up!: 57); 'Frui me!' (Enjoy me!: 90); 'Imple, copo, vinum!' (Fill me with wine, innkeeper!: 101; 'Reple!' (Refill me!: 152); 'Bibite!' (Drink!: 51); 'Reple copo!' (Refill me, innkeeper!: 153); 'Merum da, escipe vita!' (Give neat wine, and welcome, Life!: 61); 'Bene tibi sit vita!' (May life be good to you!: 40); 'Bene bibo!' (I drink well!: 43).

---

78 Dalby (2000): 137-8.

79 Dalby (2000): 247.

80 Tönnes Kleberg, *Hótels, restaurants et cabarets dans l'antiquité romaine* (Uppsala, 1957): 39-45; cf. Ray Laurence, *Roman Pompeii: Space and Society* (London, 1996): 78-87.

Shops sold a variety of wines, as is most famously indicated by the inscription at a *thermopolium* on the Via degli Augustali in Pompeii (VII.2.45) which advises that a cup of *vin ordinaire* can be purchased for an *as*, a very small sum, much better for two and Falernian for four. We are perhaps justified in being a bit skeptical whether genuine Falernian was available in a wine shop, even one in Campania where the wine was produced, for such a small sum. But even wine-bibbing in *thermopolia* was not without its refinements. In winter, Romans were wont to drink their wine heated, and at the Pompeiian Inn of Hermes (I.1.6-9) a lead cauldron encased in masonry but suspended so that a fire could be lit under it, as well as a smaller bronze vessel connected to it by a lead channel, were both intended to provide hot water for warm drinks. On the north wall of the room is depicted a man pouring wine from an amphora back into a dolium. The man is named as Hermes and is presumed to be the *caupo* (innkeeper). At another *caupona* (VIII.12.34-5), Room 12 is the associated wine-shop, and at the 'Inn of Salvius' (V.2.13), a more spacious *tabernae* offered private rooms to groups of diners as well as the open bar. Here too we see lead cauldrons for heating water for drinks.[81] Romans did not typically heat the wine itself but rather poured it into heated water. Theophrastus (in Athen. 11.782) asserts that in his time (third century BCE) Greeks poured wine into hot water but earlier had poured the water into the wine. He adds that Xenophanes (sixth century BCE) had insisted that the only correct procedure was to pour the wine into the heated water. Martial (*Ep.* 10.13) may be referring to the practice when he says that Tucca 'breaks' his crystal glasses with bumpers of Setine, but the phrase may equally refer to the ruby color of the wine showing through (*rumpere*) the glass.[82]

Upper-class Romans added refinements to their drinking rituals. One was the habit of drinking wine as an apéritif, which Pliny (*NH* 14.148), who frowns on the custom, says became fashionable during the reign of Tiberius.[83] As we have seen, the two most fashionable apéritifs were *mulsum*, wine to which honey was added immediately before service to sweeten it, and *conditum*, our prototype vermouth. Wealthy Romans drank heated wines in winter as well, but added a summer analog, chilled wine.[84] Chilled Setine seems to have been popular (Martial, *Ep.* 5.64,6.86, 9.3), made by diluting the wine with snow imported from the mountains, another practice inherited from the Greeks, as

81 J. Packer, "Inns at Pompeii" *Cronache Pompeiane* 4 (1978): 5-14; Antonio Varone, "Vini e mescite in area vesuviana. A proposito del bere a Pompei," in Di Pasquale (2010): 216-27.

82 Allen (1961): 107-09.

83 Cf. Allen (1961): 101-02.

84 Allen (1961): 120-22; Dalby (2000): 248.

so often with wine service. Wine was poured through snow in a snow strainer (*saccus vinarius*), doubtless not only to chill and dilute the wine but also to capture bits of grass and straw with which the snow had been insulated during transport and storage. Elsewhere Martial speaks of diluting and chilling Falernian by pouring the wine through snow in a *saccus vinarius* suspended over a *crater* and by putting chunks of ice in the Falernian and then pouring it through the snow (*Ep.* 5.64, 14.104; cf. Petronius 73). During Nero's reign a wine cooler was invented (Pliny *NH* 31.3) which sounds much like the modern champagne bucket except that snow was substituted for ice and a *lagoena* for the bottle.

As with Greek wine practice, so with Greek wine service pieces; practically all the utensils and vessels that the Romans used in sampling, dispensing, drinking, and storing wine derived from Greece. We have already encountered the *pithos*/dolium and the amphora for fermentation and aging of wine. For aging vintage wines the Greeks had invented a cover for the bung of the amphora. The bung itself, it will be recalled, was made of some organic material such as fennel stalk or cork and then covered over with gypsum or plaster to create a hermetic seal. But the Greeks further protected the stopper with a cover, the *krademnon*, named after a woman's cap with a strap which tied beneath the chin. The Romans had the *calantica/calautica,* otherwise the *mitra*, a leather cover over the bung, tied round the neck of the amphora or to the 'ears'. A wall painting from the House of Julia Felix in Pompeii (II.4.3-12) shows the use of the *mitra* very clearly. The lower rim of the cover is scalloped and there is a knob at the apex of the 'mitre' (the item gives us the name of the bishop's mitre) from which straps loop around the two handles of the amphora to keep it firmly in place.[85]

For sampling wine from the dolium the ancients had several options. The wine-thief siphon was known to Hipponax (Fr. 56) as early as the sixth century BCE, and he describes it as in regular use by wine tasters.[86] Egyptian reliefs from 1500 BCE show siphons being used to decant wine from a large storage vessel. The Greeks also had the ladle, *chuathos*, Latin *cyathus*, to decant wine from the dolium, but what of the amphora with its smaller neck? Warner Allen suggests the Greek *oinarasis/ephebos,* exactly like the modern wine thief or the *venancia* of Jerez, designed to plunge through the flor of a sherry to obtain a sample below without seriously disrupting the protective *velum*. Since we now

85 Allen (1961): 33.

86 Allen (1961): 33-8. But Allen's illustration from the Hubbard Amphora, a Cypriote vase of c.900 BCE, is not a true siphon, i.e., a tube for decanting liquids via hydrostatic pressure, but rather a drinking straw, probably to filter out solids from a grog.

know that both Greek and Roman vintners treasured a *flor* fermentation, surely it is not out of line to suggest they used the instrument in exactly the same way in dolia as well.

Wine for service was poured from the amphora into the *krater/crater*, the mixing pot, typically of ceramic or bronze, where it was mixed with water in various proportions. From there it was ladled with the *simpulum*, of bronze or a precious metal, into the wine flagon, *oinochoe/lagoena,* and thence into various drinking cups. As we have already seen, it is in the materials and designs of these vessels that the ancients customarily made a statement about social status. Lower-class wine bibbers from early on had to content themselves with simple earthenware pitchers, cups, beakers and goblets. These were terra cotta vessels, unglazed, but to which a slip had been applied before firing which made them somewhat waterproof and easier to clean. Wealthy gourmands, on the other hand, had a panoply of finely decorated ceramic, bronze, silver and even gold service pieces from which to choose. But by the first century CE, even fine ceramic and metal pieces were being increasingly replaced by glass, as the innovation of glass blowing, especially into molds, made mass production possible and the vessels thus produced correspondingly cheaper.[87] For Roman wine service are attested glass beakers, amphorae, *craters* and *lagoenae*. Because of their fragility, glass *craters* and amphorae will have been luxury items, but we sense that beakers made their way into the working class if not among the poor. Glass, of course has the advantages of being non-reactive and therefore not leaving a 'taint' in wine as do bronze and even silver, as well as allowing the drinker to admire the color and, to some extent, the clarity of wine, and so glass drinking vessels, some beautifully decorated, were popular among the elite as well.

Surely the most exotic of wine vessels was the beaker of fluorspar.[88] Fluorspar, otherwise fluorite or calcium fluoride, Latin *murrina*, was imported from Parthia. Carving it requires repeated heating and impregnation with resins to prevent splitting, and it is therefore difficult to work and quite fragile. Typical coloring for such vessels, so-called *maculosa murra,* 'mottled murrine', was banded purple and red, and the resin added aroma and taste. Several resins were used, but myrh, which gave its Latin name to fluorspar, was probably most often used. The use of such wine vessels first reached Rome after the conquest of Mithridates in 63 BCE. Nero paid a million sesterces for just one such

87 Allen (1961): 91-9; Stuart J. Fleming, *Roman Glass: Reflections of Everyday Life* (Philadelphia, 1997); *idem, Roman Glass: Reflections on Cultural Change* (Philadelphia, 1999).

88 Dalby (2000): 188-9.

vessel and Petronius smashed another during his extended suicide to prevent him from having a second.

Romans replaced the bung on a partially emptied amphora and sealed it, probably with plaster, then used a private signet to ensure no thieving by cellarmen or others. Martial (*Ep.* 9.88) curses his wine steward for interrupting his enjoyment of a precious Opimian vintage with rank business matters and for saying his signet is required at the very moment for resealing a wine jar. Cicero's thrifty mother (*Ep. Fam.* 16.26) even resealed empty amphorae to ensure that those emptied on the sly could not be concealed among the legitimately consumed.[89]

Oddly enough, there was one form of service in Gaul that prevented resealing of amphorae but perhaps demonstrated the height of piety, if not pretentious fashion.[90] The symbolism of wine substituting for sacrificial blood in Gallic chthonic libations was sometimes reinforced by the practice of *sabrage des cols à coup de lame,* the prototype of modern *sabrage* of champagne bottles, in which the neck of the wine amphora was struck off with a quick stroke of a sword in the same way that the necks of sacrificial victims were cut to allow blood to drain into the earth. In some cases the vaguely anthropomorphic form of the amphora ('ears', 'neck', 'shoulders') was reinforced with the addition of incised elements such as lips, eyes and mouth. In several Gallic cultic deposits, human crania have been found along with wine vessels decanted in this way, suggesting that they replicated a real sacrifice.

The Convivium

Doubtless the most exuberant expression if wine culture in Rome was its use in the banquet, and, in a sense, its expression here leads us back to the very beginnings of wine history in the ancient world. As far back as we can trace the history of wine it was not just food and intoxicant but also a prestige product. Aristocracies in rigidly hierarchical societies have always been uninhibited in flaunting their social superiority, and one way to do this is by the conspicuous consumption of luxury products, not least wine. The social venue for the display and consumption of wine and other luxury foods was the banquet. As we have seen, when wine became a staple food product in Rome, the upper classes simply responded by identifying and acquiring vintage wines. But the banquet

89 Allen (1961): 88-9.

90 Maria Paraskera and Matthieu Poux, "Libations et repas sacrés dans le monde grec et gaulois," in Brun, Poux and Tchernia (2009): 182-6.

remained, as it long had been, the proper social milieu for consumption and display of these prestige wines.[91]

The banquet as diacritical marker of social standing can be traced back, as we have seen, to the protohistory of wine. Wine features prominently in the aristocratic banquets of the Assyrians, Minoans, Mycenaeans, and Archaic Greeks. In fact, the signal elements which most clearly mark banqueting as as symbol of privilege and status are the consumption of wine with luxurious foods, and particularly the consumption of both while reclining.[92] The first iconographic representation of the fact, as noted before, is the famous relief from Nineveh, the 'Garden Feast of Assurbanipal', which shows the Assyrian monarch reclining on a high couch sipping wine from a refined cup, while his consort sits nearby on a high throne and tipples as well. Servants attend his every need and, to reinforce the idea of power and elitism, we see the severed head of an enemy bound to the branches of a tree. Banquets feature in the iconography of the Egyptians and Minoans as well. First literary evidence in the Greek world derives from Homer's accounts where it is, of course, very prominent. The Greek banquet *par excellence* in the Golden Age was the *symposion*, though strictly speaking the term refers to the drinking bout which followed the food courses and included the conversation and entertainment of the males of elite households. Sadly, our evidence for banqueting in the Hellenistic period is lacunose, especially since the Greeks abandoned the practice of figured pottery during the period, and these images are among our best evidence for Greek banqueting protocols. Still, it is clear that among Hellenistic Greek practices adopted by the Romans, in addition to the reclining position

91 Katherine M. Dunbabin, *The Roman Banquet: Images of Conviviality* (Cambridge, UK, 2003) is now the definitive work, and forms the basis for my discussion, but see also Barbara Gold and John F. Donahue, *Roman Dining* (Baltimore, 2005); Matthew B. Roller, *Dining Posture in Ancient Rome: Bodies, Values and Status* (Princeton, NJ, 2006); Oswyn Murray, ed., *Sympotica: A Symposium on the Symposion* (Oxford, 1990), especially John D'Arms, "The Roman *Convivium* and the Ideal of Equality, op. cit. (1990): 308-20; William Slater, *Dining in a Classical Context* (Ann Arbor, MI, 1991); Elke Stein-Hölkeskamp, *Das römische Gastmahl* (Munich, 2005); Dirk Schnurbusch, *Convivium: Form und Bedeutung aristokratischer Geselligkeit in der römischen Antike* (Stuttgart, 2011); Eugenia Salzu Prina Ricotti, *L'arte del convito nella Roma antica* (Rome, 1983); Luciano Landolfi, *Banchetto e società romana. Dalle origini al I sec. a. C.* (Rome, 2002); Zaccaria Ruggia, *More regio vivere. Il banchetto aristocratico e la casa romana di età arcaica* (Rome, 2003); Maurizio Zampi, *Il triclinio. Mito e politica dell'alimentazione nell'antica* Roma (Rome, 2002).

92 Dunbabin (2003): 11-35. Cf. Jean-Marie Dentzer, *Le motif du banquet couché dans le Proche-Orient et le monde grec du VIIe au IVe siècle avant J.C.* [= *BEFAR* 246] (Rome, 1982).

of men, were the use of a designated room for the purpose and many of the service implements such as the *krater* and *cyathos*.

Etruscan banqueting appears in the iconography almost as early as that of Greece, specifically in the first quarter of the sixth century BCE. During the climax of Etruscan power in the seventh and sixth centuries, 'Etruscan' culture was common, not only throughout the Etruscan homeland, but in Latium and Campania as well. Needless to say, Etruscans were enthusiastic hellenizers, and wine culture was central to Etruscan banqueting culture. Earliest evidence of banqueting in Rome herself is a small, fragmentary terracotta revetment from the Palatine Hill, dated to the sixth century, which shows the usual pattern of reclining banqueters, including women, along with wine service and a flute girl. Here as so often it is impossible to distinguish Etruscan influence from that of Greece and especially of Magna Graecia, or perhaps of Greece as mediated by the Etruscans.[93] But is is perfectly clear that the ultimate source of wine practice is Greek. At the other end of the historical spectrum, it was the Romans themselves who purveyed banqueting culture along with wine culture to most parts of the western and northeastern provinces.

Early representation in Rome show all-male gatherings, or male gatherings where females present are marked by their state of undress and unrestrained sexual behavior as *hetaerae*, both suggestive of the *symposion*, but early on we begin to see gatherings with women portrayed in dress and demeanor as respectable *matronae*, participating on roughly equal terms with men, after the Etruscan fashion.[94]

We have already met public banquets in our discussion of prestige wines; for example, those banquets detailed by Pliny (*NH* 14.97) in which Julius Caesar as dictator when celebrating his Gallic triumph of 46 BCE apportioned at the public banquet an amphora of Falernian and one of Chian per table, as he had previously done when celebrating his Spanish triumph of 60. But during his third consulship of 46, he offered Falernian, Chian, Lesbian and Mamertine, the first time ever that four different wines were offered at a Roman banquet. Most public banquets, such as those offered by Caesar, must of necessity have taken place out of doors using moveable equipment.[95] We hear of the use of

93 On the export of 'heoic' drinking to Etruria, David Ridgway, "Nestor's Cup and the Etruscans," *Oxford Journal of Archaeology* 16.3 (1997): 325-44. Etruscan antecedents: Annette Rathjes, "Il Banchetto in Italia centrale. Qulae stile di vita?," in Murray and Tecuşan (1995): 167-75; Angela Ponfrandolfo, "Simposio e *élites* sociali nel mondo etrusco e italico," op. cit. (1995): 176-96.

94 Dunbabin (2003): 67-8

95 Dunbabin (2003): 92-102.

fora, porticoes, theaters, amphitheaters, and even the Colosseum for the purpose. Semiprivate banquets might be staged at inns such as the taberna of the House of Julia Felix at Pompeii (II.iv.7), or in the two pergolate *triclinia* of the Large Vineyard (III.7) there. Such luxury accommodations may have been rented to high-status travelers, but may also have been rented to local *collegia* (guilds) for their communal dinners. In fact, such communal dinners were one of the main functions of such guilds. Though many had *curiae* (guildhalls), few such halls had special dining rooms, so couches and tables must have been set up in other locations unless *triclinia* were rented. A few guildhalls actually contain *triclinia*; for example, a building outside Pompeii in the maritime district where three sequential *triclinia* were located along one side of the still partially excavated peristyle. These *triclinia* were equipped with water features such as jets in the middle of each table and more emptying into channels around the couches. There is also the Building of the Triclinia at Ostia (I.xii.1), the seat of the *collegium* of the carpenters (*fabri tignarii*) dating to ca. 120 CE. One wing of the guildhall contains four *triclinia* with masonry couches, and each *triclinium* will have held a dozen or more. Since a communal table was impractical for such numbers, food was served on a masonry ledge at the fronts of the couches. Finally, we should mention a building at Misenum, identified by inscriptions as the *curia* of the Augustales and dating to the Flavian and Trajanic periods. Here the eastern room is a *triclinium*. In Rome itself, the *curia* of the *praecones* (heralds) on the slopes of the Palatine had numerous iconographic representations of dining, though the dining room itself has not been identified.

Communal feasts were important to mark events such as religious rites, triumphs, imperial anniversaries, and funerals of prominent citizens. There were numerous religious *collegia* which held banquets in temples, not to speak of professional religious *collegia* such as that of the flute-players, who held their public banquets in the temple of Jupiter in Rome. Oddly, much as we hear of them in the literature, representations of such feasts are rare. One exception is a fragmentary relief from a state monument showing a banquet of the Vestals. Six Vestals, identified by their regalia, recline in a *triclinium*, one holding a large *poculum* (goblet), another bearing a jar and a plate with a single cake upon it, and we also see several servants in the background. This is the only known representation of a religious feast from Rome proper, but a small sample comes from towns in northern and central Italy. For example, an altar now in the Museo Nazionale at Este, of unknown provenience, shows a crowded group of figures reclining in the standard *triclinium*, a table at the front with drinking vessels, and two servants attending at the sides. There is a huge backdrop in the form of a curtain suspended from a tree. Another representation comes from an altar block from Umbrian Sentinum now in Ancona, which shows twelve men reclining in standard fashion around a circular table upon which are drinking vessels.

In addition, large numbers of inscriptions record the donation of a *cena* or *epulum* to the public by wealthy members of the community, and Dunbabin[96] makes a convincing case that in one representation from Amiternum we are shown the *cena* at the left with reclining dignitaries, and an *epulum* at the right with seated dignitaries of lesser rank. Thus, perhaps we should imagine the town's decurions at the left and the Augustales at the right, or some such scheme. The visual scheme should be compared with the terms we find in inscriptions, namely *discumbentibus* for reclining diners, *vescentibus* for those seated, and *epulantibus* for the general public who enjoy a picnic lunch on the grounds.

In the private sphere we are told by the historians that early Romans dined, both formally and informally, seated at tables in the atrium of the *domus*, but that the Romans adopted the use of the dining room, the *triclinium*, under Greek influence, perhaps again as mediated by the Etruscans. The typical Greek dining room was located in the *andron*, 'men's quarters'. The standard room was approximately square, with a door slightly off center at the front and a raised platform, the *kline-band*, around 3 1/2 sides at the back. On this couches were placed, head to tail, to accommodate individual diners, and before each was placed a separate table. The characteristic Roman form, on the other hand, emerges in the late Republic and becomes standard in the early Empire. It is a long, narrow room divided into a forepart for service and entertainment and a rear section where three broad couches (*lecti*) were placed in a *U* shape and amidst which a large rectangular or circular table was placed to which diners on all three couches had access. The standard dining pattern was no more than nine diners, three per couch, carefully seated according to social rank. In the last years of Pompeii there is a tendency towards larger *triclinia* which could accommodate larger parties. Typically rooms exhibit floor mosaics in a T + U plan, an undecorated U-shaped band around the periphery of the room where the *lecti* will have obscured the floor, and elaborate mosaics elsewhere, a smaller square under the central table and a larger one in the forecourt where guests entered and service and entertainment took place. Walls were elaborately frescoed as well. Already by the first century CE, however, we begin to see a new arrangement to accommodate a continuous, semicircular couch, the *stibadium* or *sigma* couch, a form which obviously derives from bolsters used for outdoor dining.

96 Dunbabin (2003): 72-83.

FIGURE 36 *The Roman* cena. *After Rich,* Dictionary, *s.v.*

There is an evident adherence to Greek Hellenistic models in Roman banqueting iconography in the first two centuries of the Common Era, but by the third and fourth a distinctly Roman iconography has imposed itself. This is especially true in depictions of banquet servants.[97] Earlier depictions show one, occasionally two servers, simply clad, usually short-haired males. By late antiquity, however, we often see seven or more, elaborately decked out in short, lavishly decorated tunics and, if they are male, with long, curled and styled hair. At the 'House of Bacchus' in Complutum, Spain, a mosaic in the corridor leading to the *triclinium* shows seven such pages, each proffering a wine goblet with right hand while holding a towel draped over his left forearm. That this and other iconography reflects real practice and not just artistic fashion is attested by the proliferation in the literature of titles for such pages. We find an *a cyatho* for one who ladles wine, an *a lacuna* who handles the wine flagon, the *minister poculi* who actually proffers the cups, and the *minister vini* in charge of the entire wine service. Otherwise we hear of food bearers, distributors of garlands and perfumes, operators of the heating apparatus for wine, and even a servant who mans a fan or fly-whisk.[98] Perhaps the most specialized functionary was one whose exact title we do not know; in several of the depictions,

97 K. M. Dunbabin, "The Waiting Servant in Later Roman Art," in Gold and Donahue (2005): 115-40 and Fig. 1-25; *idem* (2003): 150-6.

98 Dunbabin (2003): 150-6.

notably an illustration in the *Vergilius Romanus*[99] of Dido's feast, we see a page proffering a specially designed water basin with a long handle with his left hand and a flagon of water with his right. Draped across his left forearm is a towel. Obviously the diners in turn put out their hands over the basin, the page poured water over them, the diners laved their hands and then dried them on the towel.

Rituals of drinking are central to banqueting ideology, and it is therefore hardly surprising that in the artistic representations the drinking apparatus occupy a prominent place.[100] Drinkers hold silver cups or bowls, servants hurry forward with freshly filled cups or a flagon of wine. Many types of cups are represented, along with the ladle for transferring wine from mixing bowl to flagon or to cups and long-handled spoons for stirring mixed wine. Several hoards of silver service containing as many as 100 objects have been discovered, such as those from the House of Menander at Pompeii (I.x.4) and from the Villa della Pisanella at Boscoreale, both of which contained these objects. Silver cups and bowls are lavishly decorated in relief with motives of birds, animals, flowers, vines, and mythological—especially Dionysiac—scenes. Similarly, the paintings of the tomb enclosure of Vestorius Priscus at Pompeii, aedile in perhaps 70-71 CE, show the young magistrate seated on a tribunal on one of the panels, a pair of gladiators on another, doubtless to commemorate the *munus* he had organized as magistrate. On a third panel is represented a banquet, perhaps one he organized as a public benefaction, and on its right we see a table laden with silver service pieces, including a large mixing bowl, four matched pairs of cups, two matching wine flagons, two drinking horns resting on stands, a row of small ladles, and two mixing spoons. This *abacus* or *mensa vasaria* was a popular form of ostentation among wealthy Romans. Beneath the *mensa vasaria* are two other vessels whose color suggests that they are bronze, identified as the basin and flagon used in the hand-washing ritual.[101]

Refinements of wine service also appear. For example, on a *loculus* slab from Isola Sacra[102] we see a tall wicker hamper with a flat bottom and straight, squared sides, with a wicker cover attached at the side. At the top of this 'fiasco' appear the characteristic 'ears' of an amphorae, and thus we see a clever way to protect the wine supply and also keep it vertical without the need for a stand. Silver wine vessels appear in both iconography and archaeology far less in late

99 *Cod. Vat. Lat.* 3867, fol. 100v [= Dunbabin (2003): Pl. XVI].

100 Dunbabin (2003): 65-7.

101 Dunbabin (2003): 85-9.

102 Tomb II, ca. 152-60 CE: Museo Ostiense Inv. 1333 [= DAI Rome 69.734 = Dunbabin (2003): Fig. 67].

antiquity than before, for reasons previously mentioned, namely, the resort to glass as the luxury material *par excellence*. Especially impressive are the so-called 'cage cups' or *diatreta* (Gr. 'perforated', 'drilled through') with their extraordinary outer shells of intricate lattice work, frequently in a contrasting color from the inner core, created by mold-blowing a blank inside another blank and then laboriously using the lathe to cut away most of the outer blank to create the lattice work.[103]

Unlike *symposion* practice, where food and wine constituted different courses, prestige wines in Roman banquets were always accompanied by prestige foods. We hear of and see in the artwork exotic fruits and fishes, seafood, game, small birds and other luxury products, elegantly presented on silver or glass tableware. An unusual representation of such foods is known as the *asaratos oikos*, 'unswept room', i.e., the representation in mosaic of food debris thrown on the floor. The fact that these are typically floor mosaics creates a clever *trompe l'oeil* realism. The most famous is from Rome, now in the Vatican Museum, signed by a Heraklitos and probably to be dated from the second century CE. The mosaic depicts bones of fish and chicken, seafood shells, lobster claws, snail shells, nuts, cherries, vegetable detritus, and even a mouse nibbling on a discarded nut. We must assume that similar detritus littered the floor of the *triclinium* in a real banquet, along with puddles of spilled wine (and perhaps worse).[104] Some of the more luxurious foods we see in the art are suckling pig, roast fowl and fish of many kinds, arrays of vegetables, and breads fashioned in elaborate shapes. We even see depicted live animals such as flamingoes, presumably to be led to the kitchen for preparation.

But the most typical 'diacritical marker' of luxurious dining for the Romans remained wine. And here as elsewhere the accouterments of heated and cooled wine were elements of ostentation. For example, a fresco from the north wall of the House of the Chaste Lovers at Pompeii (IX.xii.6) shows an outdoor banquet scene, at the right of which servants pour wine from an amphorae into a silver *crater* or *situla* placed in a large, elaborately decorated silver basin supported on three legs. We must imagine that iced water filled the basin and chilled the wine/water mixture in the crater in exactly the same fashion as the modern champagne bucket.[105] The opposite, the heating of wine, provided yet another opportunity for asserting social status via ostenta-

103 Dunbabin (2003): 163-4; S. J. Fleming, *Roman Glass: Reflections on Cultural Change* (Philadelphia, 1999): 110.

104 Dunbabin (2003): 63-6.

105 Cf. Dunbabin (2003): 54 and Fig. 26. Compare another outdoor banqueting scene, evidently based on the same archetype, now in the Naples Museum (Museo Naz. 9015,

tion, in this case in the form of an elaborate mechanism called the *authepsa*, 'self-boiler'.[106] The contraption shows up in two basic forms, one egg-shaped and the other cylindrical, each of which has internal compartments to hold the fuel as well as various mechanisms for dispensing the hot water and releasing steam from the main component, the boiler. Examples from the archaeological record show clearly that these were exotic, luxury items. In the iconography, they do not appear in banquet scenes until the second century CE, but become a standard motif in the third and fourth.

Dining and drinking were frequently accompanied by forms of entertainment.[107] The production and serving of food could form a spectacle in their own right, witness the 'ungutted' pig of Trimalchio's Feast, which is, in fact, stuffed with sausages, or Trimalchio's 'Carver the Carver' (Petronius *Sat.* 36). Admittedly this is literary parody, but doubtless it represented in exaggerated form some of the more *outré* fashions for food service as spectacle, such as Athenaeus' (IX.381F) singing cooks. Otherwise we hear of and see musicians and dancers, acrobats and jugglers.

Odd as it may seem to modern sensibilities, the Roman banquet was also intimately associated with death.[108] Not only is the banquet a constant motif at the scene of death, the tomb, but motives of death were also common at banquets of the living, a sort of visual reminder, in an era when life was fragile, that one must indeed 'seize the day'. We have Trimalchio's little articulated silver skeleton which he has a slave display to his assembled guests, again, not a purely fictional invention to judge by the numerous images of skeletons on cups, mosaics, and frescoes, not to speak of bronze and silver examples of the skeletons themselves. And not only were the dead frequently depicted in literature and art as banqueting in the afterlife, but other representations show the families of the deceased banqueting, presumably at the tomb, either in conjunction with the funeral itself, or during the Feast of the Dead, the Parentalia, or perhaps on the anniversary of the death of the person entombed. Often, in fact, the iconography is so ambiguous that it is difficult to identify the exact milieu, and that ambiguity may be quite deliberate.[109] The so-called Totenmahl represents one of the most common of all funerary motives, in fact, with a history that stretches back to Hellenistic Rome and beyond. Further-

mid-first century). Etruscsan parallels from Chiusi: F. Wiel Marin, "Due diverse associazioni di vasi nel banchetto etrusco," *RM* 104 (1997): 514-18.

106 Dunbabin (2003): 166-8.

107 Dunbabin (2003): 264.

108 Dunbabin (2003): 132-40.

109 So Dunbabin (2003): 103-40.

more, several tombs have actual *triclinia* attached, such as that of Vibrius Saturninus at Pompeii. Often the sarcophagus itself is fashioned as a dining couch, upon which a representation of the deceased and his or her consort recline to banquet. These so-called *kline* monuments often have a small sculpted *tripodium* table before them with elaborate theriomorphic feet, containing the apparatus of drinking. Additionally, often the figure of the deceased holds a cup and in frescoes is attended by pages. If a wife is depicted she is typically shown seated at the feet of her husband, but she is sometimes shown reclining with him as well.

A special type of banquet was the informal outdoor banquet, a sort of picnic, and in this context, as previously noted, the typical seating arrangement is the *stibadium* or *sigma* couch. Food is place in containers on the ground or on a low table, and shade is provided by trees or an awning draped from tree limbs. But even here we see elegant pages and/or maidservants mixing and heating or cooling wine, washing the diners' hands, and serving food and wine.[110]

Surprisingly, it now appears that the *com(m)issatio* of the handbooks, that is, a separate drinking bout after the meal, by analogy with Greek practice, is illusory.[111] The term instead seems to be consistently used in the sense of a movement from one *cena/convivium* to another in a rowdy, drunken way, and thus *ire comissatum* is roughly 'to go reveling', perhaps somewhat analogous the the modern American college students' 'pub crawl'. It is the Roman equivalent of the Greek *komos*. On the other hand, there is no doubt that the convivium could extend for several hours after food had ceased to be served, and that during this part of the banquet drinking continued while drinking games, more or less refined conversation was enjoyed, and forms of entertainment were enjoyed.

Banqueting in all its manifestations easily survived the Christian conversion. For example, the so-called Sevso Treasure from ca. 350-450 CE and of unknown provenience, contains a huge silver salver with a picnic scene in niello on the central boss, showing all the standard elements of a *stibadium* picnic. But the dedicatory inscription around the boss, with its *monogramma Christi,* indicates that Sevso's family were Christians. Other representations, including funerary paintings on the *loculi* of many catacomb burials, also show the standard elements. We do notice two modifications, however. First, although hunting as an aristocratic pastime had been a standard motif in the

110 Dunbabin (2003): 131-2.

111 Matthew B. Roller, "Convivial Wine Drinking and Comissationes," in Roller, *Dining Postures in Ancient Rome* (Princeton, NJ, 2006): 181-8.

literature and art from time immemorial, it takes a more prominent position in the iconography of late antiquity, and often the hunt and the picnic which followed it are depicted together. Second, rooms designed specifically to accommodate the *stibadium* begin to occur in the archaeology of late antiquity as, for example, at a villa from El Ruedo (Almedinilla) in Roman Baetica, dating from the late third to the early fifth centuries, and the so-called Villa of the Falconer at Argos, where a mosaic floor pattern indicates the position of the couch and of the semicircular or circular table placed in its inner curve.[112]

112 Dunbabin (2003): 141-74.

Bibliography

Abulafia, David. *The Great Sea: A Human History of the Mediterranean*. Oxford, 2011.

Ahlström, G. W. "Wine Presses and Cup-Marks of the Jenin-Megiddo Survey," *BASOR* 231 (Oct., 1978): 19-49.

Algaze, Guillermo. "Fourth-Millennium BC Trade in Greater Mesopotamia: Did It Include Wine?," in McGovern, Fleming and Katz (1996): 89-96.

Allam, Adel H. et al. "Atherosclerosis in Ancient and Modern Egyptians," *Global Heart* 9.2 (June, 2012): 197-202.

Allen, H. Warner. *A History of Wine: Great Vintage Wines from the Homeric Age to the Present Day.* London, 1961.

Alp, Sedat. *Song, Music and Dance of the Hittites: Grapes and Wine in Anatolia during the Hittite Period*, trans. Y. Eran. Ankara, 2000.

Amerine, M. A. et al. *The Technology of Winemaking.* Westport, CT, 1980, 4th ed.

Amouretti, Marie-Claire. "Oléiculture et viticulture dans la Grèce antique," in Berit Wells, ed., *Agriculture in Ancient Greece: Proceedings of the Seventh International Symposium at the Swedish Institute at Athens, 16-17 May 1990* (Stockholm, 1992): 77-86.

Amouretti, Marie-Claire and Jean-Pierre Brun, edd. *La production du vin et de l'huile au Mediterranée.* Paris, 1993.

Amouretti, Marie-Claire. "Les sous-produits de la fabrication de l'huile et du vin," in Amouretti and Brun (1993): 463-76.

Ampolo, C. "Per un studio dell'alimentazione dell'Etruria e di Roma arcaica," in Ministero dei beni culturali, *L'alimentazione nel mondo antico* (Rome, 1987): 1-22.

André, Jacques. "Contribution au vocabulaire de la viticulture: les noms de cépages," *REL* 30 (1953): 126-56.

Aranguren, di Biancamaria et al. "L'avvio della coltura della vite in Toscana: l'esempio di San Lorenzo à Greve (Firenze)," in Ciacci, Rendini and Zifferero (2007): 88-97.

Arcuri, Flaminia. "Il vino in età romana," in Luigi Rossi, ed., *Il vino nel Cilento* (Acciaroli, SA, 1994): 37-59.

Attolini, I. and P. Perkins. "The Excavation of an Etruscan Farm at Podere Tartuchino," *PBSR* 60 (1992): 1-76.

Aviram, Michael and Bianca Fuhrman. "Wine Flavanoids, LDL Cholesterol, Oxidation and Atherosclerosis," in Sandler and Pinder (2003): 140-59.

Ayalon, Etan, Rafael Frankel and Amos Kloner. *Oil and Wine Presses in Israel from the Hellenistic, Roman and Byzantine Periods.* Oxford, 2009.

Backe-Fosberg, Y. *Crossing the Bridge: An Interpretation of the Archaeological Remains in the Etruscan Bridge Complex at San Giovenale, Etruria.* Uppsala, 2005.

Badler, Virginia. "The Archaeological Evidence for Winemaking, Distribution and Consumption at Protohistoric Godin Tepe, Iran," in McGovern, Fleming and Katz (1996): 45-56.

Bakels, C. and S. Jacomet. "Access to Luxury Foods in Central Europe during the Roman Period: The Archaeological Evidence," *World Archaeology* 34 (2003): 542-57.

Bakhuizen, S. C. *Chalcis-in-Euboea: Iron and Chalcidians Abroad.* Leiden, 1976.

Bakhuizen, S. C. "Torcula Graecanica: A Note on the Archaeology of Olive and Grape Pressing," in M. Gnade, ed., *Stips Votiva: Papers Presented to C. M. Stibbe* (Amsterdam, 1991): 1-6.

Barker, Graeme. *Landscape and Society: Prehistoric Central Italy.* London, 1981.

Bartoloni, Gilda. *La cultura villanoviana: all'inizio della storia etrusca.* Rome, 1989.

Basile, Boris, Gaetano Di Pasquale, Antonella Monaco and Mauro Vella. "Sopravvivenza dell'antico paesaggio della vite in Campania," in Di Pasquale (2010): 97-105.

Beard, Mary. *The Fires of Vesuvius: Pompeii Lost and Found.* Cambridge, MA, 2010.

Beck, C. W. and C. Borromeo. "Ancient Pine Pitch: Technological Perspectives from a Hellenistic Wreck," in W. R. Biers and P. E. McGovern, edd., *Organic Contents of Ancient Vessels: Materials Analysis and Archaeological Investigation* [= *MASCA Research papers in Science and Archaeology* 7] (Philadelphia, 1990): 51-8.

Beck, C. W., Douglas R. Stewart and Elizabeth C. Stout. "Appendix D: Analysis of Naval Stores from the Late-Roman Ship," in Anna Maurite McCann and Joann Freed, edd., *Deep Water Archaeology: A Late Roman Ship from Carthage and the Ancient Trade Route Near Skerki Bank Off Northwest Sicily* [= *JRA Suppl. Series* 13] (Ann Arbor, MI, 1994): 122-6.

Beck, C. W., Elizabeth C. Stout, Karen C. Lee, Adrien A. Chase and Nicole DeRosa. "Analysis of Organic Remains in the Fabric of Minoan and Mycenaean Pottery Sherds by Gas Chromatography-Mass Spectrometry," in Martlew (2008): 12-47.

Becker, Jeffrey and Nicola Terrenato, edd. *Roman Republican Villas: Architecture, Context and Ideology.* Ann Arbor, MI, 2012.

Becker, W. "Production and Consumption of Wine: Facts, Opinions, Tendencies," in C. F. Gastineau, W. J. Darby and T. B. Turner, edd., *Fermented Food Beverages in Nutrition* (New York, 1979): 157-86.

Beltrán, Lloris. *Las ánforas romanas en España.* Zaragoza, 1970.

Bencivenga Trillmich, Clara. "Pyxous-Buxentum" *MEFRA* 106 (1988): 701-29.

Berthault, Frédéric. "*Biturica*, le vin des Bituriges," in Brun, Poux and Tchernia (2009): 236-8.

Bianco, Salvatore. "Vino e simposio nelle communità dell'Enotria," in Di Pasquale (2010): 120-31.

Billiard, Raymond. *La vigne dans l'antiquité.* Lyon, 1913.

Bird, David. *Understanding Wine Technology.* Warwickshire, UK, 2010, 3rd ed.

Bisconti, Fabrizio. *Mestieri nelle catacombe romane: appunti sul declino dell'iconografia del reale nei cimiteri cristiani di Roma.* Rome, 2000.

Bisconti, Fabrizio. "Scena di commercio del vino in un relievo inedito della regione dell'ex Vigna Chiaraviglio in S. Sebastiano," *Rivista di Archeologia Cristiana-Roma* 79 (2003): 15-44.

Bisconti, Fabrizio. "Una taverna sulla via Appia," *Archeo* 22.9 (2006): 60-5.

Bobak, Martin and Michael Marmot. "Wine and Heart Disease: A Statistical Approach," in Sandler and Pinder (2003): 108-39.

Boehm, E. W. and B. G. Coombe. "Vineyard Establishment," in Coombe and Dry (1988): 23-41.

Bömer, F. "Juppiter und die römischen Weinfeste," *Rheinisches Museum* 90.1 (1941): 30-58.

Boissinot, Philippe. "L'archéologie des vignobles antiques en France méridionale," in Ciacci, Rendini and Zifferero (2007): 35-41.

Boissinot, Philippe. "De la vigne au pressoir," in Brun, Poux and Tchernia (2009): 196-207.

Bökönyi, S., L. Costantini and J. Fitt. "The Farming Economy," in M. Gualtieri, ed., *Fourth Century B.C. Magna Graecia: A Case Study* (Partille, Sweden, 1993): 281-307.

Bonfante, Larissa, Giovannangelo Camporeale, Alexandra Carpino and John R. Clarke. *New Perspectives on Etruria and Early Rome.* Madison, WI, 2009.

Bonifay, Michel. "Le vin d'Afrique," in Brun, Poux and Tchernia (2009): 288-9.

Bonanno, Anthony. "Maltese Wine Pressing in Antiquity," *Melita Historica* 15.1 (2008): 1-18.

Boothroyd, Rodney. *Home Winemaking: Techniques and Recipes.* New York, 1986.

Botti, Aniello, David L. Thurmond and Fernando La Greca. "Un Palmento ben conservato a Novi Velia ed altri palmenti nel territorio del Cilento. Osservazioni ed ipotesi," *Annali Storici di Principato Citrà* 9.2 (2012): 5-52.

Bouvier, Michel. "Recherches sur les goûts des vins antiques," *Pallas* 53 (2000): 115-33.

Brandt, J. Rasmus and Lars Karlsson, edd. *From Huts to Houses: Transformations of Ancient Societies. Proceedings of an International Seminar Organized by the Norwegian and Swedish Institutes in Rome, 21-24 September 1997.* Stockholm, 2001.

Brears, Peter. "Pots for Potting: English Pottery and Its Role in Food Preservation in the Post-Medieval Period," in C. Anne Wilson, ed., *Waste Not, Want Not* (Edinburgh, 1991): 32-65.

Brehaut, E. *Cato the Censor On Farming.* New York, 1933.

Brun, Jean-Pierre. "Discrimination entre les installations oléicoles et vinicoles," in Amouretti and Brun (1993): 532-4.

Brun, Jean-Pierre. "Pressoirs et chais," in Tchernia and Brun (1999): 48-107.

Brun, Jean-Pierre and Fanette Laubenheimer, edd. "La viticulture en Gaule," *Gallia* 58 (2001).

Brun, Jean-Pierre. *Archéologie du vin et de l'huile dans l'Empìre romain.* Paris, 2004.

Brun, Jean-Pierre. *Archeologie de l'huile et du vin en Gaule romaine.* Paris, 2005.

Brun, Jean-Pierre. "Le techniche di spremitura dell'uva: origini e sviluppo dell'uso del torchio nel Mediterraneo occidentale," in Ciacci, Rendini and Zifferero (2007): 55-67.

Brun, Jean-Pierre, Matthieu Poux and André Tchernia. *Le vin: Nectar des dieux, génie des hommes.* Gollion, Switzerland, 2009.

Brun, Jean-Pierre and Fanette Laubenheimer. "Le vignoble gaulois," in Brun, Poux, Tchernia (2009): 217-29.

Brun, Jean-Pierre and André Tchernia. "Vendanges et vinification," in Brun, Poux and Tchernia (2009): 242-59.

Buchanan, G. A. and T. G. Amos. "Grape Pests," in Coombe and Dry (1988): 209-31.

Buxó, R. "Evidence for vines and ancient cultivation from an urban area, Lattes (Hérault), Southern France," *Antiquity* 70 (1996): 393-407.

Calabresi, A. T. "Vin Santo and Wine in a Tuscan Farmhouse," in M. Douglas, ed., *Constructive Drinking: Perspectives on Drink from Anthropology* (Cambridge, UK, 1987): 122-34.

Camin, L. and W. McCall. "Settlement Patterns and Rural Habitation in the Middle Cecina Valley Between the Hellenistic to Roman Age: The Case of Podere Cosciano," *Etruscan Studies* 9 (2002): 19-27.

Campbell, Christy. *Phylloxera: How Wine Was Saved for the World.* London, 2004. Republished in the U.S. as *The Botanist and the Vintner: How Wine Was Saved for the World.* Chapel Hill, NC, 2006.

Carandini, Andrea et al. *Schiavi e padroni nell'Etruria romana: la Villa de Settefinestre dallo scavo alla mostra.* Bari, 1979.

Carandini, Andrea. "Columella's Vineyard and the Rationality of the Roman Economy," *Opus* 2 (1983): 177-204.

Carandini, Andrea and A. Ricci, edd. *Settefinestre: Una villa schiavistica nell'Etruria romana.* Modena, 1985.

Carandini, Andrea. "La villa romana e la piantagione schiavistica," in Giulio Einaidi, ed., *Storia di Roma,* vol. 4 (Torino, 1988): 101-200.

Carandini, Andrea, ed. *La fattoria e villa dell'auditorium nel quartiere Flaminio di Roma.* Rome, 2007.

Carre, Marie-Brigette. "Transport en vrac," in Brun, Poux and Tchernia (2009): 290-1.

Carrington, R. C. "Studies in the Campanian 'Villae Rusticae'," *JRS* 21 (1931): 110-30.

Carrington, R. C. "Some Ancient Italian Country Houses," *Antiquity* 8 (1934): 261-80.

Carter, J. C. *Discovering the Greek Countryside at Metaponto.* Ann Arbor, MI, 2006.

Casson, Lionel. *Ships and Seafaring in Ancient Times.* Austin, TX, 1994.

Castelletti, L. "Rapporto preliminare sui resti vegetali della serie Neolitico-Bronzo di Pienza (Siena)," *Rivista Archeologica dell'Antica Provincia e Diocesi di Como* 156-7 (1974-5): 243-51.

Castrén, Paavo. *Ordo Populusque Pompeianus.* Rome, 1975.

Cavalieri, Duccio, Patrick E. McGovern, Daniel C. Hartl, Robert Mortimer and Mario Polsinelli. "Evidence for *S. cerevisiae* Fermentation in Ancient Wine," *Journal of Molecular Evolution* 57 (2003): S226-S232.

Cerchiai, Luca, Lorena Jannelli and Fausto Longo. *The Greek Cities of Magna Graecia and Sicily.* Los Angeles, 2004.

Chatier, François. *Taste Buds and Molecules: The Art and Science of Food, Wine and Flavor.* New York, 2009.

Chavallier, R. *La romanisation de la Celtique du Pô.* Rome, 1983.

Cheze, Catherine, Joseph Vercauteren and Robert Verpoote, edd. *Polypheols, Wine and Health* [= *Proceedings of the Phytochemical Society of Europe* 48]. Dordrecht, the Netherlands, 2001.

Ciacci, Andrea and Andrea Zifferero, edd. *Vinum: un progetto per il riconoscimento della vite silvestre nel paesaggio archeologico della Toscana e de Lazio settentrionale.* Siena, 2005.

Ciacci, Andrea. "Introduzione al 'Progetto Vinum': un progetto per il riconoscimenta della vite silvestre nel paesaggio archeologico della Toscana e del Lazio settentrionale," in Ciacci-Zifferrero (2005): 15-40.

Ciacci, Andrea, Paola Rendini and Andrea Zifferrero. *Archeologia della vite e del vino in Etruria.* Siena, 2007.

Ciacci, Andrea and Andrea Zifferero. "Il 'Progetto Vinum': Prime considerazioni conclusive," in Ciacci, Rendini and Zifferrero (2007): 249-72.

Ciacci, Andrea. "La ricostruzione del paesaggio vitivinicolo antico: l'indagine sui vitigni e la circolazione varietale," in Di Pasquale (2010): 74-9.

Ciacci, Andrea and Myriam Giannace. *Senarum vinea: Il paesaggio urbano di Siena. Forme di ricupero e valorizzazione dei vitigni storici.* Siena, 2012.

Ciampoltrini, Giulio and Poala Rendini. "Vie e porti del vino nella valle dell'Albegna in età Etrusca (VI-V secolo a.C.)," in Ciacci, Rendini and Zifferero (2007): 176-84.

Cicerelli, Caterina, Claude Albore Livadie, Lorenzo Costantini and Matteo delle Donne. "La vite a Poggiomarino, Longola: Un contesto di vinificazione dell'Età del Ferro," in Pietro G. Guzzo and Maria P. Guidobaldi, edd., *Nuove ricerche archeologiche nell'area vesuviana. Atti del Convegno Internazionale, Roma 1-3 Febbraio 2007* (Rome, 2008): 574-5.

Cogrossi, Cornelia. "Il vino nel 'Corpus iuris' e nei glossatori," in Gabriele Archetti, ed., *La civiltà del vino. Fonti, temi e produzioni vitivinicole de Medioevo al Novecento* [= *Atti del convegno, Monticelli Brusati, Antica Fratta, 5-6 Ottobre 2001*] (Brescia, 2003): 499-531.

Comba, R. *Vigne e vini nel Piemonte antico.* Alba, 1994.

Coombe, B. G. and P. R. Dry, edd. *Viticulture: Practices* (vol. 2). Ashford, South Australia, 1988, 2nd ed.

Costantini, L. "Semi e carboni del mesolitico e neolitico della Grotta dell'Uzzo, Trapani," *Quartenaria* 23 (1981): 233-47.

Costantini, L. and L. Costantini Biasini. "Cures Sabini-paletnobotanica. Nota preliminare," *Archeologia Laziale* 7.1 (1985): 86-90.

Costantini, L. "The Origin of the Mediterranean Diet in Italy," *Rivista di Antropologia* 76 Suppl. (1988): 7-15.

Costantini, L. "Archeologia della vitivinicoltura in Basilicato: un bilancio delle ricerche archeobotaniche a Pizzica Pantanello, Metaponto," in Ciacci, Rendini and Zifferero (2007): 98-105.

Cotton, M. Aylwin. *The Late Republican Villa at Posto Francolise.* London, 1979.

Cotton, M. Aylwin and G. Metraux. *The San Rocco Villa at Francolise.* London, 1985.

Cox, Jeff. *From Vines to Wines: The Complete Guide to Growing Grapes and Making Your Own Wine.* North Adams, MA, 1999.

Creasy, Glen L. and Leroy L. Creasy. "Grape-derived Flavanoids and Stilbenes," in Sandler and Pinder (2003): 199-227.

Crispino, Luigi. *Virtu terapeutiche del vino.* Acciaroli, SA, 1997.

Crispino, Luigi. *Alimentazione mediterranea e stile di vita.* Acciaroli, SA, 2008.

Cristofani, M. "Duo sunt liquores," in *L'alimentazione nel mondo antico* (Rome, 1987): 37-40.

Cunliffe, Barry, ed. *The Illustrated History of Prehistoric Europe.* Oxford, 1994.

Cunliffe, Barry. "Iron Age Societies in Western Europe and Beyond, 800-140 BC," in Cunliffe (1994): 336-72.

Curtel, G. *La vigne et le vin chez les Romans*. Paris, 1903.

Curtis, Robert I. *Ancient Food Technology.* Leiden, 2001.

Curtis, Robert I. "Storage and Transport," in Wilkins and Nadeau (2015): Ch. 16.

Dalby, Andrew. *Empire of Pleasurers: Luxury and Indulgence in the Roman World.* London, 2000.

Dalby, Andrew. *Food in the Ancient World from A to Z.* London, 2003.

D'Arms, John H. *Romans on the Bay of Naples.* Cambridge, MA, 1970.

D'Arms, John H. "The Roman *Convivium* and the Ideal of Equality," in Murray (1990): 308-20.

D'Arms, John H. "Ville rustiche e ville di 'otium'," in F. Zevi, ed., *Pompeii 79: Raccolta di studi per il decimonono centenario dell'eruzione vesuviana* (Naples, 1979): 65-86.

D'Arms, John H. and E. C. Kopff. *The Seaborne Commerce of Ancient Rome: Studies in Archaeology and History* [= *MAAR* 36]. Rome, 1980.

Day, J. "Agriculture in the Life of Pompeii," *Yale Classical Studies* 3 (1932): 166-208.

de Blij, H. J. *Wine: A Geographic Appreciation.* Totowa, NJ, 1983.

De Caro, Stefano. "Anfore per pece del Bruzio," *Klearchos* 105-08 (1985): 21-32.

De Caro, Stefano. *La villa rustica in località villa Regina a Boscoreale.* Rome, 1994.

De Cazanove, Olivier. "Rituels romains dans les vignobles," in Murray and Tecuşan (1995): 214-23.

Della Corte, M. "Altra villa rustica, esplorata da Signore Giovanni di Palma, nel fondo 'Agricoltura' di sua proprietà in contrada Pisanella, commune di Boscoreale, l'anno 1906 (giorni 15) e l'anno 1908 (mesi 6)," *NS* (1921): 46-7.

Delpino, Filippo. "L'ellenizzazione dell'Etruria villanoviana: sui rapporti tra Grecia e Etruria fra il IX e VII secolo a.C.," in *Atti del 2° Convegno Internazionale Etrusco* (Rome, 1985): 105-16.

Delpino, Filippo. "Viticoltura, produzione e consumo del vino nell'Etruria protostorica," in Ciacci, Rendini and Zifferero (2007): 133-46.

Dentzer, Jean-Marie. *Le motif du banquet couché dans le Proche-Orient et le monde grec du VIIe au IVe siècle avant J.C.* [= *BEFAR* 246]. Rome, 1982.

De Romanis, F. *Crossings: Early Mediterranean Contacts with India.* New Delhi, 1997.

De Rossi, Gianfranco. "Indicatori archeologici della produzione e diffusione del vino della Baia di Napoli in età altomedievale," *insulae Diomedeae* 4 (2003): 541-9.

Desbat, Armand. "Le commerce des vins à Lugdunum," in Brun, Poux and Tchernia (2009): 264-73.

de St. Denis, E. "Falx vinitoria," *Revue Archéologique* 41 (1953): 163-76.

De Vos, Arnold. *Pompei, Ercolano, Stabia.* Rome, 1982.

Dietler, M. "Driven by Drink: The Role of Drinking in the Political Economy and the Case of Early Iron-Age France," *Journal of Anthropological Archaeology* 9 (1990): 352-406.

Dietler, M. "Feasts and Commensal Politics in the Political Economy: Food, Power and Status in Prehistoric Europe," in Wiessner and Schiefenhovel (1996): 87-125.

Dion, Roger. "La viticulture dans l'antiquité grecque," *Revue des deux Mondes* (Feb, 1952): 465-86.

Dion, Roger. *L'histoire de la vigne et du vin en France.* Paris, 1959.

Di Pasquale, Gaetano and Elda Russo Ermolì. "L'ambiente della vite nell'area tirrenica e le evidenze della sua coltivazione in epoca etrusca," in Di Pasquale (2010): 62-5.

Di Pasquale, Giovanni, ed. *Vinum nostrum. Arte, scienza e miti del vino nelle civiltà del Mediterraneo antico.* Florence, 2010.

Drachmann, A. G. *Ancient Oil Mills and Presses.* Copenhagen, 1932.

Dufkoá, M. and J. Pecírka. "Excavations of Farms and Farmhouses in the Chora of Chersonesos in the Crimea," *Eirene* 8 (1970): 123-74.

Dunbabin, Katherine M. "Wine and Water at the Roman Convivium," *JRA* 6 (1993): 116-41.

Dunbabin, Katherine M. *The Roman Banquet: Images of Conviviality.* Cambridge, UK, 2003.

Dunbabin, Katherine M. "The Waiting Servant in Later Roman Art," in Gold and Donahue (2005): 115-40 and Fig. 1-25.

Dunbabin, T. J. *The Western Greeks: The History of Sicily and South Italy from the Foundation of the Greek Colonies to 480 B. C.* Oxford, 1948.

Dyson, Stephen. *Archaeology and History in Sardinia from the Stone Age to the Middle Ages: Shepherds, Sailors and Conquerors.* Philadelphia, 2007.

Eisinger, Josef. "Lead and Wine: Eberhard Gockel and the Colica Pictonum," *Medical History* 26 (1982): 279-302.

Eistreicher, Stefan K. *Wine: From Neolithic Times to the 21st Century.* New York, 2006.

Etienne, Robert and Françoise Mayet. *Le vin hispanique.* Paris, 2000.

Farkas, T. *Technology and Biochemistry of Wine.* New York, 1988.

Fasciato, Micheline. "'Ad quadrigam fori vinarii': autour du port au vin d'Ostie," *Mélanges d'archéologie et d'histoire* 59.1 (1947): 65-81.

Fatucchi, A. "Palmenti romani da uva dell'Etruria nord-orientale," *Rivista di Storia dell'Agricoltura* 27.2 (1987): 9-29.

Fayer, C. *Aspetti di vita quotidiana nella Roma arcaica: dalle origini all'età monarchica.* Rome, 1982.

Figueiral, I. et al. "Archaeology, vine growing and wine production in Roman Southern France: The site of Gasquinoy (Béziers, Hérault)," *Journal of Archaeological Science* 37 (2010): 139-49.

Fitton Brown, A. D. "Black Wine," *CR* 12 (1962): 192-5.

Fleming, Stuart J. *Roman Glass: Reflections of Everyday Life.* Philadelphia, 1997.

Fleming, Stuart J. *Roman Glass: Reflections on Cultural Change.* Philadelphia, 1999.

Flouretzos, Pavlos. "La production de vin chypriote à l'époque préhistorique," in Brun, Poux and Tchernia (2009): 38-9.

Forbes, Patrick. *Champagne: The Wine, the Land, and the People.* London, 1967.

Forbes, R. J. *Studies in Ancient Technology, vol 3.* Leiden, 1965, 2nd ed.

Formenti, Françoise and J. M. Duthel. "The Analysis of Wine and Other Organics Inside Amphoras of the Roman Period," in McGovern, Fleming and Katz (1996): 79-85.

Forni, Gaetano. "Tappe e tracce della domesticazione della vite," in Ciacci-Zifferrero (2005): 41-6.

Forni, Gaetano. "Quando e come sorse la viticoltura in Italia," in Ciacci, Rendini and Zifferrero (2007): 69-81.

Fracchia, Helena. "The Romanization of the Ager Buxentinus," in E. Lo Cascio and A. Storchi Marino, edd. *Modalità insediative e strutture agrarie nell'Italia meridionale in età romana* (Bari, 1994): 55-73.

François, Paul, ed. *Le vin de Rome* [= *Pallas: Revue d'études antiques* 53] (2000).

Frazer, J. Edwin, ed. *The Roman Villa. Villa Urbana.* Philadelphia, 1998.

Freeman, B. M., E. Tassie and M. D. Rebbechi. "Training and Trellising," in Coombe and Dry (1988): 42-65.

Fregoni, M. *Origini della vite e della viticoltura. Contributi dei populi antichi.* Rome, 1991.

Fugazzola Delpino, Maria Antonietta, Andrea Pessina and Vincenzo Tiné, edd. *Civiltà dell'argilla. Le prime communità del Neolitico.* Dragoni, CE, 2004.

Fugelsang, Kenneth C. *Wine Microbiology.* New York, 1997.

Gagneux-Granade, Marguerite. "Retica, le vin des Rhétes," in Brun, Poux and Tchernia (2009): 239-40.

Galli, E. "Alla ricerca di Sibari," *Atti e Memorie della Società Magna Grecia* (Rome, 1929): 46-8.

Galli, E. "Due ville romane in agro Sibaritano," in *Atti del II Congresso Nazionale di Studi Romani*, vol. 1 (Rome, 1931): 267.

Gazzetti, Gianfranco. *La villa dei Volusii a Fiano romano.* Rome, 1994.

Giannace, Myriam. "Archeologia della viticoltura: metodi di coltivazione e lavorazione dell'uva nell'antichità preromana," in Ciacci and Zifferero (2005): 75-81.

Giannace, Myriam. "I comprensori indagati nel 2005. l'alta Valle dell'Ombrone (SI), l'alta Valdese (SI), la Val d'Orcia (SI), la bassa Val di Cornia (LI), l'alta Valle dell'Albegna," in Ciacci, Rendini and Zifferero (2007): 201-05.

Giannace, Myriam, A. Masi and F. Vallelonga. "Il 'Progetto Vinum': vite silvestri e siti archeologici," in Ciacci-Zifferero (2005): 39-45.

Gilles, Karl-Josef. "Die erste Blüte des Moselweins. In der Spätantike boomt die Arbeit in der Kelteranlagen der Region," *Antike Welt* 36.4 (2005): 29-35.

Gold, Barbara and John F. Donahue, edd. *Roman Dining.* Baltimore, 2005.

Goldberg, David M., Joseph Yan and George J. Soleas. "Absorption of Three Wine-Related Polyphenols in Three Different Matrices by Health Subjects," *Clinical Biochemistry* 36 (1) (2003): 79-87.

Goode, Jamie. *The Science of Wine: From Vine to Glass.* Berkeley, 2005.

Gorny, Ronald L. "Viniculture and Ancient Anatolia," in McGovern, Fleming and Katz (1996): 133-74.

Grace, Virginia R. *Amphoras and the Ancient Wine Trade* [= *Picture Book 6, Excavations of the Athenian Agora*]. Princeton, 1979.

Gras, Michel. "Vin et société à Rome et dans le Latium à l'epoque archaîque," in *Modes de contact et processus de transformation dans les sociétés anciennes: Actes du Colloque de Cortona (24-30 Mai 1981)* (Pisa-Rome, 1983): 1067-75.

Gras, Michel. *Trafics tyrrhéniens archaïques.* Rome, 1985.

Greco, E. "Poseidonia entre le VIe et la IVe siècle avant J.C.: quelques problèmes de topographie historique," *Revue Archéologique, Nouvelle Série, Fasc. 2* (1979.2): 219-34.

Green, Carin M. C. "The Shepherd of the People: Varro on Herding for the Villa Publica in *De re rustica* 2," in Becker and Terrenato (2012): 32-44.

Greene, Joseph A. "The Beginnings of Grape Cultivation and Wine Production in Phoenician/Punic North Africa," in McGovern, Fleming and Katz (1996): 311-22.

Guadagno, Giuseppe. *Campania antica: storia, archeologia, vita quotidiana.* Benevento, 2009.

Gualtieri, M. *Fourth Century BC Magna Graecia: A Case Study.* Partille, Sweden, 1993.

Guasch-Jané, M. R. et al. "First Evidence of White Wine in Ancient Egypt from Tutankhamon's Tomb," *Journal of Archaeological Science* 33.8 (2006): 1075-80.

Guasch-Jané, M. R. *Wine in Ancient Egypt: A Cultural and Analytical Study* [= *BAR International Series* 1851]. Oxford, 2008.

Gulli, Claudio. "L'iconografia della vite nella produzione figurativa sienese dalle origini al XIX secolo," in Ciacci and Gianacce (2012): 133-53.

Guzzo, Pietro Giovanni et al. *Casali di ieri, casali di oggi.* Naples, 2000.

Halliday, James and Hugh Johnson. *The Art and Science of Wine.* Buffalo, NY, 2007, 2nd ed.

Hayden, Brian. "Feasting in Prehistoric and Traditional Societies," in Wiessner and Schiefenhovel (1996): 127-48.

Heltzer, Michael. "Vineyards and Wine in Ugarit (Property and Distribution)," *Ugarit Forschungen* 22 (1990): 119-35.

Hesnard, A. and Ch. Lemoine. "Les amphores du Cécube et du Falerne. Prospections, typologie, analyses," *MÉFRA* 93.1 (1981): 243-395.

Hesnard, A. "Aire de production des gréco-italiques et des Dr. 1," in *Amphores romaines et histoire économique. Dix ans de recherche. Actes du colloque de Sienne (22-24 mai 1986)* [= *Coll. de l'École française de Rome* 114] (1989).

Heurgon, J. "Les Lasii pompéiens et l'importation des vins en Gaul," *Parola del Passato* 7 (1952): 113-8.

Hodos, Tamar. "Wine Wares in Protohistoric Eastern Sicily," in Christopher Smith and John Serrati, edd., *Sicily from Aeneas to Augustus: New Approaches in Archaeology and History* (Edinburgh, 2000): 41-54.

Holloway, R. Ross. *The Archaeology of Ancient Sicily.* London, 2002.

Hornsey, Ian. *The Chemistry and Biology of Winemaking.* Cambridge, UK, 2007.

Horsnaes, Helle W. *The Cultural Development in North Western Lucania c. 600-273 BC.* Rome, 2002.

Hue, Michel and Abbaye de Flaran. *Lestagnac, un chai gallo-romain: La viticulture en Gascogne antique.* Valence-sur-Baise (Gers), 2003.

Imazio, S. and M. Labro. "I marcatori molecolari e il loro impiego nello studio della domesticazione della vite, attraverso il confronto con la vite selvatica," in Ciacci-Zifferrero (2005): 57-66.

Imazio, S., M. Labra, A. Scienza and O. Failla. "Chloroplast Microsatellites to Investigate the Origin of the Grapevine," *Genetic Resources and Crop Evolution* 53 (2006): 1003-11.

Imazio, S. et al. "'Progetto Vinum': Metodi analisi del genoma e primi risultati," in Ciacci, Rendini and Zifferrero (2007): 238-48.

Immerwahr, H. R. "New Wine in Ancient Wineskins: The Evidence from Attic Vases," *Hesperia* 61 (1992): 121-32.

Jackisch, Philip. *Modern Winemaking.* Ithaca, NY, 1985.

Jaquat, Christiane and Danièle Martinoli. "*Vitis vinifera* L.: Wild or Cultivated? Study of the Grape Pips Found at Petra, Jordan; 150 B.C.-A.D. 40," *Vegetation History and Archaeobotany* 8 (1999): 25-30.

Jackson, Ron S. *Wine Science: Principles, Practice, Perception.* San Diego, 2000, 2nd. ed.

James, T. G. H. "The Earliest History of Wine and Its Importance in Ancient Egypt," in McGovern, Fleming and Katz (1996): 197-213.

Jashemski, Wilhelmina F. "The Caupona of Euxinus at Pompeii," *Archaeology* 20 (1967): 36-44.

Jashemski, Wilhelmina F. "Large Vineyard Discovered in Ancient Pompeii," *Science* 180, No. 4088 (25 May 1973): 821-30.

Jashemski, Wilhelmina F. "The Discovery of a Market-Garden Orchard at Pompeii: The Garden of the 'House of the Ship Europa'," *AJA* 78 (1974): 391-404.

Jashemski, Wilhelmina F. *The Gardens of Pompeii, Herculaneum and the Villas Destroyed by Vesuvius.* New Rochelle, NY, 1979-1993.

Jeffs, Julian. *Sherry.* London, 1992, 4th ed.

Johannowsky, W. "Appunti su Pyxous-Buxentum," *Atti e Memorie della Società Magna Grecia* 4 (1992): 173-83.

Joncheray, J. P. "L'Epare grecque, ou etrusque, de Bon Porte," *Cahiers d'Archéologie Subaquatique* 5 (1976): 5-53.

Johnson, Hugh. *The Story of Wine.* London, 2004.

Jouanna, Jacques and Laurence Villard. *Vin et Santé en Grèce ancienne. Actes du colloque organisé à l'Université de Rouen et à Paris (Université de Paris IV Sorbonne et ENS) 28-30 septembre 1998.* Paris, 2002.

Jones, G. D. B. "Capena and the Ager Capenas," *PBSR* 30 (1962): 116-208.

Jones, G. D. B. "Capena and the Ager Capenas," *PBSR* 31 (1963): 100-58.

Jones, G. D. B. "Il Tavoliere romano. L'agricoltura romana attraverso l'aerofotografia e lo scavo," *Archeologia Classica* 3.2 (1980): 85-100.

Jurisic, Mario. *Ancient Shipwrecks of the Adriatic: Maritime Transport During the First and Second Centuries AD.* Oxford, 2013.

King, Anthony. *Roman Gaul and Germany.* Berkeley, 1990.

Kinsella, J. E., E. Frankel, J. B. German and J. Kanner. "Possible Mechanisms for the Protective Role of Antioxidants in Wine and Plant Foods," *Food Technology* 47 (1993): 85-9.

Kirker, K. *Die sakrale Bedeutung des Weines in Altertum.* Giessen, 1910.

Klatsky, Arthur L., Gary D. Friedman, Mary Anne Armstrong and Harald Kipp. "Wine, Liquor, Beer, and Mortality," *American Journal of Epidemiology* 158 (6) (2003): 585-95.

Kleberg, Tönnes. *Hótels, restaurants et cabarets dans l'antiquité romaine.* Uppsala, 1957.

Kobayashi, Shozo, Nami Goto-Yamamoto and Hirohiko Hirochika. "Retrotransposon-Induced Mutations in Grape Skin Color," *Science* 304 (14 May 2004): 982.

Koehler, Carolyn G. "Wine Amphoras in Ancient Greek Trade," in McGovern, Fleming and Katz (1996): 323-37.

Lachiver, M. *Vins, vignes et vignerons: histoire des vignobles français.* Paris, 1988.

La Greca, Fernando. "Ville romane nel Cilento," *Annali Storici di Principato Citrà* 4.2 (2006): 5-18.

Lamboglia, Nino. "La nave romana di Albenga," *Rivista di Studi Liguri* 18 (1952): 131-236.

Landi, Maurizio. "Così i Romani scoprivano il vino annacquato," *Storica* 20 (Oct. 2010): 14.

Landolfi, Luciano. *Banchetto e società romana. Dalle origini al I sec. a. C.* Rome, 2002.

Laubenheimer, Fanette. *La production des amphores en Gaule Narbonnaise sous le haut-empire.* Paris, 1985.

Laubenheimer, Fanette. *Le temps des amphores en gaule: vins, huiles et sauces.* Paris, 1990.

Laubenheimer, Fanette. "Amphores et dolia," in Brun, Poux and Tchernia (2009): 278-87.

Laurence, Ray. *Roman Pompeii: Space and Society.* London, 1996.

Lauter, H. "Hellenistische Vorläufer der römischen Villa," in Frazer (1998): 211-7.

Lemaître, Séverine. "Du Chio au *passum*: les vins grecs en Gaule," in Brun, Poux and Tchernia (2009): 274-5.

Lenger, Dincer Savas. "Il mercato del vino nell'antichità in Asia Minore," in *Archeologia e Cibo* [= *Nutrirsi: Rivisti online di scienza della nutrizione umana*] (2006): 1-14.

Lenormant, F. *La Grande-Grèce, vol. 1.* Paris, 1881.

Leonard, Albert Jr. "'Canaanite Jars' and the Late Bronze Age Aegeo-Levantine Wine Trade," in McGovern, Fleming and Katz (1996): 233-54.

Lequément, R. "Le vin africaine à l'époque impérial," *Antiquités africaines* 16 (1980): 185-93.

Lesko, Leonard H. *King Tut's Wine Cellar.* Berkeley, 1977.

Lesko, Leonard H. "Egyptian Wine Production during the New Kingdom," in McGovern, Fleming and Katz (1996): 215-30.

Levadoux, L. "Les populations sauvages et cultivées de *Vitis vinifera* L." *Annales de l'Amelioration des Plantes* 1 (1956): 59-118.

Lightfoot, Dale R. and Frank W. Eddy. "The Agricultural Utility of Lithic-Mulch Gardens: Past and Present," *GeoJournal* 34.4 (1994): 425-37.

Lissarrague, François. *The Aesthetics of the Greek Banquet: Images of Wine and Ritual.* Princeton, NJ, 1990.

Locatelli, Daniela, ed. *Banchetto e simposio in Etruria. Simboli e immagini del potere.* Modena, 2008.

Loeschke, S. *Denkmäler von Weinbau aus der Zeit der Römer-Herrschaft am Mosel, Saar und Ruwer.* Trier, 1933.

Longo, Oddone. "Techniche di vinificazione nel mondo greco-romano," *Quaderni di Storia* 57 (2003): 197-214.

Lutz, H. F. *Viticulture and Brewing in the Ancient Orient.* Leipzig, 1922.

Manacorda, Daniele. "Nel calice dell'Imperatore," *Archeo* 259 (2006): 104-07.

Mangana, M. and K. Kotsakis. "A New Method for the Identification of Wild and Cultivated Charred Grape Seeds," *Journal of Archaeological Science* 23 (1966): 409-18.

Manzi, Luigi. *La viticoltura e l'enologia presso i Romani.* Rome, 1883.

Marazzi, M., C. Giardino and C. Pepe. "Vivara, Settlement of Punta d'Alaca, Naples, Italy Late Helladic I-II, 16th-15th Centuries B.C.," in Martlew (2008): 273-80.

Marcone, Arnaldo. *Storia dell'agricoltura romana: dal mondo arcaico all'età imperiale.* Rome, 1997.

Margalit, Yair. *Concepts in Wine Technology.* San Francisco, 2004.

Mariotti Lippi, Marta and Miria Mori Secci. "L'archeobotanica, il paleoambiente e la storia della vite in Toscana," in Ciacci, Rendini and Zifferrero (2007): 82-7.

Mariotti Lippi. "The Massaciuccoli Holocene Pollen Sequence and the Vegetation History of the Coastal Plains of the Mar Ligure," *Vegetation History and Archaeobotany* 16 (2007): 267-77.

Marlière, Élise. *L'outre et le tonneau dans l'Occident romain.* Montagnac, 2002.

Martin-Kilcher, S. "Amphorae in the West: Discussion and Research since 1989," in J. Eiriny and J. Lund, edd., *Transport Amphorae and Trade in the Eastern Mediterranean* (Athens, 2004): 263-72.

Martlew, Holley, ed. *Archaeology Meets Science: Biomolecular Investigations in Bronze Age Greece; The Primary Scientific Evidence, 1997-2003.* Oxford, 2008.

Marzano, Annalisa. *Roman Villas in Central Italy: A Social and Economic History.* Leiden, 2007.

Masi, Andrea. "Un esempio di archeologia dell'agricoltura: i palmenti," in Ciacci-Zifferero (2005): 83-95.

Masi, Andrea. "'Progetto Vinum': Archeologia dei siti campionati e primi risultati," in Ciacci, Rendini and Zifferero (2007): 201-16.

Mauné, Stéphane. "La villa de Vareilles," in Brun, Poux and Tchernia (2009): 212-5.

Mayerson, Philip. "Jar Stoppers and the Sealing of Winejars," *Zeischrift für Papyrologie und Epigraphik* 136 (2001): 219-20.

McCarthy, M. G., L. D. Jones and G. Due. "Irrigation—Principles and Practices," in Coombe and Dry (1988): 104-28.

McDonald, A. H. *Republican Rome.* London, 1966.

McDonald, A. H. *Houses, Villas and Palaces in the Roman World.* London, 1975.

McGovern, Patrick E., Donald L. Glusker, Lawrence J. Exner and Mary M. Voigt. "Neolithic Resinated Wine," *Nature* 381.6 (1996): 480-81.

McGovern, Patrick E., Stuart Fleming and Solomon H. Katz, edd. *The Origins and Ancient History of Wine*. Amsterdam, 1996.

McGovern, Patrick E. and Rudolph H. Michel. "The Analytical and Archaeological Challenge of Detecting Ancient Wine: Two Case Studies from the Ancient Near East," in McGovern, Fleming and Katz (1996): 57-65.

McGovern, Patrick E., U. Hartung, V. R. Badler, D. L. Glusker and L. J. Exner. "The Beginnings of Winemaking and Viniculture in the Ancient Near East and Egypt," *Expedition* 39 (1) (1997): 3-21.

McGovern, Patrick E. *Ancient Wine: The Search for the Origins of Viniculture*. Princeton, NJ, 2003.

McGovern, Patrick E.. "The Archaeological and Chemical Hunt for the Origins of Viniculture in the Near East and Etruria," in Ciacci, Rendini and Zifferrero (2007): 108-22.

McGovern, Patrick E., Donald L. Glusker, Lawrence J. Exner and Gretchen R. Hall. "The Chemical Identification of Resinated Wine and a Mixed Fermented Beverage in Bronze-Age Pottery Vessels of Greece," in Martlew (2008): 169-218.

McGovern, Patrick E., Armen Mirzoian and Gretchen R. Hall. "Ancient Greek Herbal Wines," *Proceedings of the National Academy of Science USA* 106 (2009): 7361-66.

McGovern, Patrick E.. *Uncorking the Past: The Quest for Wine, Beer and Other Alcoholic Beverages.* Berkeley, 2009.

McGovern, Patrick E., A. Mirzoian and G. R. Hall. "Ancient Egyptian Herbal Wines," *Proceedings of the National Academy of Sciences USA* 106 (2009): 7361-66.

McGrew, John R. "Basic Guide to Pruning," in *Growing Wine Grapes* (Ann Arbor, MI, 1993): 25-41.

McKay, A. G. *Ancient Campania, vol 2*. Hamilton, ON, 1972.

McKay, A. G. *Houses, Villas and Palaces in the Roman World.* Baltimore, 1998.

Merangou-Lerat, A. "Les vins et les amphores de Crète," *Etudes Crètoises* 30 (1995).

Merangou-Lerat, A. "Vin et amphores de Crète en Campanie," in *Eukrata: Mélanges offertes à Claude Vatin* (Aix-en-Provence, 1994): 137-43.

Meredith, Carole P. "Science as a Window into Wine History," *Bulletin of the American Academy of Arts and Sciences* 56.2 (Winter, 2003): 54-70.

Mielsch, H. *Die römische Villa: Architektur und Lebensform.* Munich, 1987.

Mielsch, H. *La villa romana.* Florence, 1990.

Morel, Jean-Paul. "Greek Colonization in Italy and in the West," in T. Hackens, Nancy D. Holloway and R. Ross Holloway, edd. *Crossroads of the Mediterranean* (Providence, RI, 1981): 123-61.

Morel, Jean-Paul. "Les trafics maritimes de la Grande Grèce," in Francesco Prontera, ed., *La Magna Grecia e il mare: studi di storia marittima* (Taranto, 1996): 143-72.

Moretti, M. and A. Sgubini Moretti. *La Villa Volusii a Lucus Feronia.* Rome, 1977.

Mori Secci, Miria. "Archeobotanica della vite," in Ciacci-Zifferrero (2005): 67-74.

Murray, Mary Anne, Neil Boulton and Carl Heron. "Viticulture and Wine Production," in P. T. Nicholson and I. Shaw, edd., *Ancient Egyptian Materials and Technology* (Cambridge, 2000): 577-608.

Murray, Oswyn, ed. *Sympotica: A Symposium on the Symposion.* Oxford, 1990.

Murray, Oswyn and Manuela Tecuşan. *In Vino Veritas.* London, 1995.

Myles, Sean et al. "Genetic Structure and the Domestication History of the Grape," *Proceedings of the National Academy of Sciences USA* (2010): 3530-5.

Negrul, A. M. "Evolution of Cultivated Forms of Grapes," *Comptes Réndus (Doklady) de l'Academie des Sciences de l'USSR* 18.8 (1938): 585-88.

Negrul, A. M. "Proles of Cultivated Grapevines and their Classification," *Ampelography: URSS* 1 (1946): 133-216.

Neyses, A. "Drei neuentdeckte gallo-römische Weinkelterhäuser im Moselgebiet," *Antike Welt* 10.2 (1979): 56-8.

Nicholas, P. R., A. P. Chapman and R. M. Cirami. "Grapevine Propagation," in Coombe and Dry (1988): 1-22.

Nriagu, Jerome O. *Lead and Lead Poisoning in Antiquity.* New York, 1983.

Nuñez, Diego Rivera and Michael J. Walker. "A Review of Paleobotanical Findings of Early *Vitis* in the Mediterranean and of the Origins of Cultivated Grapevines, with Special Reference to the New Pointers to Prehistoric Exploitation in the Western Mediterranean," *Review of Palaeobotany and Palynology* 61 (1989): 205-37.

Nutton, V. "Medicine and the Roman Army," *Medical History* 13.3 (1969): 260-70.

Obisesan, Thomas O. "Wine: Protective in Macular Degeneration," in Sandler and Pinder (2003): 285-98.

Odiot, Thierry. "Le site du Molard à Donzère (Drôme)," in Brun, Poux and Tchernia (2009): 208-9.

Olmer, Fabienne. "La vin hispanique," in Brun, Poux and Tchernia (2009): 276-7.

Olmo, H. P. "The Origin and Domestication of the *Vinifera* Grape," in McGovern, Fleming and Katz (1996): 31-43.

Östenberg, C. E. "Villa Sambuco," in Axel Boëthius, *Etruscan Culture. Land and People* (Mälmo, 1962): 313-20.

Packer, J. "Inns at Pompeii," *Cronache Pompeiane* 4 (1978): 5-53.

Pain, Stephanie. "Grog of the Greeks," *New Scientist* 2214 (1999): 54-7.

Pallottino, Massimo. *A History of Earliest Italy.* Ann Arbor, 1991.

Panella, Clementine. "Retroterra, porti e mercati: l'essempio dell'ager Falernus," in J. H. D'Arms and E. C. Kopff, edd., *The Seaborne Commerce of Ancient Rome: Studies in Archaeology and History* [=*MAAR* 36] (1980): 251-60.

Paolucci, Giulio. "L'occupazione dello spazio agrario in Etruria: vici e fattorie," in Consiglio regionale della Toscana, *Il mondo rurale etrusco* (Florence, 2009): 39-55.

Papàsogli, Giorgio. *L'agricoltura degli Etruschi e dei Romani.* Rome, 1942.

Paraskera, Maria and Matthieu Poux. "Libations et repas sacrés dans le monde grec et gaulois," in Brun, Poux and Tchernia (2009): 169-87.

Parker, A. J. *Ancient Shipwrecks of the Mediterranean and Roman Provinces.* Oxford, 1992.

Parker, Grant. *The Making of Roman India.* Cambridge, 2008.

Pasqui, Angiolo. "La villa pompeiana della Pisanella presso Boscoreale," *Monumenti Antichi* 7 (1897): 397-554.

Patroni, G. "L'origine della *domus*," *Rendiconti Reale Accademia dei Lincei, Classe di Scienze* 11 (1902): 467-507.

Paul, Harry W. *Bacchic Medicine: Wine and Alcohol Therapies from Napoleon to the French Pardox.* Amsterdam, 2001.

Peacock, David. "Recent Discoveries of Roman Amphora Kilns in Italy," *AntJ* 57 (1977): 262-9.

Peacock, David and D. F. Williams, edd. *Amphorae and the Roman Economy.* New York, 1986.

Peacock, David and D. F. Williams, edd. *Food for the Gods: New Light on the Ancient Incense Trade.* Oxford, 2006.

Peña, J. Theodore. *Roman Pottery in the Archaeological Record.* Cambridge, 2007.

Percival, J. *The Roman Villa.* London, 1976.

Perkins, Philip. *Etruscan Settlement, Society and Material Culture in Central Coastal Etruria* [= BAR *Int Ser.* 788]. Oxford, 1999.

Perkins, Philip. "Production and Distribution of Wine in the Etruscan Albegna Valley," in Ciacci, Rendini and Zifferero (2007): 185-90.

Pesce, Angelo. "Un testimone d'eccezione: Axel Boëthius," in Guzzo (2000): 105-11.

Petit-Aupert, Catherine and Pierre Sillières. "L'installation viticole de Lestagnac," in Brun, Poux and Tchernia (2009): 210-11.

Piccaluga, G. "Numa e il vino," SMSR 33 (1962): 99-103.

Piccoli, Francesca. *Il vino nel nord Italia in epoca romana.* Verona, 2004.

Pikoulas, Yanis. "Oinou istoro (Narro il vino). La ricerca in Grecia (1990-2006)," in Ciacci, Rendini and Zifferero (2007): 42-7.

Platon, Leftéris and Katérina Kopaka. "Ληνοί Μινωικοί. Installations minoennes de traitement des produits liquides," *Bulletin de correspondence hellénique* 117.1 (1993): 35-101.

Pomarèdes, Hervé. "Viticulture et équipements de production dans la cité de Lodève durant le haut Empire: apports des découvertes récentes autour de Clermont-L'Hérault," *Revue archéologique de Narbonnaise* 41 (2008): 7-41.

Pomeroy, Patrice. "Le navire romaine de la Madrague de Giens," *Comptes-rendus* 126 (1982): 133-54.

Ponfrandolfo, Angela. "Simposio e *élites* sociali nel mondo etrusco e italico," in Murray and Tecuşan (1995): 176-96.

Potter, Timothy W. *The Changing Landscape of South Etruria.* London, 1979.

Potter, Timothy W. and K. M. Dunbabin. "A Roman Villa at Crocicchie, Via Clodia," *PBSR* 47 (1979): 19-26.

Potter, Timothy W. *Roman Italy.* Berkeley, 1987.

Potter, Timothy W. and A. C. King. *Excavations at the Mola di Monte Gelato: A Roman and Medieval Settlement in South Etruria.* London, 1997.

Poux, Matthieu. "Le 'vin du triumvir', premier cru lyonnais," in Brun, Poux, Tchernia (2009): 232-3.

Powell, Marvin A. "Wine and the Vine in Ancient Mesopotamia: The Cuneiform Evidence," in McGovern, Fleming and Katz (1996): 97-122.

Puglisi, S. *La valle dei palmenti. Archeologia vitivinicola e rupestre in Sicili*a. Messina, 1990.

Purcell, N. "Wine and Wealth in Ancient Italy," *JRS* 75 (1985): 1-19.

Py, F. and M. Py. "Les amphores étrusques du Vannage et de Villevielle, Gard," *MEFRA* 86 (1974): 141-254.

Py, F. and R. Buxó i Capdevila. "La viticulture en Gaule à l'Âge du Fer," *Gallia* 58.1 (2001): 29-43.

Quilici, L. "Opifici rupestri dell'Italia centrale in età antica e medioevale," in *Atti del XXI Convegno di Studi Maceratesi (Matelica, 1985)* (Macerata, 1988): 41-66.

Quilici, L. "Segni del paesaggio agrario nell'Etruria rupestre. Impianti per la viticoltura," in Marina Martelli, ed., *Tyrrhenoi Philotechnoi* (Rome, 1994): 183-93.

Rathbone, D. "Italian Wines in Roman Egypt," *Opus* 2.1 (1983): 81-98.

Rathjes, Annette. "Il banchetto in Italia centrale. Quale stile di vita?," in Murray and Tecuşan (1995): 167-75.

Rauh, Nicholas K. et al. "Viticulture, Oleoculture and Economic Development in Roman Rough Cilicia," *Münstersche Beiträge zur antiken Handelgeschichte* 25.1 (2006): 49-98.

Reade, Julian Edgeworth. "The Symposium in Ancient Mesopotamia: Archaeological Evidence," in Murray and Tecuşan (1995): 35-56.

Reay, Brandon. "Cato's *De agri cultura* and the Spectacle of Expertise," in Becker and Terrenato (2012): 61-8.

Renfrew, Colin. *The Emergence of Civilization: The Cyclades and the Aegean in the Third Millennium B.C.* London, 1972.

Renfrew, Jane M. *Paleobotany: The Prehistoric Food Plants of the Near East and Europe.* London, 1973.

Renfrew, Jane M. "Paleobotanical Finds of *Vitis* from Greece," in McGovern, Fleming and Katz (1996): 255-67

Renfrew, Jane M. "Archaeology and the Origins of Wine Production," in Sandler and Pinder (2003): 56-69.

Reutti, F., ed. *Die römische Villa.* Darmstadt, 1990.

Rheinisches Landesmuseum Trier. *2000 Jahre Weinkultur am Mosel-Saar-Ruwer.* Trier, 1987.

Ricci, Clotilde. *La coltura della vite e la fabbricazione del vino nell'Egitto Greco-Romano.* Milan, 1924.

Riddle, J. R. *Dioscorides on Pharmacy and Medicine.* Austin, TX, 1985.

Ridgway, David. *The First Western Greeks.* Cambridge, 1992.

Ridgway, David. "Nestor's Cup and the Etruscans," *Oxford Journal of Archaeology* 16.3 (1997): 325-44.

Roller, Matthew B. *Dining Posture in Ancient Rome: Bodies, Values and Status.* Princeton, NJ, 2006.

Roller, Matthew B. "Convivial Wine Drinking and *Comissationes*," in Roller (2006): 181-8.

Rossi, Luigi, ed. *Il vino nel Cilento.* Acciaroli, SA, 1994.

Rossiter, J. J. "Wine and Oil Processing at Roman Farms in Italy," *Phoenix* 35 (1981): 344-61.

Rossiter, J. J. and E. Haldenby. "A Winemaking Plant in Pompeii Insula II.5," *Echos du Monde Classique/Classical Views* 33, NS 8 (1989): 229-39.

Rostovtzeff, Michael I. *The Social and Economic History of the Roman Empire.* Oxford, 1957, 2nd. ed.

Ruggia, Zaccaria. *More regio vivere. Il banchetto aristocratico e la casa romana di età arcaica.* Rome, 2003.

Russo Tagliente, Alfonsina. *Edilizia domestica in Apulia e Lucania: ellenizzazione e società nella tipologia abitativa indigena tra VIII e il III secolo a.C.* Lecce, 1992.

Salza Prina Ricotti, Eugenia. *L'arte del convito nella Roma antica.* Rome, 1983.

Sandbach, F. H. "Vergil, *Georgics* II. 277," *CR* 42 (1928): 59-60.

Sandler, Merton and Roger Pinder, edd. *Wine: A Scientific Exploration.* New York, 2003.

Sannino, Silvestro. *Civiltà agricola vesuviana.* Salerno, 2009.

Santangeli Valenzani, Riccardo and Rita Volpe. "La restituzione del paesaggio agrario della vitivinicoltura a Rome e nel suburbio," in Ciacci, Rendini and Zifferero (2007): 48-54.

Savay-Gueraz, Hugues. "*Allobrogica*, le vin du Allobroges," in Brun, Poux and Tchernia (2009): 234-5.

Scagliarini Cortaita, D. "Le 'villae' romane nell'Italia settentrionale," in E. Roffia, ed., *Ville romane sul Lago di Garda* (Brescia, 1997): 53-81.

Scandone Matthiae, Gabriella. "Il vino nell'antico Egitto," in Murray and Tecuşan (1995): 57-61.

Schilling, R. "La signification du culte des Vinalia," in *La religion romaine de Vénus: depuis les orìgines jusqu'au temps d'Auguste.* Paris, 1982.

Schnurbusch, Dirk. *Convivium: Form und Bedeutung aristokratischer Geselligkeit in der römischen Antike.* Stuttgart, 2011.

Scienza, Attilio, A. O. Failla and L. Valenti. "Grapevine Germoplasm Diversity and Conservation," in G. H. Bunce, L. Ryszowski and M. G. Paoletti, edd., *Landscape Ecology and Agrosystems* (Boca Raton, LA, 1993): 183-95.

Scienza, Attilio and M. Boselli, edd. *Vini e vitigni della Campania. Tremila anni di storia.* Naples, 2003.

Scienza, Attilio, Osvaldo Failla and Roberto Anzani. "Consistenza e valutazione del patrimonio viticolo spontaneo," in Ciacci-Zifferrero (2005): 47-55.

Scienza, Attilio. "Interesse e attualità dello studio della vite selvatica nella Toscana e nel Lazio settentrionale: una comune area di ricerca tra archeologia e biologia," in Ciacci and Zifferero (2005): 11-14.

Scienza, Attilio. "L'origine dei vitigni coltivati: Una storia interdisciplinare della cultura europea," in Di Pasquale (2010): 25-31.

Scienza, Attilio. "La storia della civiltà europea racconta attraverso l'origine dei suoi vitigni," in Giovanni Negri and Elisibetta Petrini, *Roma, Caput Vini* (Milan, 2011): 181-202.

Scullard, H. H. *Festivals and Ceremonies of the Roman Republic.* London, 1981.

Sculli, O. *I palmenti di Ferruzzano. Archeologia del vino e testimonianze di cultura materiale in un territorio della Calabria meridionale.* Florence, 2002.

Sculli, O. *I vitigni autoctoni della Locride.* Cosenza, 2004.

Sereni, Emilio. "Per la storia delle più antiche tecniche e della nomenclatura della vite e del vino in Italia," *Atti e Memorie dell'Academia toscana di scienze e letture La Columbaria* 29 (1964): 75-204.

Sereni, Emilio. *Terra nuova e buoi rossi.* Torino, 1981.

Sereni, Emilio. *History of the Italian Agricultural Landscape.* Princeton, NJ, 1997.

Setari, Elisabetta. "La cultura del vino nel mondo italico. Forme, immagini, rituali," in Di Pasquale (2010): 110-19.

Singer, Caroline. "The Incense Kingdoms of Yemen," in Peacock and Williams (2006): 10-25.

Singleton, Vernon R. "An Enologist's Commentary on Ancient Wines," in McGovern, Fleming and Katz (1996): 67-77.

Skurlatidou, Eudokia. "Techniques and Methods of Viticulture in Ancient Times: The Examples Found in Greece and the Black Sea," *Eirene* 42 (2006): 73-88.

Slane, Kathleen Warner. "Observations on Mediterranean Amphoras and Tablewares Found in India," in V. Begley and R. de Pluma, edd., *Rome and India: The Ancient Sea Trade* (Madison, WI, 1991): 204-15.

Slater, William. *Dining in a Classical Context.* Ann Arbor, MI, 1991.

Small, A., ed. *The Excavations at San Giovanni di Ruoti.* Toronto, 1994.

Small, A. "Changes in the Pattern of Land Settlement and Land Use Around Gravina and Monte Irsi," in Elio Lo Cascio and Alfredina Storchi Marino, *Modalità insediative e strutture agrarie nell'Italia meridionale in età romana* (Bari, 2001): 35-53.

Smith, H. and G. Jones. "Experiments on the Effects of Charring on Cultivated Grape Seeds," *Journal of Archaeological Science* 17 (1990): 317-27.

Smith, J. T. *Roman Villas: A Study in Social Structure.* London, 1997.

Stanco, Enrico Angelo. "La localizzazione di Statonia: nuove considerazioni in base alle antiche fonti," *MEFR* 106 (1994): 247-58.

Stefani, Greta. "Le ville rustiche del territorio Vesuviano," in Guzzo (2000): 13-56.

Stefani, Greta. "La villa del Fondo Prisco in località Civita," in Guzzo (2000): 45-7.

Stefani, Greta. "La villa del fondo Imperiali in località Civita di Nitto," in Guzzo (2000): 49-53.

Stefani, Greta. "La villa del fondo Agricoltura in località Pisanella," in Guzzo (2000): 59-64.

Stefani, Greta. "La Villa della Pisanella," in Guzzo (2000): 65-8.

Stefani, Greta. "La villa in località Congiani," in Guzzo (2000): 69-71.

Stefani, Greta. "La Villa 2 di Terzigno," in Guzzo (2000): 75-8.

Stein-Hölkeskamp, Elke. *Das römische Gastmahl.* Munich, 2005.

Stevenson, A. C. "Studies in the Vegetational History of S. W. Spain. Spain II. Palynological Investigations at Laguna de las Madres, S. W. Spain," *Journal of Biogeography* 12 (1985): 293-314.

Stronach, David. "The Imagery of the Wine Bowl: Wine in Assyria in the Early First Millennium BC," in McGovern, Fleming and Katz (1996): 175-95.

Talbert, Richard, ed. *Barrington Atlas of the Greek and Roman World.* Princeton, NJ, 2000.

Tassie, E. and B. M. Freeman. "Pruning," in Coombe and Dry (1988): 66-84.

Tchernia, André. "Il vino: produzione e commercio," in Fausto Zevi and Elena Fiorentini, edd., *Pompei 79: Raccolta di studi per il decimonono centenario dell'eruzione vesuviana* (Naples, 1979): 87-96.

Tchernia, André. *Le vin de l'Italie romaine. Essai d'histoire économique d'après les amphores.* Rome, 1986.

Tchernia, André. "Le vignoble italien du 1er siècle avant notre ére au IIIe siècle de notre ère: répartition et évolution," in Amouretti and Brun (1993): 282-7.

Tchernia, André. "The Dromedary of the Peticii and Trade with the East," in F. De Romanis and A. Tchernia, edd., *Crossings: Early Mediterranean Contacts with India* (New Delhi, 1997): 238-49.

Tchernia, André. "Le cercle de L. Licinius Crassus et la naissance de la hiérarchie des vins à Rome," *Comptes-rendus* 141 (1997): 1247-59.

Tchernia, André and Jean-Pierre Brun. *Le vin romain antique.* Grenoble, 1999.

Tchernia, André. "La vinification au début de notre ère et le goût des vins romains," in Tchernia and Brun (1999): 118-9.

Terrenato, Nicola. "The Auditorium Site in Rome and the Origins of the Villa," *JRS* 14.1 (2001): 5-32.

Terrenato, Nicola. "The Enigma of 'Catonian' Villas: The *De agri cultura* in the Context of Second-Century BC Italian Architecture," in Becker and Terrenato (2012): 69-93.

Thurmond, David L. *A Handbook of Food Processing in Classical Rome: For Her Bounty No Winter.* Leiden, 2006.

Torelli, Mario. "The Early Villa: Roman Contributions to the Development of a Greek Prototype," in Becker and Terrenato (2012): 18-31.

Unwin, Tim. *Wine and the Vine: An Historical Geography of Viticulture and the Wine Trade.* London, 1991.

Vandermersch, Christian. *Vins et amphores de grande Grèce et de Sicilie, IVe-IIIe s. avant J.-C.* Naples, 1994.

Vandermersch, Christian. "Vigne, vin et économie dans l'Italie du Sud grecque à l'époque archaïque," *Ostraka* 5.1 (1996): 155-85.

Vandiver, P. and C. G. Koehler. "Structure, Processing, Properties and Styles of Coirnthian Amphoras," in W. D. Kingery, ed., *Ceramics and Civilization, vol. 2* (Columbus, OH, 1986): 173-215.

van Dommelen, P., C. Gomez Bellard and G. Pérez Jordà. "Produzione agraria nella Sardegna punica fra cereali e vino," *L'Africa Romana* 18 (2010): 1187-1202.

Vernou, Christian. "Le vin des *Parisii* et des cités septentrionales," in Brun, Poux and Tchernia (2009): 241.

Vernou, Christian. "À la recherche des outils antiques des vignerons," in Brun, Poux and Tchernia (2009): 260-3.

Vernou, Christian. "La vente au détail. Un monument attestant de la commercialisation du vin chez les Lingons," in Brun, Poux and Tchernia (2009): 316-7.

Versari, A., G. P. Parpinello and M. Cattaneo. "*Leuconostoc oenos* and Malolactic Fermentation: A Review," *Journal of Industrial Microbiology and Biotechnology* 23.6 (Dec., 1999): 447-55.

Vine, Richard P., Ellen Harkness and Sally J. Linton. *Winemaking: From Grapegrowing to Marketplace.* New York, 2002, 2nd ed.

Vitali, Daniele, Fanette Laubenheimer and Laurence Benquet. "La produzione e il commercio del vino nell'Etruria romana: Le fornaci di Albinia (Orbetello, GR)," in Ciacci, Rendini and Zifferero (2007): 191-200.

Volpe, Rita. "Republican Villas in the *Suburbium* of Rome," in Becker and Terrenato (2012): 94-110.

von Blanckenhagen, P., M. A. Cotton and J. B. Ward-Perkins. "Two Roman Villas at Francolise, Prov. Caserta," *PBSR* 33 (1965): 55-69.

Vouillamoz, José, Anna Schneider and M. Stella Grando. "Microsatellite Analysis of Alpine Grape Cultivars (*Vitis vinifera* L.): Alleged Descendants of Pliny the Elder's *Raetica* Are Genetically Related," *Genetic Resources and Crop Evolution* 54 (2007): 1095-1104.

Walsh, Carey Ellen. *Fruit of the Vine: Viticulture in Ancient Israel* [= *Harvard Semitic Monographs* 60]. Cambridge, MA, 2000.

Waltzing, Jean Pierre. *Étude historique sur les corporations professionnelles chez les Romains, vol 1.* Louvain, 1895.

Wardle, K. A. "The Palace Civilizations of Minoan Crete and Mycenaean Greece, 800-140 BC," in Cunliffe (1994): 202-43.

Weaver, Robert J. *Grape Growing.* New York, 1976.

Wecowski, Marek. *The Rise of the Greek Aristocratic Banquet.* Oxford, 2014.

Weisse, Martin E. and Renée S. Moore. "Antimicrobial Effects of Wine," in Sandler and Pinder (2003): 299-313.

Wells, Berit, ed. *Agriculture in Ancient Greece: Proceedings of the Seventh International Symposium at the Swedish Institute at Athens.* Stockholm, 1990.

White, K. D. *Agricultural Implements of the Roman World.* Cambridge, 1967.

White, K. D. *Roman Farming.* Ithaca, NY, 1970.

White, K. D. *Farm Equipment of the Roman World.* Cambridge, UK, 1975.

Widemann, F. et al. "Analytical and Typological Studies of Gallo-Roman Workshops Producing Amphorae in the Area of Narbonne," *Archaeophysika* 10 (1978): 317-41.

Widemann, F., F. Laubenheimer and J. Leblanc. "Amphorae Workshops in Western Narbonensis: The Non-resolution Space Problems," in *XIXth Symposium on Archaeometry and Archaeological Prospection* (London, 1979): 57-71.

Wiel-Marin, F. "Due diverse associazioni di vasi nel banchetto etrusco," *RM* 104 (1997): 514-18.

Wiessner, P. and W. Schiefenhovel, edd. *Food and the Status Quest: An Interdisciplinary Perspective.* New York, 1996.

Wightman, Edith Mary. *Roman Trier and the Treveri.* New York, 1971.

Wilkins, John M. and Shaun Hill, edd. *Food in the Ancient World.* Oxford, 2006.

Wilkins, John M. and Robin Nadeau, edd. *A Companion to Food in the Ancient World.* Oxford, 2015.

Will, Elizabeth Lyding. "The Mediterranean Shipping Amphoras from Arikamedu," in V. Begley and R. De Puma, *Rome and India: The Ancient Sea Trade* (Madison, WI, 1991): 151-56.

Wilson, James E. *Terroir.* Berkeley, 1998.

Winkler, A. J., James A. Cook, W. M. Kliewer and Lloyd A. Lider. *General Viticulture.* Berkeley, 1974.

Winter, W., ed. *Indo-European and the Indo-Europeans: A Reconstruction and Historical Analysis of a Proto-Language and a Proto-Culture,* trans. J. N. Nichols, 2 vols. Berlin, 1995.

Zampi, Maurizio. *Il triclinio. Mito e politica dell'alimentazione nell'antica.* Rome, 2002.

Zemer, Avashalom. *Storage Jars in the Ancient Sea Trade.* Haifa, 1978.

Zettler, Richard L. and Naomi F. Miller. "Searching for Wine in the Archaeological Record of Ancient Mesopotamia of the Third and Second Millennium BC," in McGovern, Fleming and Katz (1996): 123-31.

Zifferero, Andrea. "La produzione e il commercio del vino in Etruria," in Ciacci and Zifferero (2005): 97-120.

Zohary, Daniel and Maria Hopf. *Domestication of Plants in the Old World.* Oxford, 1988.

Zohary, Daniel. "The Domestication of the Grapevine *Vitis Vinifera* L. in the Near East," in McGovern, Fleming and Katz (1996): 23-30.

Index